Applications of Microsoft® Excel in Analytical Chemistry

Second Edition

Stanley R. Crouch
Michigan State University

F. James Holler
University of Kentucky

BROOKS/COLE
CENGAGE Learning·

Australia • Brazil • Japan • Korea • Mexico • Singapore • Spain • United Kingdom • United States

For product information and technology assistance, contact us at **Cengage Learning Customer & Sales Support, 1-800-354-9706**

For permission to use material from this text or product, submit all requests online at **www.cengage.com/permissions** Further permissions questions can be emailed to **permissionrequest@cengage.com**

ISBN-13: 978-1-285-08795-5
ISBN-10: 1-285-08795-X

Brooks/Cole
20 Davis Drive
Belmont, CA 94002-3098
USA

Cengage Learning is a leading provider of customized learning solutions with office locations around the globe, including Singapore, the United Kingdom, Australia, Mexico, Brazil, and Japan. Locate your local office at: **www.cengage.com/global**

Cengage Learning products are represented in Canada by Nelson Education, Ltd.

To learn more about Brooks/Cole, visit **www.cengage.com/brookscole**

Purchase any of our products at your local college store or at our preferred online store **www.cengagebrain.com**

Printed in the United States of America
1 2 3 4 5 6 7 16 15 14 13 12

CONTENTS

Chapter 1

Excel Basics

The spreadsheet is one of the most useful personal computer tools. Spreadsheets were first used in business applications for financial analysis, modeling, forecasting, database management and similar tasks. In analytical chemistry, spreadsheets are particularly useful for statistical and graphical analysis, computing and plotting titration curves, performing equilibrium calculations, curve fitting, simulations, matrix manipulations, and solving equations.[1] Throughout this book, we present many examples to illustrate the construction and applications of spreadsheets for performing tasks in quantitative analytical chemistry. Because of its widespread availability and general utility, we have chosen to illustrate our examples using Microsoft® Excel 2010. Because there are few changes from Excel 2007, most of the worksheets and instructions also apply to Excel 2007. When significant differences occur, they are noted in the text or in the appropriate figure. Because Excel 2010 and Excel 2007 are very different from previous versions, we refer you to our previous book for illustrations and examples using Office 1997-2003.[2] Unless otherwise noted, we assume that Excel 2010 is configured with the default options as delivered from the manufacturer. Although the syntax and commands for other spreadsheet software are somewhat different from those in Excel, the general concepts and principles of operation are similar. If a program other than Excel is used, the precise instructions must be modified in the examples presented.

---

[1] For more information on the use of spreadsheets in chemistry, see R. de Levie, *How to Use Excel in Analytical Chemistry and in General Scientific Data Analysis,* (New York: Cambridge University Press, 2001); .E. J. Billo, *Excel for Chemists: A Comprehensive Guide,* 2nd ed. (New York, Wiley-VCH, 2001)D. Diamond and V. C. A. Hanratty, *Spreadsheet Applications in Chemistry Using Microsoft Excel,* (New York: Wiley, 1997); H. Freiser, *Concepts & Calculations in Analytical Chemistry: A Spreadsheet Approach,* (New York: CRC Press, 1992);.

[2] S. R. Crouch and F. J. Holler, *Applications of Microsoft® Excel in Analytical Chemistry.* Belmont, CA: Brooks/Cole, 2004.

1

It is our conviction that we learn best by doing, not by reading about doing. Although software producers have made great strides in generating manuals and other materials for their products, it is still generally true that when we know enough to read a software manual efficiently, we no longer need the manual. With this in mind, we have tried to provide a series of spreadsheet exercises that evolve in the context of analytical chemistry, not a software manual. Excel commands and syntax are introduced only when they are needed to accomplish a particular task. If more detailed information is needed, please consult the help screens. Help is available at the click of a mouse button from within Excel by clicking on the Help icon ⊙ or by pressing F1.

Getting Started

In this book, we will assume that you are familiar with *Windows™*. If you need assistance with Windows, please consult the Windows guide *Getting Started* or use the on-line help facility available. To start Excel, double click on the Excel icon or use the **Start** button and click on **Start/All Programs/Microsoft Office/Microsoft Office Excel 2010**. The window shown in Figure 1-1 then appears on your computer screen.

Versions of Excel prior to Excel 2007 contained menus such as File, Edit, View, Insert, Format, Tools, etc. The menus and toolbars have been completely removed from Excel 2007 and 2010 and replaced by the *ribbon*, a two-dimensional layout of icons and words. Each tab, such as Home, Insert, Page Layout, Formulas, Data, Review, and View, brings up a different ribbon with its own set of icons and descriptions. Although the ribbon takes up space, it can be minimized by clicking the minimize ribbon arrow, entering Ctrl+F1, or by right clicking anywhere on the ribbon and selection Minimize the Ribbon from the list that appears. To maximize workspace, you may operate with the ribbon minimized. Excel 2007 does not have the File button. Instead commands such as Save, Print, Open, Close, and Send are located in the Office button to the left

of the Home tab.

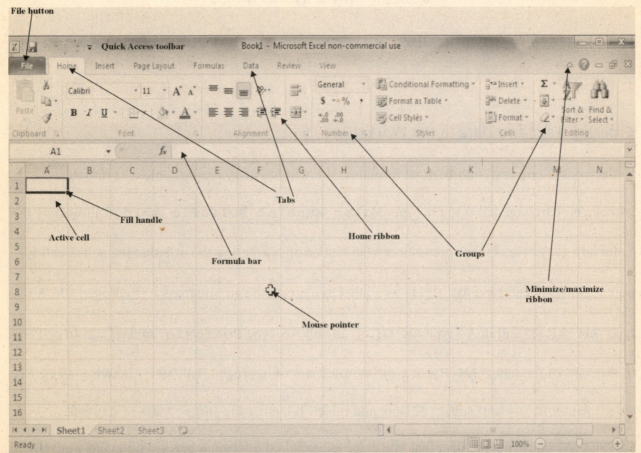

Figure 1-1 The opening window in Microsoft Excel. Note the location of the File button, the Quick Access Toolbar, the active cell, and the mouse pointer.

Below the ribbon in Figure 1-1 is the *worksheet* consisting of a grid of *cells* arranged in

rows and columns. The rows are labeled 1, 2, 3, and so on, and the columns are labeled A, B, C,

and so on. Each cell has a unique location specified by its address. For example, the *active cell,*

which is surrounded by a dark outline as shown in Figure 1-1, has the address A1. The address of

the active cell is always displayed in the box just above the first column of the displayed worksheet in the *formula bar*. You may verify this display of the active cell by clicking on various cells of the worksheet.

Calculating a Molar Mass

In our first exercise we will use Excel as a basic calculator to find the molar mass of sulfuric acid, H_2SO_4.

Entering Text and Data in the Worksheet.

Cells may contain text, numbers, or formulas. We will begin by typing some text into the worksheet. Click on cell A1, and type **Molar Mass of Sulfuric Acid** followed by the Enter key **[↵]**. This is the spreadsheet title. Notice after entering the title that the active cell is now A2. In this cell type **AM H[↵]** as a label to indicate the atomic mass of hydrogen. In A3 type **AM S[↵]**, and in A4 type **AM O[↵]**. In cell A6, type **Sulfuric Acid[↵]**. In cell B2 to the right of the label **AM H**, enter the atomic mass of hydrogen, 1.00794. Likewise in cell B3, enter the atomic mass of sulfur, 32.066, and in cell B4, enter the atomic mass of oxygen, 15.9994. As you type, the data that you enter appears in the formula bar. If you make a mistake, just click the mouse in the formula bar, and make necessary corrections. Because the title of the spreadsheet should be easily distinguished from the body, it is appropriate to format the title in boldface font. This can be done by selecting cell A1. In the formula bar select the entire title by dragging the mouse over the words **Molar Mass of Sulfuric Acid**. When the title has been selected, click the Bold button in the Font group on the Home tab. This action will make the title appear in boldface font.

Entering an Equation

In cell B6 we will enter the formula that we want Excel to use to calculate the molar mass of

sulfuric acid. Type into cell B6, the following:

$$=2*B2+B3+4*B4[\downarrow]$$

The expression just typed is called a *formula*. In Excel, formulas begin with an equal sign **[=]**

followed by the desired numerical expression. This formula will calculate the molar mass of

H_2SO_4 by summing twice the atomic mass of hydrogen (cell B2), the atomic mass of sulfur (cell

B3), and four times the atomic mass of oxygen (cell B4). The result should be as shown in Figure

1-2.

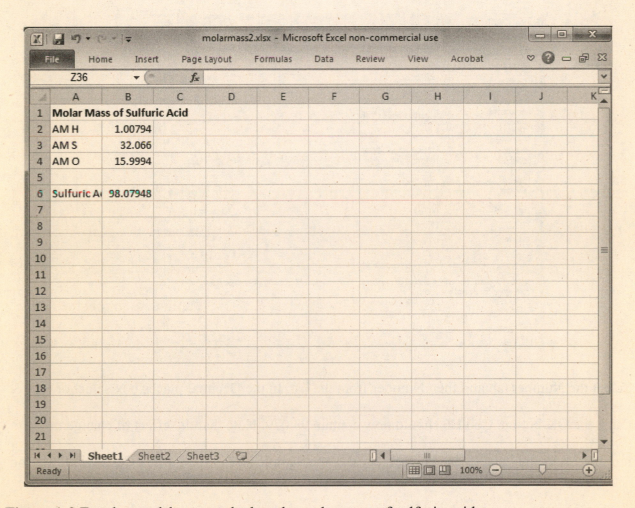

Figure 1-2 Excel spreadsheet to calculate the molar mass of sulfuric acid.

Note in Figure 1-2, that Excel has expressed the molar mass of sulfuric acid to five digits past the decimal point. In Section 6D of *Fundamentals of Analytical Chemistry, 9th edition* (FAC9) and Section 6D of *Analytical Chemistry: An Introduction, 7th edition* (AC7), reference to the significant figure convention indicates that the molar mass should be expressed to only three digits beyond the decimal point since the atomic mass of sulfur is only known to this number of digits. Hence, a more appropriate result would be 98.079 for the molar mass of H_2SO_4. To change the number of displayed digits, bring up the Home ribbon and then click on cell B6. From the Cells group, select the Format command and Format Cells… from the pull down menu. The Format Cells window shown below then appears on the screen.

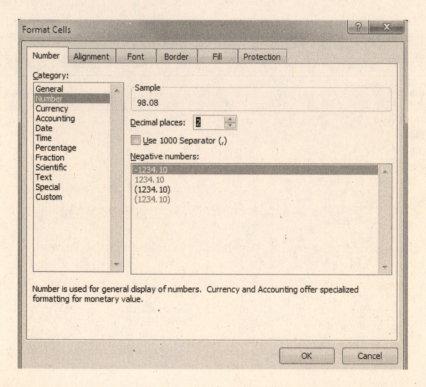

Select the Number tab and then Number from the list. In the Decimal places box select or type 3. Click the OK button. Cell B6 should now contain 98.079. Note that the effect of change the number of decimal places can be previewed in the Sample box in the Format Cells window. You can also increase or decrease the number of decimal places by clicking the Increase or Decrease

Decimal button in the Number group on the Home ribbon.

Documenting the Worksheet

Since the spreadsheet does not display the equations entered or indicate which cells contained

data, it is important to document all cells that contain an equations or data. There are several

different documentation schemes, but we shall introduce a method that is quite easy to

implement. Make cell A9 the active cell and type **Documentation[↵]**. Make the font for this

cell boldface. In cell A10, type

Cells B2:B4=user entries[↵]

The colon between B2 and B4 specifies a range. Thus, B2:B4 means the range of cells B2

through B4.

In cell A11, type

Cell B6=2*B2+B3+4*B4[↵]

The spreadsheet should now appear as shown in Figure 1-3. This documentation indicates the

data entered by the user and shows the formula entered in cell B6 to calculate the molar mass of

sulfuric acid. In many cases, it is apparent which cells contain user entered data. Hence, often the

documentation section will contain only formulas.

	A	B	C
1	**Molar Mass of Sulfuric Acid**		
2	AM H	1.00794	
3	AM S	32.066	
4	AM O	15.9994	
5			
6	Sulfuric A(	98.079	
7			
8			
9	**Documentation**		
10	Cells B2:B4=user entries		
11	Cell B6=2*B2+B3+4*B4		

Figure 1-3 Final spreadsheet for calculating the molar mass of sulfuric acid including a documentation section. Note that we have omitted the Excel ribbon and formula bar for clarity in this figure.

If desired, you can save your file to the hard disk by clicking on the File (Office in Excel 2007) button and choosing **Save As**. You can save as an Excel Workbook and various other formats including a format compatible with Excel 97-2003. Choose **Excel Workbook** and enter a location and a file name such as **molarmass**. Excel will automatically append the file extension **.xlsx** to the file name so that it will appear as **molarmass.xlsx**. Choosing to save in a format compatible with Excel 97-2003 appends the file extension **.xls** to the file.

Laboratory Notebook Example

As another illustration of spreadsheet use, we will use Excel to carry out some functions of the laboratory notebook illustrated in FAC9, Figure 2-24 and AC7, Figure 2-24. We begin by typing some text into the worksheet. Click on cell A1 and type **Gravimetric Determination of Chloride** followed by the Enter key **[↵]**. Notice that the active cell is now A2, so you may now type **Samples[↵]**. As you type, the data that you enter appears in the formula bar. If you make a mistake, just click the mouse in the formula bar, and make necessary corrections.

Continue to type text into the cells of column A as shown below.

```
Mass of bottle plus sample, g[↵]

Mass of bottle less sample, g[↵]

Mass of sample, g[↵]

[↵]

Crucible masses, with AgCl, g[↵]

Crucible masses, empty, g[↵]

Mass of AgCl, g[↵]

[↵]

%Chloride[↵]

Mean %Chloride[↵]

Standard Deviation, %Chloride[↵]

RSD, parts per thousand[↵]
```

When you have finished entering the text, the worksheet should appear as shown in Figure 1-4.

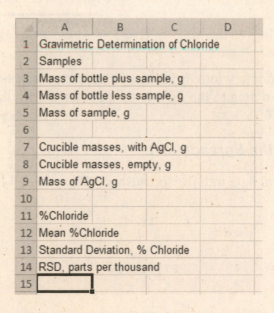

	A	B	C	D
1	Gravimetric Determination of Chloride			
2	Samples			
3	Mass of bottle plus sample, g			
4	Mass of bottle less sample, g			
5	Mass of sample, g			
6				
7	Crucible masses, with AgCl, g			
8	Crucible masses, empty, g			
9	Mass of AgCl, g			
10				
11	%Chloride			
12	Mean %Chloride			
13	Standard Deviation, % Chloride			
14	RSD, parts per thousand			
15				

Figure 1-4 The appearance of the worksheet after entering the labels.

Changing the Width of a Column

Notice that the labels that you typed into column A are wider than the column. You can change

the width of the column by placing the mouse pointer on the boundary between column A and

column B in the column head as shown in Figure 1-5a and dragging the boundary to the right so

that all of the text shows in the column as in Figure 1-5b .

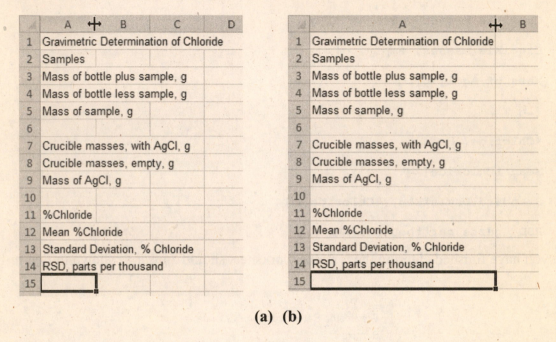

(a) (b)

Figure 1-5 Changing the column width. (a) Place the mouse pointer on the boundary between column A and column B, and drag to the right to the position shown in (b).

Entering Numbers into the Spreadsheet.

Now let's enter the numerical data into the spreadsheet. Click on cell B2 and type

1 [↵]

27.6115 [↵]

27.2185 [↵]

At this point, we wish to calculate the difference between the data in cells B3 and B4, so we type

$$=\texttt{b3-b4}\,[\hookleftarrow]$$

Notice that the difference between the contents of cell B3 and cell B4 is displayed in cell B5.

	A	B	C	D
1	Gravimetric Determination of Chloride			
2	Samples	1	2	3
3	Mass of bottle plus sample, g	27.6115	27.2185	26.8105
4	Mass of bottle less sample, g	27.2185	26.8105	26.4517
5	Mass of sample, g	0.393		
6				
7	Crucible masses, with AgCl, g			
8	Crucible masses, empty, g			
9	Mass of AgCl, g			

Figure 1-6 Sample data entry.

Filling Cells Using the Fill Handle.

The formulas for cells C5 and D5 should be identical to the formula in cell B5 except that the cell references for the data are different. In cell C5, we want to compute the difference between the contents of cells C3 and C4, and in cell D5, we want the difference between D3 and D4. We could type the formulas in cells C5 and D5 as we did for cell B5, but Excel provides an easy way to duplicate formulas, and it automatically changes the cell references to the appropriate values for us. To duplicate a formula in cells adjacent to an existing formula, simply click on the cell containing the formula, which is cell B5 in our example, then click on the fill handle (see Figure 1-1), and drag the corner of the rectangle to the right so that it encompasses the cells where you want the formula to be duplicated. Try it now. Click on cell B5, click on the fill handle, and drag to the right to fill cells C5 and D5. When you let up on the mouse button, the spreadsheet should appear like Figure 1-7. Now click on cell B5, and view the formula in the formula bar. Compare

the formula to those in cells C5 and D5.

	A	B	C	D
1	Gravimetric Determination of Chloride			
2	Samples	1	2	3
3	Mass of bottle plus sample, g	27.6115	27.2185	26.8105
4	Mass of bottle less sample, g	27.2185	26.8105	26.4517
5	Mass of sample, g	0.393	0.408	0.3588

Figure 1-7 Use of the fill handle to copy formulas into adjacent cells of a spreadsheet. In this example, we clicked on cell B2, clicked on the fill handle, and dragged the rectangle to the right to fill cells C5 and D5. The formulas in cells B5, C5, and D5 are identical, but the cell references in the formulas refer to data in columns B, C, and D, respectively.

Now we want to perform the same operations on the data in rows 7, 8, and 9 shown in

Figure 1-8, so enter the remaining data into the spreadsheet now.

	A	B	C	D
1	Gravimetric Determination of Chloride			
2	Samples	1	2	3
3	Mass of bottle plus sample, g	27.6115	27.2185	26.8105
4	Mass of bottle less sample, g	27.2185	26.8105	26.4517
5	Mass of sample, g	0.393	0.408	0.3588
6				
7	Crucible masses, with AgCl, g	21.4296	23.4915	21.8323
8	Crucible masses, empty, g	20.7926	22.8311	21.2483
9	Mass of AgCl, g			

Figure 1-8 Entering the data into the spreadsheet in preparation for calculating the mass of dry silver chloride in the crucibles.

Now click on cell B9, and type the following formula:

$$=b7-b8 \; [↵]$$

Again click on cell B9, click on the fill handle, and drag through columns C and D to copy the

formula to cells C9 and D9. The mass of silver chloride should now be calculated for all three

crucibles.

Making Complex Calculations with Excel

The equation for finding the % chloride in each of the samples is (see FAC9, Chapter 12, or

AC7, Chapter 8)

$$\%\text{chloride} = \frac{\dfrac{\text{mass AgCl}}{\text{molar mass AgCl}} \times \text{molar mass Cl}}{\text{mass sample}} \times 100\% = \frac{\dfrac{\text{mass AgCl}}{143.321\,\text{grams/mol}} \times 35.4527\,\text{grams/mol}}{\text{mass sample}} \times 100\%$$

Our task is now to translate this equation into an Excel formula and type it into cell B11 as

shown below.

$$\texttt{=B9*35.4527*100/143.321/B5[↵]}$$

Once you have typed the formula, click on cell B11, and drag on the fill handle to copy the

formula into cells C11 and D11. The %chloride for samples 2 and 3 should now appear in the

spreadsheet as shown in Figure 1-9.

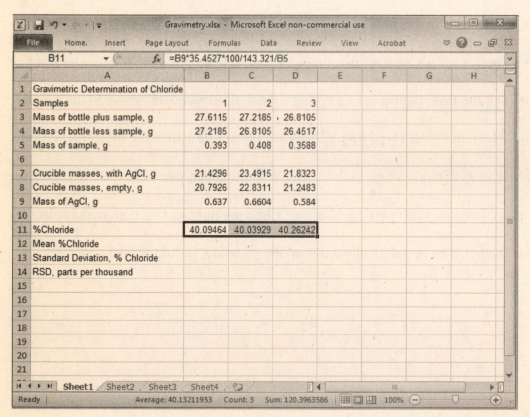

Figure 1-9 Completing the calculation of percent chloride. Type the formula in cell B11, click on the fill handle, and drag to the right through cell D11.

We will complete and document the spreadsheet in Chapter 2 after we have explored some of the important calculations of statistical analysis. For now, click on the File (Office) button and choose **Save As** and **Excel Workbook**. Enter a file name such as `grav_chloride`, and save the spreadsheet for retrieval and editing later.

An Example From Gravimetric Analysis

Let's now use some of the basics we have learned to solve a problem of gravimetric analysis. In Example 12-2 of FAC9 or 8-2 of AC7, we calculate the percentage of Fe and Fe_3O_4 in a 1.1324-g sample of an iron ore. After precipitation as $Fe_2O_3 \cdot xH_2O$, the residue was ignited to give 0.5394 g of pure Fe_2O_3.

More Cell Formatting

First, select cell A1 and type a title such as `Gravimetric Analysis Example` in bold. You can either do as before by typing the title in regular font, selecting it and clicking the Bold button, or you can click the Bold button before typing so that all subsequent typing in the active cell appears in bold. Next, in cell A2 type `mppt`. Now we'll learn how to make the abbreviation ppt appear as a subscript as in m_{ppt}. Select cell A2. In the formula bar, use the mouse to highlight (select) the `ppt` part of `mppt`. With `ppt` highlighted, Click the right mouse button and select **Format Cells** from the list. The Format Cells window shown below should appear.

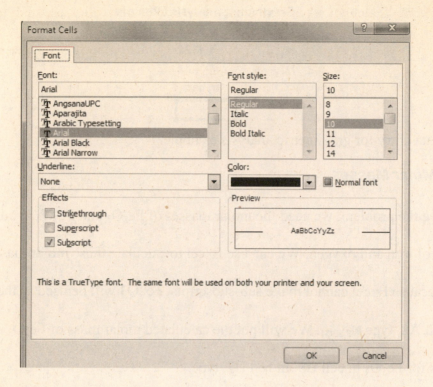

Note that since cell A2 contained only text, only the **Font** tab appears in the window.

Recall from the molar mass example that when there are numbers in the cell, the Format Cells

window contains tabs for **Number**, **Alignment**, **Font**, **Border**, **Fill**, and **Protection**. Select

Subscript in the Effects box so that a checkmark appears as shown. Click on the OK button and

note that cell A2 now contains m_{ppt} as a label for mass of precipitate. Similarly in cells A3, A4

and A5 type m_{samp}, M_{Fe}, and M_O as labels for sample mass, atomic mass of iron, and atomic mass

of oxygen.

Entering the Data

In cell B2, enter the number 0.5394 for the mass of the precipitate. In cell B3, type the number

1.1324 for the sample mass. In cells B4 and B5, type the atomic masses of iron (55.847) and

oxygen (15.9994). Your spreadsheet should now look as shown in Figure 1-10.

	A	B	C
1	Gravimetric Analysis Example		
2	m_{ppt}	0.5394	
3	m_{samp}	1.1324	
4	M_{Fe}	55.847	
5	M_O	15.9994	
6			

Figure 1-10 Data entry for gravimetric analysis example.

Calculating Molar Masses

In order to solve the problem, we need the molar masses of Fe_2O_3 and Fe_3O_4 in addition to the atomic masses of iron and oxygen. We can use Excel to calculate these molar masses. In cell A7 type M_{Fe2O3}. Because Excel cannot make sub-subscripts, Fe2O3 will be used as the subscript. Likewise in cell A8, type M_{Fe3O4}. We will put the calculated molar mass of Fe_2O_3 in cell B7 and the molar mass for Fe_3O_4 in cell B8. In cell B7, type

$$=2*B4+3*B5 \,[↵]$$

and in B8 type

$$=3*B4+4*B5 \,[↵]$$

The molar masses of Fe_2O_3 (159.692) and Fe_3O_4 (231.539) should appear in cells B7 and B8. If more than three digits beyond the decimal point are displayed, change the number format to show three digits.

Calculating the Percentages

Our final task is to use the mass of the sample, the mass of the precipitate, the molar masses, and stoichiometric information to calculate the desired percentages. Type into cells A10 and A11 the labels %Fe and %Fe$_3$O$_4$. For Fe, the following equation allows us to calculate the percentage.

$$\% \text{ Fe} = \frac{\dfrac{m_{ppt}}{\mathcal{M}_{\text{Fe}_2\text{O}_3}} \times 2\,\mathcal{M}_{\text{Fe}}}{m_{samp}} \times 100\%$$

Type into cell B10, the formula

=B2/B7*2*B4/B3*100[↵]

The calculation should return the result 33.32 for % Fe. Again adjust the number of significant

figures if too many digits are displayed.

For Fe_3O_4, the equation for the percentage is

$$\% \text{ Fe}_3\text{O}_4 = \frac{\dfrac{m_{ppt}}{\mathcal{M}_{\text{Fe}_2\text{O}_3}} \times \dfrac{2}{3} \times \mathcal{M}_{\text{Fe}_3\text{O}_4}}{m_{samp}} \times 100\%$$

Type into cell B11, the formula

=B2/B7*2/3*B8/B3*100[↵]

This should return the result 46.04 for % Fe_3O_4. Note that because these calculations involve

only multiplications and division, it is not necessary to tell Excel the order in which to do the

calculations. This *hierarchy of operations* is necessary when there is a combination of

multiplications or divisions and additions or subtractions. We will see later how to add

parentheses to tell Excel how to carry out the calculation.

Documenting the Spreadsheet

Now let's add the documentation to the spreadsheet. We will assume that cells B2 through B5

contain user-entered values and so will omit these from the documentation. It should be apparent

to someone looking at the spreadsheet and the problem that these are values to be entered. We

mainly want to document the calculations done in cells B7, B8, B10 and B11. Select cell A13 and type **Documentation[↵]** into this cell in bold. Instead of retyping the formulas entered

in cells B7, B8, B10 and B11 from scratch, there is an easy way to copy them into the

documentation cells. This shortcut also prevents typing errors in entering the formulas in the

documentation. To illustrate, select cell A14, and type **Cell B7[↵]** in this cell. Now select

cell B7 and highlight the formula displayed in the formula bar. Click on the Copy icon in the

Clipboard group on the Home tab. To prevent Excel from copying the formula and changing the

cell references, hit the Escape key on the keyboard to cancel the operation. The text copied,

however, is still in the Windows clipboard. Now select cell A14 and position the cursor after the

B7 in the formula bar. Click on the Paste icon. This will copy the formula for the molar mass of

Fe_2O_3 into cell A14 as a text string. Repeat these copy operations for cells A15 through A17.

You can also use the keyboard shortcuts **Ctrl-c** for copy and **Ctrl-v** for paste instead of

using the mouse. Your final spreadsheet should be as shown in Figure 1-11.

	A	B	C
1	**Gravimetric Analysis Example**		
2	m_{ppt}	0.5394	
3	m_{samp}	1.1324	
4	M_{Fe}	55.847	
5	M_O	15.9994	
6			
7	M_{Fe2O3}	159.692	
8	M_{Fe3O4}	231.539	
9			
10	%Fe	33.32	
11	%Fe_3O_4	46.04	
12			
13	**Documentation**		
14	Cell B7=2*B4+3*B5		
15	Cell B8=3*B4+4*B5		
16	Cell B10=B2/B7*2*B4/B3*100		
17	Cell B11=B2/B7*2/3*B8/B3*100		

Figure 1-11 Final spreadsheet for the gravimetric analysis example.

Save your spreadsheet to the disk with a file name such as **grav_analysis.xls**.

Importing and Using Molar Mass Data

In this exercise, we will learn how to import data from an external data source, to manipulate the data to obtain the desired numerical values for molar masses of the elements, to look up the appropriate values of the molar masses of the elements, and finally, to calculate molar masses of compounds.

Importing Data from Web Pages

The development of the Internet and the resulting capability for widespread storage and retrieval of textual and numerical data have made it quite easy to gain access to the latest values for molar masses, universal constants of nature, and raw data from the scientific literature. The most important potential advantage of direct import of numerical data is the elimination of human transcription errors. In addition, if you require more than a few values, importing them may save considerable time. If only a few values need to be imported, the easiest way to get the data into a spreadsheet is to use the basic editing features of all Windows programs. For example, you may simply highlight the desired number or string of characters, and click on the copy icon (or the keyboard shortcut `Ctrl-c`), and then place the cursor in the desired location in your spreadsheet, and click on the paste icon (`Ctrl-v`) or on **Paste <u>S</u>pecial…**in the Paste menu in the Clipboard group. Another way is to use the Data Import function in Excel. We will test this approach by importing atomic weight data from the Web. With a blank worksheet and the cursor in cell A1, click on the **Data** tab in Excel. At the far left in the **Data Ribbon** is a group titled **Get External Data**. Click on the **From Web** icon. Note that this opens up a New Web Query window inside Excel. Type into the address bar, the URL

http://www.chem.qmul.ac.uk/iupac/AtWt/. Go down the page to the Table of Atomic Weights

listed in atomic number order. Click on the yellow arrow next to the table to obtain.

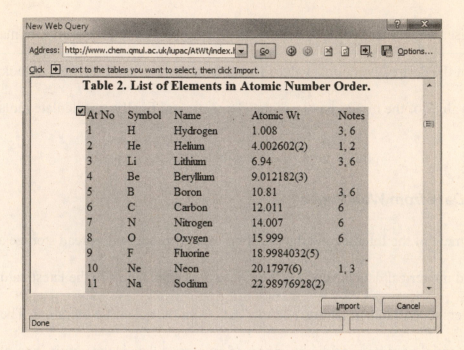

Click on **Import** and select **Existing Worksheet** and make certain =$A$1 is inserted into the

range bar. Click on **OK**. The data should now appear formatted as shown in Figure 1-12.

	A	B	C	D	E	F
1	At No	Symbol	Name	Atomic Wt	Notes	
2	1	H	Hydrogen	1.008	3, 6	
3	2	He	Helium	4.002602(2)	1, 2	
4	3	Li	Lithium	6.94	3, 6	
5	4	Be	Beryllium	9.012182(3)		
6	5	B	Boron	10.81	3, 6	
7	6	C	Carbon	12.011	6	
8	7	N	Nitrogen	14.007	6	
9	8	O	Oxygen	15.999	6	
10	9	F	Fluorine	18.9984032(5)		
11	10	Ne	Neon	20.1797(6)	1, 3	
12	11	Na	Sodium	22.98976928(2)		
13	12	Mg	Magnesium	24.3050(6)		
14	13	Al	Aluminium	26.9815386(8)		
15	14	Si	Silicon	28.085	6	
16	15	P	Phosphorus	30.973762(2)		
17	16	S	Sulfur	32.06	6	
18	17	Cl	Chlorine	35.45	3, 6	
19	18	Ar	Argon	39.948(1)	1, 2	
20	19	K	Potassium	39.0983(1)		
21	20	Ca	Calcium	40.078(4)		
22	21	Sc	Scandium	44.955912(6)		
23	22	Ti	Titanium	47.867(1)		
24	23	V	Vanadium	50.9415(1)		
25	24	Cr	Chromium	51.9961(6)		
26	25	Mn	Manganese	54.938045(5)		
27	26	Fe	Iron	55.845(2)		
28	27	Co	Cobalt	58.933195(5)		
29	28	Ni	Nickel	58.6934(4)	2	
30	29	Cu	Copper	63.546(3)	2	
31	30	Zn	Zinc	65.38(2)	2	

Figure 1-12 First 30 elements from the table of atomic weight data imported into Excel using a **Web Query**.

Scroll down the worksheet, and notice that each column is now exactly the correct width to accommodate the largest number of characters in the column and that no cells have wrapped text. Text in the cells is left justified, and numerical data are right justified.

Dealing with Character Strings

Generally, Excel is able to recognize the type of data that has been entered or imported into its cells. For example, in cell A2 of Figure 1-12, Excel has recognized the number 1, and so, it is right justified in the cell. In fact, all of the atomic numbers in column A are correctly recognized as numerical data. In cell C2, Excel recognizes that the cell contains only alphabetical characters, which are left justified. Numbers in column E without commas are interpreted as numeric while those with commas are interpreted as text.

Let us now focus on the atomic masses in column D of Figure 1-12 and note some important characteristics of the data. First, Excel has interpreted some of the data in column D as text rather than numeric data. This is because there is a digit at the end of many lines in parentheses. This digit is the uncertainty in the last place of the atomic mass. For example, we might write the atomic mass of helium as 4.002602 ± 0.000002 rather than $4.002602(2)$. Uncertainties in atomic masses can be used to compute the uncertainty in any results that are derived from atomic masses, such as molar masses of compounds. Although it is a relatively simple matter to cut (**Ctrl-x,** or **Cut** in the Clipboard group) and paste the uncertainty into another cell, it could just be deleted from each cell if it is not to be used. To see how Excel interprets the atomic masses without the parenthetical uncertainty, click on D3, copy the data, click on cell G3, and paste the data into the cell. Then click in the formula bar, and use the backspace or delete key to remove the characters "(2)", and depress the **[Enter]** key. Notice that now Excel interprets the atomic mass of helium as numeric data, and the number 4.002602 is right justified in the cell. It would be straightforward but rather tedious to perform these operations on all of the 113 atomic masses in the table, and furthermore, there would be many opportunities to delete the wrong characters and create errors in the table. Fortunately, Excel has many built-in functions that allow us to deal with situations such as the IUPAC table. Delete the number in cell G3.

Find and Replace. One way to remove, or strip, the parenthetical uncertainties in the table of

atomic masses is to use Excel's built in **Find/Replace** function. We will illustrate this approach

with a few of the entries. Copy the atomic masses from hydrogen through copper, including the

uncertainties, to Column F, cells F2:F30. Now highlight cells F2:F30. Go to the **Editing** group

on the **Home Ribbon** and click on the **Find & Select** icon. Select **Replace…** in the menu that

appears. This should bring up the **Find and Replace** window. Click on Options to display the

entire dialog box as shown below. Make sure the Replace tab is selected as shown.

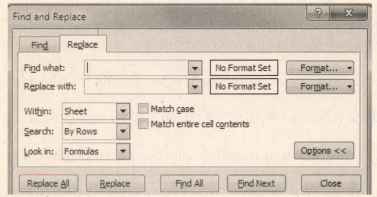

Figure 1-13 Find and Replace window.

Type **(*)** in the Find what: box and leave the Replace with: box blank. Here the

asterisk is a wild card. Choosing **(*)** as your search text means that the parentheses and

anything enclosed will be found and replaced with nothing in this case. In the Search box, choose

By Columns. Now click on Replace All. Note that entries with parentheses have been stripped of

the parenthetical characters. Now delete the 29 entries in Column F since we will not explore this

approach further.

While the Find/Replace function works well for these data, it is not as generally useful as

the built-in functions for manipulating strings of alphabetical characters and numbers. These functions are called **string functions.** We will use string functions to strip the parenthetical

uncertainties from the data in column D and produce a column of numerical atomic masses.

The Find Function. The Excel worksheet function **FIND(find_text,within_text,start_num)**

permits us to find the position in a text string of any alphanumeric character that we specify. For

example, in our table of atomic masses, it would be useful to know where the parentheses are in

each of the text strings of column D. Consider once again the atomic mass of helium represented

in cell D3 as the text string "4.002602(2)". We surround the string with quotation marks because

Excel recognizes characters in quotation marks to be strings. If we count the characters in the

string from left to right, we find that the left parenthesis is in the ninth position and the right

parenthesis is in the eleventh. Excel permits us to automatically find the position of the left

parenthesis by using the function FIND("(","4.002602(2)",1) where the string "(" is **find_text**,

the string "4.002602(2)" is **within_text,** and the number 1 is **start_num**, which is the character

position in the string where we'd like Excel to begin counting. If **start_num** is omitted, it is

assumed to be the first character of the string. Let's test the function by clicking on cell G3 and

typing

```
=FIND("(","4.002602(2)",1) [↵]
```

which produces the number 9 in cell G3 indicating that the left parenthesis occurs in the ninth

character position of the atomic mass. Now click on cell G4, and type

```
=FIND(")","4.002602(2)",1) [↵]
```

and the number 11 appears in cell G4 indicating the position of the right parenthesis. We can use

the FIND function to locate the position of any character in any string. Now instead of typing the string, we may use its cell reference, which in the case of the atomic mass of helium is D3.

Delete the results in column G. Click on cell G3 and type

```
=FIND("(",D3) [↵]
```

and once again, the number 9 appears in the cell. Note that **start_num** has been omitted because

we wish to begin the search with the first character in the string. Now use copy command to

copy cell G3 to the other cells with parenthetical characters through G14, and your worksheet

should appear as shown in Figure 1-14. After you have observed the results of finding the

parentheses in cells G3:G14, check that the resulting numbers correspond to the positions of the

left parentheses in column D. Then clear column G.

	A	B	C	D	E	F	G
1	At No	Symbol	Name	Atomic Wt	Notes		
2	1	H	Hydrogen	1.008	3, 6		
3	2	He	Helium	4.002602(2)	1, 2		9
4	3	Li	Lithium	6.94	3, 6		
5	4	Be	Beryllium	9.012182(3)			9
6	5	B	Boron	10.81	3, 6		
7	6	C	Carbon	12.011	6		
8	7	N	Nitrogen	14.007	6		
9	8	O	Oxygen	15.999	6		
10	9	F	Fluorine	18.9984032(5)			11
11	10	Ne	Neon	20.1797(6)	1, 3		8
12	11	Na	Sodium	22.98976928(2)			12
13	12	Mg	Magnesium	24.3050(6)			8
14	13	Al	Aluminium	26.9815386(8)			11

Figure 1-14 Worksheet showing the results of using the FIND function to locate the position of
the left parenthesis in each of the atomic masses of cells D3 through D14.

The MID Function. Now that we have learned how to find characters within strings, we can use

the Excel function **MID(text,start_num,num_chars)** to extract the numeric atomic mass data

from the strings of column D. The variable **text** is the string of interest, **start_num** is the

character position where we'd like the extraction to begin, and **num_chars** is the number of

characters that we'd like extracted from the string. In our example, the starting position is always

1 because the strings all begin with the first digit of the atomic mass. The number of characters

will be determined by the FIND function, which will locate the right parenthesis for us as before.

Let's try it by clicking on cell F3 and typing

`=MID(D3,1,FIND("(",D3)) [↵]`

You will notice that the atomic mass of helium appears in cell F3, but we don't quite

have it right yet because the left parenthesis appears at the end of the string. This difficulty is

easily fixed by typing **-1** at the end of the FIND function, which subtracts one from the

character position of the left parenthesis to give the last character position of the atomic mass.

Click on cell F3, then click in the formula bar at the end of the FIND function, and change the

cell contents to the following.

`=MID(D3,1,FIND("(",D3)-1) [↵]`

Now the atomic mass of helium appears as 4.002602 in the cell. All that remains is to

copy the formula in cell F3 to the other cells with parenthetical characters. If you copy into cells

in which there no parentheses exist in column D, the notation #VALUE! appears, which

indicates an error since Excel is trying to find non-existent parentheses. For cells without

parentheses, you can just copy the data values from column D and paste them in column F. Your

worksheet should now appear as shown in Figure 1-15. Scan column F, and you will see that

some atomic masses are still not displayed correctly. Problems at the end of the chapter ask you

to devise formulas to make these conversions automatically. Excel has functions that permit you

to perform checks on results of functions so that when errors occur, you can perform automatic

corrections. Note that the right-left justification is different for those atomic masses extracted and

those directly imported. Select column F and right justify all the numbers.

	A	B	C	D	E	F
1	At No	Symbol	Name	Atomic Wt	Notes	
2	1	H	Hydrogen	1.008	3, 6	1.008
3	2	He	Helium	4.002602(2)	1, 2	4.002602
4	3	Li	Lithium	6.94	3, 6	6.94
5	4	Be	Beryllium	9.012182(3)		9.012182
6	5	B	Boron	10.81	3, 6	10.81
7	6	C	Carbon	12.011	6	12.011
8	7	N	Nitrogen	14.007	6	14.007
9	8	O	Oxygen	15.999	6	15.999
10	9	F	Fluorine	18.9984032(5)		18.9984032
11	10	Ne	Neon	20.1797(6)	1, 3	20.1797

Figure 1-15 Extracting atomic masses from strings. Note that the atomic masses for the elements through Ne now appear correctly in column F.

Using VLOOKUP to Locate Data. The ultimate goal in this exercise is to calculate molar masses of compounds in a relatively straightforward and automatic fashion. Since we have the symbols for all of the elements in column B of our worksheet and the corresponding atomic masses of the elements in column F, it would be quite useful if there were a way to look up a given atomic mass by just specifying its symbol. Excel provides a convenient means to accomplish this task.

The function **VLOOKUP(lookup_value,table_array,col_index_num,range_lookup)** finds **lookup_value** in the first column of a section of a worksheet specified by **table_array** and returns the corresponding contents in the column indicated by **col_index_num**. Let's now use this function to look up the atomic mass of fluorine. Begin by clicking in cell G1 and typing

```
Element[→]
```

where [→] indicates the Right cursor key.

```
No. Atoms[→]
```

```
At. Mass[↵]
```

```
=VLOOKUP("F",$B$2:$F$119,5,FALSE) [↵]
```

Your worksheet should look like the one shown in Figure 1-16 with the molar mass of fluorine displayed in cell I2. Excel has looked up the atomic mass of fluorine (specified by its symbol "F" as the **lookup_value)** in the rectangular region of the worksheet specified by the variable **table_array,** which in this example is B2:F119. This region, or **array,** contains the atomic symbols in the first column of the array (column B in the worksheet) and the extracted atomic masses in the fifth column (column F in the worksheet). Hence, **col_index_num** is set to 5 in the function to indicate that we want the atomic mass in the fifth column of the array. Excel assumes that the lookup value is contained in the first column of the array. The logical variable **range_lookup,** which is set to FALSE here, tells Excel that the match between the atomic symbol being sought and the result must be *exact.* If this variable is set to TRUE, VLOOKUP will find an approximate match. If no match is found, an error results. Try several different element symbols in the VLOOKUP function in cell I2, and note the results. Note that the dollar signs in the array tell Excel that this range is absolute and should not change when we change rows or columns. Absolute and relative references are further discussed in Chapter 2.

	A	B	C	D	E	F	G	H	I	J
1	At No	Symbol	Name	Atomic Wt	Notes		Element	No. atoms	At. Mass	Mass
2	1	H	Hydrogen	1.008	3, 6	1.008			18.9984032	
3	2	He	Helium	4.002602(2)	1, 2	4.002602				
4	3	Li	Lithium	6.94	3, 6	6.94				
5	4	Be	Beryllium	9.012182(3)		9.012182				
6	5	B	Boron	10.81	3, 6	10.81				
7	6	C	Carbon	12.011	6	12.011				
8	7	N	Nitrogen	14.007	6	14.007				
9	8	O	Oxygen	15.999	6	15.999				
10	9	F	Fluorine	18.9984032(5)		18.9984032				

Figure 1-16 Using VLOOKUP to look up and display the atomic mass of fluorine.

Now we will generalize the lookup function so that we can look up the atomic mass of any element by simply typing its symbol into a cell. Click on cell I2, click in the formula bar, and edit the contents to read as follows:

```
=VLOOKUP(G2,$B$2:$F$119,5,FALSE) [↵]
```

The error condition #N/A then appears in cell I2 because G2 is blank and thus contains no element symbol. Click on cell G2, and type **Fe**. The atomic mass of iron now appears in cell I2. Try typing several other element symbols in cell G2, and note the results. When you are satisfied that the LOOKUP function is working properly, click on cell I2, and copy the contents into cell I3 using the fill handle. Then type various element symbols in cell G3, and your worksheet should look something like the one shown in Figure 1-17.

G	H	I	J
Element	No. atoms	At. Mass	Mass
Fe		55.845	
S		32.06	

Figure 1-17 The atomic masses of any two elements can be looked up by typing their symbols in cells G2 and G3.

Making the Calculation

The last step in our exercise is to create formulas that will calculate the molar mass of a compound from the atomic masses looked up by the functions in cells G2 and G3. We will confine ourselves to binary compounds for now and leave more complex cases for the problems at the end of the chapter. Let's calculate the molar mass of NaCl. Begin by clicking on cell G2, and type

```
Na [→]
```

```
1 [↵]
```

    1[←]

    Cl[↵]

Your worksheet should now display the atomic masses of Na and Cl in cells I2 and I3,

respectively, and the number 1 in cells H2 and H3 to indicate the number of atoms of each

element in the formula of NaCl. Now click on cell J1 and enter the following.

    Mass[↵]

    =H2*I2[↵]

Copy the formula from J2 to J3 using the fill handle, and enter the following equation in cell J4.

    =J2+J3[↵]

This formula adds the contents of cells J2 and J3, which contain the total mass of Na and Cl, and

displays the molar mass of NaCl in cell J4. Your worksheet should now be similar to the one

shown in Figure 1-18. Test this worksheet with several binary compounds in the table of molar

masses in the inside back cover of this textbook. Check the molar masses from the worksheet

against those that you find in the table. Note that Excel has no automatic way to keep track of

significant figures, and so molar masses calculated using your worksheet must be rounded to

reflect only those digits that are significant. To finish this activity, save your worksheet to a disk

or flash drive for future activity using a descriptive file name.

G	H	I	J
Element	No. atoms	At. Mass	Mass
Na	1	22.98976928	22.98977
Cl	1	35.45	35.45
			58.43977

Figure 1-18 Calculating the molar mass of NaCl. The worksheet is now general for binary compounds. Type the symbol for the first element in cell G1 and the number of atoms of the element in H1. Type the symbol and the number of atoms of the second element in G2 and H2. The molar mass of the compound is displayed in cell J4.

Summary

In this chapter, we have learned some of the basics of spreadsheet operation. We have typed text and data into a spreadsheet, formatted text and numbers, changed column widths, duplicated cells with the fill handle, entered formulas, and learned how to incorporate documentation. We have also learned how to import data from external sources such as the Web and use string functions to treat the imported data. We have developed a spreadsheet for calculating molar masses of compounds that will be useful in many applications. The remaining chapters build on the skills we have learned here and introduce many additional Excel functions and operations useful in analytical chemistry.

Problems

1.   Describe the use of the following Excel functions after reading about them in the Excel

   help facility.

   (a)   SQRT

   (b)   SUM

   (c)   PI

   (d)   FACT

(e)   EXP

(f)   LOG

2.   Use the Excel help facility to look up the use of the COUNT function. Use the function to

determine the number of data in each column of the spreadsheet of Figure 1-9. The count

function is quite useful for determining the number data entered into a given area of a

spreadsheet.

3.   Write an Excel FIND function to eliminate the square from the atomic mass of

technetium in the Atomic Mass table and display the numeric characters of the atomic

mass.

4.   Devise an Excel formula for elements 43, 61, 84-89, and 93-118 that will automatically

remove the square brackets from the table of atomic masses and eliminate the #VALUE!

error described in conjunction with Figure 1-15.

5.   Devise an Excel formula that will automatically convert the uncertainties in parentheses

for many elements into numerical values like 0.000002 for helium.

6.   Use the worksheet of Figure 1-18 to calculate the molar masses of the following

compounds.

(a) HCl

(b) NH_3

(c) ZnS

(d) $AgCl$

(e) $PbCl_2$

(f) Bi_2O_3

(g) Al_2O_3

7.   Modify the Excel worksheet of Figure 1-18 to compute the molar mass of compounds

      containing (a) three elements and (b) five elements.

8.   Modify the worksheet of Figure 1-18 to calculate the molar masses of the following

      compounds.

   (a) Na_2SO_4

   (b) $Ba(IO_3)_2$

   (c) CaC_2O_4

   (d) $KMnO_4$

   (e) $K_4Fe(CN)_6$

   (f) $Na_2S_2O_3 \cdot 5H_2O$

Chapter 2

Basic Statistical Analysis with Excel

Excel is an extremely versatile tool for statistical analysis. With a set of data, the user can readily calculate such descriptive statistics as the mean, the standard deviation, the median, and the range. We show here how Excel can be used to find these basic statistical quantities and to obtain confidence intervals for the means of data sets. Chapter 3 illustrates statistical tests with Excel and Analysis of Variance (ANOVA).

Calculating a Mean with Excel

In this example, we learn to calculate the mean of a set of data. First, we define formulas to calculate the mean, and then we use the built-in functions of Excel to accomplish the task. The data used are from six replicate determinations of iron (Example 5-1 of FAC9 and Example 5-1 of AC7). The results (x_i values) were 19.4, 19.5, 19.6, 19.8, 20.1, and 20.3 ppm iron(III).

Entering the Data

Let's begin by starting Excel with a clean spreadsheet. In cell B1 enter the heading **Data[↵]** and make it boldface. Now enter in column B under the heading the results given above. Click on cell A11, and type

Total[↵]

N[↵]

Mean[↵]

Make these labels boldface. Your worksheet should now look like that shown in Figure 2-1.

	A	B	C
1		Data	
2		19.4	
3		19.5	
4		19.6	
5		19.8	
6		20.1	
7		20.3	
8			
9			
10			
11	Total		
12	N		
13	Mean		
14			

Figure 2-1 Data entry for statistical calculations.

Finding the Mean

The mean is calculated according to the equation

$$\bar{x} = \frac{\sum_{i=1}^{N} x_i}{N}$$

Click on cell B11, and type

$$\text{=SUM(B2:B7) } [\hookleftarrow]$$

This formula calculates the sum of the values in cells B2 through B7 and displays the result in cell B11. Now, we will obtain the number of data points N in cell B12 by typing

$$\text{=COUNT(B2:B7) } [\hookleftarrow]$$

The **COUNT** function counts the number of nonzero cells in the range B2:B7 and displays the result in cell B12. Since we have found the sum of the values and the number N of data points, we can find the mean $\bar{x}$ of the data set by typing the following formula in cell B13.

$$\text{=B11/B12} [\hookleftarrow]$$

At this point in the exercise, your worksheet should appear as shown in Figure 2-2.

	A	B	C
4		19.6	
5		19.8	
6		20.1	
7		20.3	
8			
9			
10			
11	Total	118.7	
12	N	6	
13	Mean	19.7833	
14			

Figure 2-2 Spreadsheet after computing the sum, count and mean values.

Note that the mean contains many more figures after the decimal point than any of the data values. It is appropriate, as discussed in FAC9, Section 6D and AC7, Section 6D, to round the mean to 19.8 either by invoking the **Format Cells** window or by clicking on the Decrease Decimal button (see Chapter 1).

Using Excel's Built-in Functions

Excel has built-in functions to compute many important and useful quantities. Now we shall see how to use them to calculate the mean, or in Excel's syntax, the average. Click on cell C13 and type

$$=\text{AVERAGE (B2:B7)} \ [\leftarrow]$$

Notice that the mean determined using the built-in **AVERAGE** function is identical to the value in cell B13 that you determined by typing a formula. Again format the result to contain only 1 figure after the decimal point. Before proceeding or terminating your Excel session, save your file to a storage device as `average.xls`.

Finding the Deviations from the Mean

We can now use Excel to determine the deviations d_i of each of the data values from the mean using the definition

$$d_i = |x_i - \bar{x}|$$

Click on cell C1 and type the boldface label

Deviation[↵]

With cell C2 active, type

=ABS(B2-$B$13) [↵]

This formula computes the absolute value **ABS()** of the difference between our first value in B2 and the mean value in B13. Note that this formula is a bit different than those that we have used previously. We have typed a dollar sign, $, before the B and before the 13 in the second cell reference. This type of cell reference is called an *absolute reference.* This syntax means that no matter where we might copy the contents of the cell C2, the reference will always be to cell B13. The other type of cell reference that we consider here is the *relative reference*, which is exemplified by B2. The reason that we use a relative reference for B2 and an absolute reference for B13 is that we want to copy the formula in C2 into cells C3-C7, and we want the mean referred to by **$B$13** to be subtracted from each of the successive values in column B. Let's now copy the formula by clicking on cell C2, clicking on the fill handle, and dragging the rectangle through C7. When you release the mouse button, your worksheet should look like that shown in Figure 2-3.

	A	B	C
1		Data	Deviation
2		19.4	0.383333
3		19.5	0.283333
4		19.6	0.183333
5		19.8	0.016667
6		20.1	0.316667
7		20.3	0.516667
8			
9			
10			
11	Total	118.7	
12	N	6	
13	Mean	19.8	19.8

Figure 2-3 Worksheet with deviations computed.

Now click on cell C3, and notice that it contains the formula **=ABS(B3-$B$13)**.

Compare this formula with the one in cell C2 and C4-C7. The absolute cell reference **$B$13**

appears in all of the cells. As you can see, we have accomplished our task of calculating the

deviation from the mean for all of the data. Now we will edit the formula in cell C13 to find the

mean deviation of the data.

Editing Formulas

To edit the formula to calculate the mean deviation of the data, click on C13, and then click on

the formula in the formula bar. Use the arrow keys, **[←]** and **[→]**, and either the

[Backspace] or the **[Delete]** key to replace both Bs in the formula with Cs so that it reads:

=AVERAGE(C2:C7). Finally, type **[↵]**, and the mean deviation will appear in cell C13. Your

worksheet should appear as in Figure 2-4. Save the file by clicking on the save icon in the Quick

Access toolbar, by clicking the File (Office) button and saving, or by typing **[Ctrl+S]**.

	A	B	C	D
1		Data	Deviation	
2		19.4	0.383333	
3		19.5	0.283333	
4		19.6	0.183333	
5		19.8	0.016667	
6		20.1	0.316667	
7		20.3	0.516667	
8				
9				
10				
11	Total	118.7		
12	N	6		
13	Mean	19.8	0.3	
14				

Figure 2-4 Worksheet showing deviations and mean deviation.

Computing the Standard Deviation

In this exercise, we will calcuate the sample standard devation, the sample variance, and the relative standard deviation of two sets of data. We begin with the spreadsheet and data from the previous example. The sample standard deviation *s* is given by the equation

$$s = \sqrt{\frac{\sum_{i=1}^{N}(x_i - \bar{x})^2}{N-1}}$$

and the variance is s^2.

Finding the Variance

If you are continuing the previous exercise, begin with the worksheet shown in Figure 2-4 on your computer screen. Otherwise, retrieve the file **average.xls** from your disk by clicking on the Office button and opening the file. Make cell D1 the active cell, and type the label

```
Deviation^2[↵]
```

Make the label boldface. With cell D2 the active cell, type

$$\text{=C2^2 [↵]}$$

and the square of the deviation in cell C2 appears in D2. Copy this formula into the other cells in

column D by once again clicking on cell D2, clicking on the fill handle, and dragging the fill

handle through cell D7. You have now calculated the squares of the deviations of each of the

data from the mean value in cell B13.

A Shortcut for Performing a Summation

To find the variance, we must first find the sum of the squares of the deviations, so now click on

cell D11, and then click on the AutoSum icon in the Editing group on the Home ribbon. The

worksheet should now appear as shown in Figure 2-5

	A	B	C	D	E	F
1		Data	Deviation	Deviation^2		
2		19.4	0.383333	0.146944		
3		19.5	0.283333	0.080278		
4		19.6	0.183333	0.033611		
5		19.8	0.016667	0.000278		
6		20.1	0.316667	0.100278		
7		20.3	0.516667	0.266944		
8						
9						
10						
11	Total	118.7		=SUM(D2:D10)		
12	N	6		SUM(number1, [number2], ...)		
13	Mean	19.8	0.3			

Figure 2-5 Worksheet with AutoSum function selected.

The dashed box shown above now surrounds the column of data in cells D2-D10, which appear

as arguments of the SUM function in cell D11 and in the formula bar. Note that Excel assumes

that you want to add all of the numerical data above the active cell and automatically completes

the formula. When you type [↵], the sum of the squares of the deviations appears in cell D11.

Since cells D8 through D10 are blank, they contribute nothing to the sum, and so there is no harm in leaving the references to blank cells in the formula. Be aware, however, that references

to blank cells could pose difficulty under certain circumstances.

The final step in calculating the variance is to divide the sum of the squares of the

deviations by the number of degrees of freedom, which is $N - 1$. We'll type the formula for

carrying out this last calculation in cell D12. Before proceeding, type the label **Variance** in

F12 in boldface font. Now click on D12, and type

$$\texttt{=D11/(B12-1) [↵]}$$

Excel calculates the variance and it appears in the cell. Notice that you must enclose the

difference $B12 - 1$ in parentheses so that Excel computes the number of degrees of freedom

before the division is carried out. If we had not enclosed the number of degrees of freedom, B12

$- 1$, in parentheses, Excel would have divided D11 by B12 and then subtracted one, which is

incorrect. To illustrate this point, suppose $D11 = 12$ and $B12 = 3$. If we leave off the parentheses,

$D11/B12 - 1 = 3$, but if we put them in, $D11/(B12 - 1) = 6$. The order of mathematical operations

in Excel is extremely important. Remember that just as in algebra, Excel performs

exponentiation before multiplication and division, and it performs multiplication and division

before addition and subtraction. As in the present example, we can change the order of

operations by properly placing parentheses. The order that Excel uses in evaluating various

mathematical and logical operations is shown in the table that follows.

Order of Operations		
Order	Operator	Description
1	–	Negation
2	%	Percent
3	^	Exponentiation
4	* and /	Multiplication and Division
5	+ and -	Addition and subtraction
6	= , <, >, <=, >=, <>	Comparison

Finding the Standard Deviation

Our next step is to calculate the standard deviation by extracting the square root of the variance.

Click on D13, and type

$$=SQRT(D12)\ [\lrcorner]$$

Then click on F13, and type in bold

$$Standard\ Devation[\lrcorner]$$

Your worksheet should then appear similar to Figure 2-6.

	A	B	C	D	E	F	G
1		Data	Deviation	Deviation^2			
2		19.4	0.383333	0.146944			
3		19.5	0.283333	0.080278			
4		19.6	0.183333	0.033611			
5		19.8	0.016667	0.000278			
6		20.1	0.316667	0.100278			
7		20.3	0.516667	0.266944			
8							
9							
10							
11	Total	118.7		0.628333			
12	N	6		0.125667		Variance	
13	Mean	19.8	0.3	0.354495		Standard Deviation	
14							

Figure 2-6 Worksheet after computing variance and standard deviation.

Notice that we have deliberately left cells E12 and E13 blank. We will now use the built-in variance and standard deviation functions of Excel to check our formulas.

The Built-in Statistical Functions of Excel

Click on cell E12, and then type

=VAR (

Now click in cell B2, and drag the mouse into cell B7 so that the worksheet appears as shown in Figure 2-7.

	A	B	C	D	E	F	G
1		Data	Deviation	Deviation^2			
2		19.4	0.383333	0.146944			
3		19.5	0.283333	0.080278			
4		19.6	0.183333	0.033611			
5		19.8	0.016667	0.000278			
6		20.1	0.316667	0.100278			
7		20.3	0.516667	0.266944			
8							
9							
10							
11	Total	118.7		0.628333			
12	N	6		0.125667	=VAR(B2:B7		
13	Mean	19.8	0.3	0.354495	VAR(**number1**, [number2], ...) n		

Figure 2-7 Defining the range for the variance (VAR) function.

Notice that the cell references B2:B7 appear in cell E12 and in the formula bar. Now, let up on the mouse button, and type **[↵]**, and the variance appears in cell E12. If you have performed these operations correctly, the values displayed in cells D12 and E12 are identical.

The active cell should now be E13. If it is not, click on it, and type

=STDEV (

and click and drag to highlight cells B2:B7 as you did previously. Let up on the mouse button,

type [↵], and the standard deviation appears in cell E13. The computed values in cells D13 and

E13 should be equal although they may show different numbers of decimal places. It is important

to note that the Excel STDEV and VAR functions calculate the *sample standard deviation* and

the *sample variance,* not the corresponding population statistics. These built-in functions are

quite convenient since your sample will generally be sufficiently small that you will want to

calculate sample statistics rather than population statistics. Excel 2010 has the functions VAR,P,

VAR,S, STDEV,P and STDEV,S for the population and sample variances and standard

deviations, respectively. The functions VAR and STDEV are retained for compatability with

Excel 2007 and earlier versions.

Up to this point, we haven't paid much attention to the number of decimal places

displayed in cells D12, E12, D13, and E13. To control the number of decimal places in a cell or

range of cells, highlight the target cell(s), and click on the Increase Decimal icon. Highlight cells

D13 and E13 now, and try it. Then click on the Decrease Decimal icon to reverse the process.

Now decrease the number of decimal places until only one significant figure is displayed. Note

that Excel conveniently rounds the data for us.

The Coefficient of Variation, or Percent Relative Standard Deviation

We will now calculate the coefficient of variation (CV), also known as the percent relative

standard deviation (%RSD). The CV is given by

$$CV = \frac{s}{\bar{x}} \times 100\%$$

Click in cell E14, and type

=E13*100/B13[↵]

Then click in cell F13 and type the label **CV, %[↵]**. Note that we have multiplied the ratio of E13 to B13 by 100 so that the relative standard deviation is expressed as a percentage. Move the decimal point to indicate only two significant figures in the CV.

Standard Error of the Mean

We will now calculate the standard error of the mean s_m from the equation

$$s_m = \frac{s}{\sqrt{N}}$$

Click in cell F15 and type the label **Standard Error of the Mean[↵]** in bold. In cell E15, type

$$\texttt{=E13/SQRT(B12) [↵]}$$

The standard error of the mean should now appear in cell E15. Format this result to contain 1 digit after the decimal place. Your worksheet should now look like that shown in Figure 2-8.

	A	B	C	D	E	F	G
1		Data	Deviation	Deviation^2			
2		19.4	0.383333	0.146944			
3		19.5	0.283333	0.080278			
4		19.6	0.183333	0.033611			
5		19.8	0.016667	0.000278			
6		20.1	0.316667	0.100278			
7		20.3	0.516667	0.266944			
8							
9							
10							
11	Total	118.7		0.628333			
12	N	6		0.1	0.1	Variance	
13	Mean	19.8	0.3	0.4	0.4	Standard Deviation	
14					1.8	CV, %	
15					0.1	Standard Error of Mean	
16							

Figure 2-8 Final worksheet for computing basic statistics.

We have now constructed a general-purpose spreadsheet that you may use to make basic statistical calculations. To complete this part of the exercise, select a convenient location, construct a formula to display the number of degrees of freedom, and then add a label in an adjacent cell to identify this important variable. Save the file for future use in problems and laboratory calculations. You can use the spreadsheet for other data sets by deleting or overtyping the data in cells B2:B7. To clear the data from your worksheet, just click and drag to highlight cells B2:B7, and strike **[Delete]** or choose Cle<u>a</u>r All from the Editing group on the Home ribbon. Alternatively, you may simply click on B2, and begin typing the data. Terminate each data entry with **[↵]**. Note that room is provided for 9 data points in the worksheet. If you need more rows, select cell A10 and then click on the Insert command in the Cells group on the Home ribbon. From the pull-down menu, select Insert Sheet <u>R</u>ows. You will have to extend any formulas from row 7 to include any added values.

As a final example of standard deviations, retrieve the spreadsheet that we created in Chapter 1 for the gravimetric determination of chloride, which we called **grav_chloride.xls**. Enter formulas into cells B12-B14 to compute the mean, standard deviation, and the RSD in parts per thousand of the percent chloride in the samples. In this example, multiply the relative standard deviation by one thousand in cell B14. Adjust the decimal point in the results to display the proper number of significant figures. Figure 2-9 shows the results. You can also document your worksheet by the method described in Chapter 1. Save your worksheet so that you can use it as a model for making laboratory calculations.

	A	B	C	D
1	Gravimetric Determination of Chloride			
2	Samples	1	2	3
3	Mass of bottle plus sample, g	27.6115	27.2185	26.8105
4	Mass of bottle less sample, g	27.2185	26.8105	26.4517
5	Mass of sample, g	0.393	0.408	0.3588
6				
7	Crucible masses, with AgCl, g	21.4296	23.4915	21.8323
8	Crucible masses, empty, g	20.7926	22.8311	21.2483
9	Mass of AgCl, g	0.637	0.6604	0.584
10				
11	%Chloride	40.0946	40.0393	40.2624
12	Mean %Chloride	40.1321		
13	Standard Deviation, %Chloride	0.12		
14	RSD, parts per thousand	2.90		
15				

Figure 2-9 Worksheet for gravimetric determination of chloride begun in Chapter 1.

Computing Descriptive Statistics with the Analysis ToolPak

There is yet another way to compute the mean, the standard deviation and other statistical quantities. This method takes advantage of the Descriptive Statistics package in Excel's Analysis ToolPak. The advantage of this method is that the formulas and calculations are done automatically. You will need the file **average.xls** for this exercise.

The Analysis ToolPak

On the Data ribbon, select Data Analysis in the Analysis group. The window shown below should appear if the ToolPak has been installed and enabled. Select Descriptive Statistics and click OK.

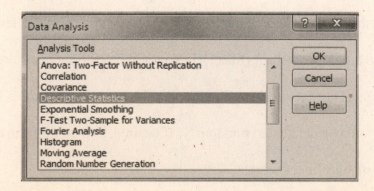

If the Analysis ToolPak has not been installed, the Excel Setup program must be run to install it. If it has been installed, but not enabled, the ToolPak must be enabled by clicking on Excel Options under the File (Office) button. In the Excel options menu, click on **Add-Ins** and select the Analysis ToolPak. After installation and enabling, the Analysis group should appear on the Data ribbon.

With the Descriptive Statistics window open, click and drag to highlight cells B2:B7 as shown in Figure 2-10. Note that these cells appear in the Input Range: box.

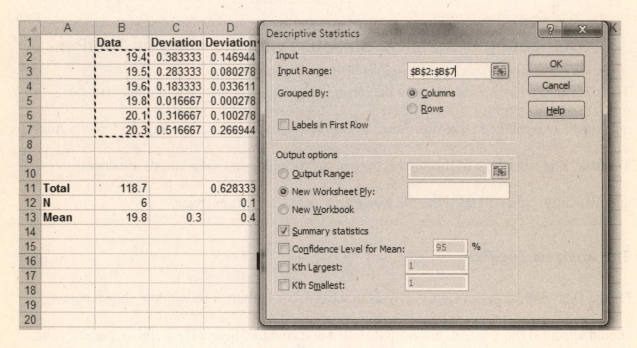

Figure 2-10 Worksheet with Descriptive Statistics Window

Click on New Worksheet Ply:, select Summary Statistics and click on OK. This action inserts the statistical results into a new worksheet (Sheet 4). You will have to increase the width of column A to read all the labels. The statistical results should be as shown in Figure 2-11. Note that the results are identical to those calculated previously. If desired, you could format the cells with the appropriate number of significant figures.

	A	B	C
1	*Column1*		
2			
3	Mean	19.78333	
4	Standard Error	0.144722	
5	Median	19.7	
6	Mode	#N/A	
7	Standard Deviation	0.354495	
8	Sample Variance	0.125667	
9	Kurtosis	-1.35279	
10	Skewness	0.568672	
11	Range	0.9	
12	Minimum	19.4	
13	Maximum	20.3	
14	Sum	118.7	
15	Count	6	

Figure 2-11 Descriptive statistics.

Note that Excel produces some additional statistics such as the median, the range, and the minimum and maximum values. Excel also produces some measures of the symmetry of the data distribution called the *kurtosis* and the *skewness*. The skewness is negative for a distribution that tails to the left and positive for one that tails to the right. Kurtosis is a measure of the amount of peakedness in the distribution. The detailed descriptions of kurtosis and skewness are beyond the scope of this discussion.[1]

Calculating the Pooled Standard Deviation

In this example, we calculate the pooled standard deviation of the four sets of glucose results from Example 6-2 of FAC9. We'll use the Equation 6-7 given in Feature 6-4 of FAC9.

$$s_{pooled} = \sqrt{\frac{\sum_{i=1}^{N_1}(x_i - \bar{x}_1)^2 + \sum_{j=1}^{N_2}(x_j - \bar{x}_2)^2 + \sum_{k=1}^{N_3}(x_k - \bar{x}_3)^2 + \cdots}{N_1 + N_2 + N_3 + \cdots - N_t}}$$

Enter the data and column and row headings as shown in Figure 2-12.

---

[1] See http://en.wikipedia.org/wiki/Kurtosis and http://en.wikipedia.org/wiki/Skewness for more detail.

	A	B	C	D	E	F	G
1		Month 1	Month 2	Month 3	Month 4		
2	x_1	1108	992	788	799		
3	x_2	1122	975	805	745		
4	x_3	1075	1022	779	750		
5	x_4	1099	1001	822	774		
6	x_5	1115	991	800	777		
7	x_6	1083			800		
8	x_7	1100			758		
9	Average						
10	N						No. Sets
11	DF						Total DF
12	Dev^2						Total Dev^2
13					S_{pooled}		

Figure 2-12 Data entry for glucose example.

You can find the mean (average) values of the first data set by clicking on cell B9 and typing

$$\texttt{=AVERAGE (B2:B8) [↵]}$$

You can find the number of measurements N in cell B10 and the number of degrees of freedom

in cell B11 by typing

$$\texttt{=COUNT (B2:B8) [↵]}$$

$$\texttt{=B10-1 [↵]}$$

Finding the Sum of the Squares of the Deviations

To compute the pooled standard deviation, it is necessary to find the sum of the squares of the

deviations from the mean for each of the four data sets. There are two ways to do this. We can

take advantage of the VARIANCE function and use it to find the sum of the squares as follows.

$$s^2 = \frac{\sum\limits_{i=1}^{N}(x_i - \bar{x})^2}{N-1}$$

$$\sum\limits_{i=1}^{N}(x_i - \bar{x})^2 = s^2(N-1)$$

So, for each of the data sets, we could compute the variance and multiply it by the number of degrees of freedom $N - 1$, which we found by using the COUNT function.

The second way takes advantage of Excel's built-in function DEVSQ() to calculate the sum of the squares of the deviations directly. With the cursor in cell B12, type

=DEVSQ(B2:B8) [↵]

The first two columns of your worksheet should look like those shown in Figure 2-13 after adjusting the number of decimal places displayed.

	A	B
1		**Month 1**
2	x_1	1108
3	x_2	1122
4	x_3	1075
5	x_4	1099
6	x_5	1115
7	x_6	1083
8	x_7	1100
9	Average	1100.3
10	N	7
11	DF	6
12	Dev^2	1687.43

Figure 2-13 First two columns of pooled standard deviation worksheet.

Now, click and drag from B9 to B12 to highlight those cells, and click on the fill handle, and

drag to the right to copy the formulas into columns C through E. When you let up on the mouse

button, your worksheet should appear similar to Figure 2-14.

	A	B	C	D	E	F	G
1		Month 1	Month 2	Month 3	Month 4		
2	x_1	1108	992	788	799		
3	x_2	1122	975	805	745		
4	x_3	1075	1022	779	750		
5	x_4	1099	1001	822	774		
6	x_5	1115	991	800	777		
7	x_6	1083			800		
8	x_7	1100			758		
9	Average	1100.3	996.2	798.8	771.9		
10	N	7	5	5	7	No. Sets	
11	DF	6	4	4	6	Total DF	
12	Dev^2	1687.43	1182.80	1086.80	2950.86	Total Dev^2	
13					S_{pooled}		

Figure 2-14 Pooled standard deviation worksheet after computing averages and sums of squares.

Note that the sums of the squares of the deviations (**Dev^2**) from each of the four means are

displayed in row 12.

Finding the Total Sum of Squares

To find the total sum of squares, click on cell F12, and then click on the AutoSum icon in the

Editing group on the Home tab to produce the following on the screen.

	A	B	C	D	E	F	G
9	Average	1100.3	996.2	798.8	771.9		
10	N	7	5	5	7	No. Sets	
11	DF	6	4	4	6	Total DF	
12	Dev^2	1687.43	1182.80	1086.80	2950.86	=SUM(B12:E12)	^2
13					S_{pooled}	SUM(**number1**, [number2], ...)	

Figure 2-15 Finding the total sum of squares.

Notice that Excel correctly guesses that you want to find the sum of the data in cells B12:E12, so type [↵], and the total sum of squares appears in cell F12. Now click on F12, and drag the fill handle up to copy the sum formula into cell F11. When you let up on the mouse button, the total number of degrees of freedom appears in F11. You could also have found the total number of degrees of freedom by summing the values in cells B10:E10 and subtracting the number of data sets (4) as in the Equation in Feature 6-4 of FAC9 or Feature 6-4 of AC7.

You have now made all of the intermediate calculations necessary to compute the pooled standard deviation. Now click on cell F13, and type

$$=\text{SQRT(F12/F11)} \text{ [↵]}$$

and the pooled standard deviation appears in the cell. For completeness, add a formula to cell F10 to find the number of data sets in the analysis. This is a good time to save your worksheet to the disk under the name **pooled_std_dev.xls**. The completed worksheet is shown in Figure 2-16 after documentation.

	A	B	C	D	E	F	G
1		Month 1	Month 2	Month 3	Month 4		
2	x_1	1108	992	788	799		
3	x_2	1122	975	805	745		
4	x_3	1075	1022	779	750		
5	x_4	1099	1001	822	774		
6	x_5	1115	991	800	777		
7	x_6	1083			800		
8	x_7	1100			758		
9	Average	1100.3	996.2	798.8	771.9		
10	N	7	5	5	7	4	No. Sets
11	DF	6	4	4	6	20.00	Total DF
12	Dev^2	1687.43	1182.80	1086.80	2950.86	6907.89	Total Dev^2
13					s_{pooled}	18.58479	
14							
15	Documentation						
16	Cell B9=AVERAGE(B2:B8)			Cell F10=COUNT(B10:E10)			
17	Cell B10=COUNT(B2:B8)			Cell F11=SUM(B11:E11)			
18	Cell B11=B10-1			Cell F12=SUM(B12:E12)			
19	Cell B12=DEVSQ(B2:B8)			Cell F13=SQRT(F12/F11)			

Figure 2-16 Completed pooled standard deviation worksheet.

In this exercise, we have constructed a worksheet to calculate the pooled standard deviation of several sets of data. As extensions of this exercise, you may use the worksheet to solve other pooled standard deviation problems. You may also wish to expand the worksheet to accommodate more data points within data sets and larger numbers of sets.

Confidence Intervals

Excel allows us to calculate confidence intervals both for situations where the population standard deviation σ is known and for those where we must use the sample standard deviation s as an estimate of σ.

Confidence Intervals When σ is Known

Excel has a built-in statistical function **CONFIDENCE.NORM() [CONFIDENCE()** in Excel 2007] that we can use when we know the population standard deviation σ. We will use the Month 1 glucose data from the pooled standard deviation example and do the calculations of Example 7-1 of FAC9. Open the file **pooled_std_dev.xls**. Copy the data from Month 1 by selecting cells B2:B8 and clicking on the Copy icon in the Clipboard group on the Home ribbon. Open a new worksheet, by clicking on the Sheet2 tab as shown in Figure 2-17.

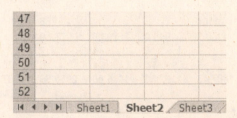

Figure 2-17 Opening a new worksheet by clicking on the Sheet2 tab.

Now paste the values into the new worksheet by selecting cell B2 and clicking on the Paste icon in the Clipboard group. This should insert the seven values into cells B2:B8 of Sheet2.

Select cell B10 and click on the Insert Function icon in the formula bar. This opens the Insert

Function window, shown in Figure 2-18. Select Statistical from the category: list.

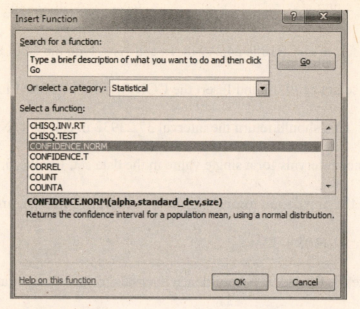

Figure 2-18 Insert Function window.

Select CONFIDENCE,NORM from the function: list (just CONFIDENCE in Excel 2007). Click

on OK and this opens the Function Arguments window shown in Figure 2-19.

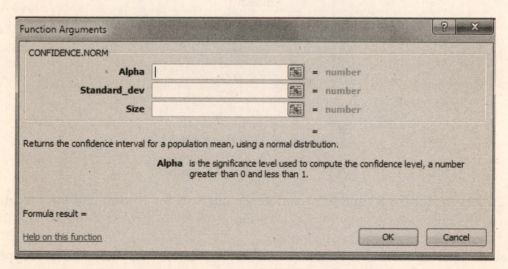

Figure 2-19 Function Arguments window for CONFIDENCE,NORM.

We will now do the first part of Example 7-1 of FAC9 in which we calculate the 80% and 95% confidence intervals for the first entry 1108 mg/L glucose. In the Function Arguments window choose 0.20 for alpha, 19 for the Standard_dev, and 1 for the Size (size here means sample size). Click on OK. The result in cell B10 is 24.34948. In cell A10, type the label **80% CI**. In cell A11 type **95% CI**. Now select cell B11 and Insert the CONFIDENCE,NORM function as before, but chose 0.05 for alpha. This should return the interval 37.23932 in cell B11. Note that these are the 80 and 95% confidence intervals for a single value in the data set. For the first value, there is 80% probability that the population mean μ lies in the interval 1108 ± 24, and 95% probability that μ lies in the interval 1108 ± 37.

Now compute the 80 and 95% confidence intervals for the mean of all seven measurements (1100.3) in cells C10 and C11. Your worksheet should appear as in Figure 2-20 after labeling the cells. Save your work as the file **pooled_std_dev.xls**.

	A	B	C
1		**Month 1**	
2		1108	
3		1122	
4		1075	
5		1099	
6		1115	
7		1083	
8		1100	
9			
10	**80% CI**	24.34948	9.203238
11	**95% CI**	37.23932	14.07514
12		**1 result**	**7 results**

Figure 2-20 Results for the 80 and 95% confidence intervals for 1 result and for 7 results.

Instead of using the Insert Function icon on the formula bar, you could have typed into cell B10 the function **=CONFIDENCE,NORM(0.20,19,1) [↵]**. However, it is relatively easy to make mistakes in typing in complicated expressions and usually more reliable to use Excel's built-in facility.

Confidence Intervals When s Must be Used

When σ is unknown and s must be used, the t statistic is used to calculate the confidence interval instead of the z value. Let's take the same example from the file **pooled_std_dev.xls** as in the previous exercise. Choose Sheet2 from the workbook. You can find the confidence interval from

$$CI = \bar{x} \pm \frac{ts}{\sqrt{N}}$$

Find the mean, sample standard deviation, t value, and 95% confidence interval in cells B15 through B19. The value of t can be found from the Excel function TINV(0.05,6), where 0.05 is the probability level (95% confidence level) and 6 is the number of degrees of freedom. The result of the CI calculation is shown below.

15	Calculation of CI	
16	Mean	1100.286
17	Std. Dev.	16.77015
18	t value	2.446912
19	95% CI	15.50981
20		
21	Documentation	
22	Cell B16=AVERAGE(B2:B8)	
23	Cell B17=STDEV(B2:B8)	
24	Cell B18=TINV(0.05,6)	
25	Cell B19=B18*B17/SQRT(7)	

In Excel 2010 you can find the confidence interval by using the function CONFIDENCE,T with alpha = 0.05, Std. Dev. = 16.77015, and a sample size of 7. This should return exactly the same value as found in cell B19. Another way to find the confidence interval is to use the Analysis ToolPak. Select Data Analysis from the Analysis group on the Data Ribbon and choose Descriptive Statistics. Enter cells B2:B8 as the Input Range:, select New Worksheet Ply:, check Summary Statistics and Confidence Level for Mean:. Enter 95% for the confidence level as shown in the Descriptive Statistics window in Figure 2-21.

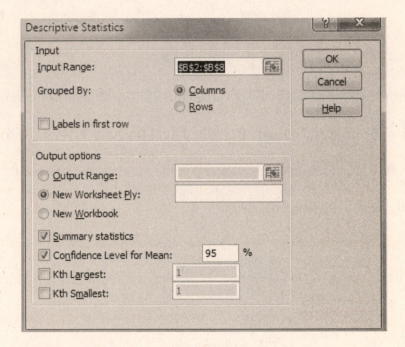

Figure 2-21 Adding the confidence level for the mean to the descriptive statistics.

Click on OK and note that a new worksheet opens with the descriptive statistics and the confidence interval as shown in Figure 2-22. You will have to increase the width of column A to read all the labels. Note that this confidence interval is identical to that calculated manually in cell B19. The interval is larger than the 95% confidence interval for the case when σ was known even though the known σ is larger than the estimated value of s. Your new worksheet should

appear as shown in Figure 2-22.

	A	B
1	*Column1*	
2		
3	Mean	1100.286
4	Standard Error	6.338523
5	Median	1100
6	Mode	#N/A
7	Standard Deviation	16.77015
8	Sample Variance	281.2381
9	Kurtosis	-0.85214
10	Skewness	-0.3686
11	Range	47
12	Minimum	1075
13	Maximum	1122
14	Sum	7702
15	Count	7
16	Confidence Level(95.0%)	15.50981

Figure 2-22 Descriptive statistics including 95% confidence interval based on s as an estimate of σ.

Save your workbook as **pooled_std_dev.xls**.

Summary

In this chapter, we learned how to perform basic statistical operations such as calculating the

mean and standard deviations of a data set. In addition, we learned how to compute the pooled

standard deviation of several sets of data. Finally, we used Excel to calculate confidence

intervals for mean values when the population standard deviation is known and when it is

unknown.

Problems

1.   Find the mean and median of each of the following sets of data. Determine the deviation

from the mean for each data point within the sets and find the mean deviation for each

set.

(a) 0.0110 0.0104 0.0105

(b) 24.53 24.68 24.77 24.81 24.73

(c) 188 190 194 187

(d) 4.52×10^{-3} 4.47×10^{-3}

4.63×10^{-3} 4.48×10^{-3}

4.53×10^{-3} 4.58×10^{-3}

(e) 39.83 39.61 39.25 39.68

(f) 850 862 849 869 865

2.   For each of the data sets in problem 1, find the standard deviation, the variance, the

coefficient of variation and the standard deviation of the mean.

3.   Analysis of several plant-food preparations for potassium ion yielded the following data:

Sample	Percent K^+
1	5.15, 5.03, 5.04, 5.18, 5.20
2	7.18, 7.17, 6.97
3	4.00, 3.93, 4.15, 3.86
4	4.68, 4.85, 4.79, 4.62
5	6.04, 6.02, 5.82, 6.06, 5.88

The preparations were randomly drawn from the same population.

(a)   Find the mean and standard deviation s for each sample.

(b)   Obtain the pooled value s_{pooled}.

4.   Six bottles of wine of the same variety were analyzed for residual sugar content with the

following results:

Bottle	Percent (w/v) Residual Sugar
1	0.99, 0.84, 1.02
2	1.02, 1.13, 1.17, 1.02
3	1.25, 1.32, 1.13, 1.20, 1.12
4	0.72, 0.77, 0.61, 0.58
5	0.90, 0.92, 0.73
6	0.70, 0.88, 0.72, 0.73

(a)   Evaluate the standard deviation s for each set of data.

(b)   Pool the data to obtain an absolute standard deviation for the method.

5.   Nine samples of illicit heroin preparations were analyzed in duplicate by a gas

chromatographic method. The samples can be assumed to have been drawn randomly

from the same population. Pool the following data to establish an estimate of σ for the

procedure.

Sample	Heroin, %	Sample	Heroin, %
1	2.24, 2.27	6	1.07, 1.02
2	8.4, 8.7	7	14.4, 14.8
3	7.6, 7.5	8	21.9, 21.1
4	11.9, 12.6	9	8.8, 8.4
5	4.3, 4.2		

6.   Calculate a pooled estimate of σ from the following spectrophotometric determination of

NTA (nitrilotriacetic acid) in water from the Ohio River:

Sample	NTA, ppb
1	12, 17, 15, 8
2	32, 31, 32
3	25, 29, 23, 29, 26

7.   Consider the following sets of replicate measurements:

A	B	C	D	E	F
3.5	70.24	0.812	2.7	70.65	0.514
3.1	70.22	0.792	3.0	70.63	0.503
3.1	70.10	0.794	2.6	70.64	0.486
3.3		0.900	2.8	70.21	0.497
2.5			3.2		0.472

Calculate the mean and the standard deviation for each of these six data sets. Use the

sample standard deviation and find the 95% confidence intervals for the means of each

data set. What does this interval mean?

8.   Calculate the 95% confidence interval for the means of each set of data in Problem 7 if σ

is known from other experiments and has values of: set A, 0.20; set B, 0.070; set C,

0.0090; set D, 0.30; set E, 0.15; set F, 0.015.

9.   An atomic absorption method for the determination of the amount of iron present in used

jet engine oil was found, from pooling 30 triplicate analyses, to have a standard deviation

$s = 2.4$ µg Fe/mL. If s is a good estimate of σ, calculate the 80 and 95% confidence

intervals for the result, 18.5 µg Fe/mL, if it was based on (a) a single analysis, (b) the

mean of two analyses, (c) the mean of four analyses.

10.   An atomic absorption method for determination of copper in fuels yielded a pooled

standard deviation of $s = 0.32$ µg Cu/mL ($s \rightarrow \sigma$). The analysis of an oil from a

reciprocating aircraft engine showed a copper content of 8.53 µg Cu/mL. Calculate the 90

and 99% confidence intervals for the result if it was based on (a) a single analysis, (b) the

mean of 4 analyses, (c) the mean of 16 analyses.

11.	A volumetric calcium analysis on triplicate samples of the blood serum of a patient

	believed to be suffering from a hyperparathyroid condition produced the following data:

	meq Ca/L = 3.15, 3.25, 3.26. What is the 95% confidence limit for the mean of the data,

	assuming

	(a)	no prior information about the precision of the analysis?

	(b)	From prior experience it is believed that $\sigma = 0.056$ meq Ca/L?

12.	A chemist obtained the following data for percent lindane in the triplicate analysis of an

	insecticide preparation: 7.47, 6.98, 7.27. Calculate the 90% confidence interval for the

	mean of the three data, assuming that

	(a)	the only information about the precision of the method is the precision for the

		three data points.

	(b)	on the basis of long experience with the method, hat σ is judged to be 0.28%

		lindane.

Chapter 3

Statistical Tests with Excel

In this chapter we describe several statistical tests that can be done with Excel. We include here various t-tests for comparing means, the F-test for comparing variances and Analysis of Variance (ANOVA) for multiple comparisons. We also briefly discuss distribution-free tests.

t Tests

Excel has built-in features that allow you to perform t tests to compare two sample means with equal variances, two means with unequal variances, and paired results.

Comparing Means Assuming Equal Variances

We will use an example here to compare two sample means assuming that both samples have equal variances. In this example, two different analytical methods are compared for determining the mass of protein in randomly drawn 10.00 mg samples. We would like to know whether the two means are the same or different at the 95% confidence level. The null hypothesis is H_0: $\mu_1 = \mu_2$ and the alternative is H_a: $\mu_1 \neq \mu_2$, a two-tailed test. The results by the two methods in mg protein are given in Figure 3-1

	A	B	C
1		Method 1	Method 2
2		2.5	2.6
3		3.1	2.1
4		2.4	2.4
5		2.5	2
6		3	2.4
7		2.8	2.1
8			

Figure 3-1 Results for two analytical methods in mg protein/10.00 mg sample.

Manual Calculation. We will first manually calculate the value of *t* and compare it to the

critical value. Later, we will use the built-in procedures in Excel. Let us first obtain the means for

both methods, the standard deviations, the variances, the number of samples, and the number of

degrees of freedom by methods discussed in Chapter 2. The results of these calculations are

shown in Figure 3-2

	A	B	C
1		Method 1	Method 2
2		2.5	2.6
3		3.1	2.1
4		2.4	2.4
5		2.5	2
6		3	2.4
7		2.8	2.1
8			
9	Average	2.716667	2.266667
10	STDEV	0.292689	0.233809
11	VAR	0.085667	0.054667
12	*N*	6	6
13	DF	5	5

Figure 3-2 Basic statistical calculations for the two methods.

To calculate the *t* value, we need to compute the pooled standard deviation as discussed

in Chapter 2. To do that, we need the sum of the squares of the deviations from the means. We

will use Excel's function **DEVSQ()** for this purpose as we did in Chapter 2. In cell A14, type the

label **Dev^2** in bold. In cell B14, type

$$\texttt{=DEVSQ(B2:B7) [↵]}$$

Alternatively, you can click on the Insert Function icon and choose the DEVSQ function in the

Insert Function window. Copy the cell B14 formula into cell C14. Your worksheet should now

look like that in Figure 3-3.

	A	B	C
1		Method 1	Method 2
2		2.5	2.6
3		3.1	2.1
4		2.4	2.4
5		2.5	2
6		3	2.4
7		2.8	2.1
8			
9	Average	2.716667	2.266667
10	STDEV	0.292689	0.233809
11	VAR	0.085667	0.054667
12	N	6	6
13	DF	5	5
14	DEV^2	0.428333	0.273333

Figure 3-3 Worksheet after calculating the squares of the deviations from the mean.

Now use the AutoSum function to calculate the total degrees of freedom in cell D13 and the total sum of the squares of the deviations in cell D14. Add labels in cells E13 and E14. In cell D15, compute the pooled variance by typing

$$=D14/D13[↵]$$

In cell D16, calculate the pooled standard deviation by typing

$$=SQRT(D15)[↵]$$

Add labels for these new values.

Next, we will compute the value of t by using the equation

$$t = \frac{\bar{x}_1 - \bar{x}_2}{s_{\text{pooled}}\sqrt{\dfrac{N_1 + N_2}{N_1 N_2}}}$$

Type in cell B17,

$$=(B9-C9)/(D16*SQRT((B12+C12)/(B12*C12)))[↵]$$

Type **Value of** **t** as a label in cell A17. Your worksheet should now look similar to Figure 3-4 except for rows 18 and 19. In cell B18, we have added the critical value of t. You can obtain the value either by looking it up in a t table such as Table 7-3 in FAC9 or Table 7-2 in AC7 or by

using the Excel 2010 function T.INV.2T(probability,deg_freedom). In our case, we typed into

cell B18

$$\texttt{=T.INV.2T(0.05,10)[↵]}$$

In Excel 2007, the corresponding function is TINV(probability,deg_freedom), where the two-

tailed value is assumed. Note that Excel 2010 also allows the earlier function to be used for

compatibility. In cell B19, we have calculated the probability of getting a t value of 2.942441 by

random chance from the Excel 2010 function T.DIST.2T(x,deg_freedom), where x is the text

value of t given in cell B17 (2.942441), and deg_freedom is 10 for our case.. This function

produces a two-tailed test. You can also use the Excel 2007 function

TDIST(x,deg_freedom,tails), where the number of tails (2) must be specified. Excel 2010 allows

the earlier function to be used for compatibility. Hence, we typed into cell B19 the formula

$$\texttt{=T.DIST.2T(B17,10,2) [↵]}$$

	A	B	C	D	E
1		Method 1	Method 2		
2		2.5	2.6		
3		3.1	2.1		
4		2.4	2.4		
5		2.5	2		
6		3	2.4		
7		2.8	2.1		
8					
9	Average	2.716667	2.266667		
10	STDEV	0.292689	0.233809		
11	VAR	0.085667	0.054667		
12	N	6	6		
13	DF	5	5	10	Total DF
14	DEV^2	0.428333	0.273333	0.701667	Total DEV^2
15				0.070167	VAR Pooled
16				0.26489	s_{pooled}
17	Value of t	2.942441			
18	Critical t	2.228139			
19	Probability	0.014724			

Figure 3-4 Worksheet after calculating pooled variance, pooled standard deviation, value of t, critical value of t, and probability of obtaining a t value of 2.942441.

Since $t > 2.228$, we reject the null hypothesis and conclude that there is a difference in the means at the 95% confidence level. Note from cell B19, there is only a 1.5% probability of obtaining a t value this large because of random error.

Using the Excel Built-In Functions. We will first use the Excel 2010 function T.TEST(array1,array2,tails,type) to find the probability of obtaining a difference in means of the magnitude obtained. In Excel 2007, the function is TTEST(array1,array2,tails,type). Again, Excel 2010 allows the earlier function to be used. In cell A21, type the label `ttest`. Select cell B21. Click on the Insert (Paste) Function button. Choose the Statistical Category and select T.TEST from the Insert Function Window. This opens the T.TEST Function Arguments window shown in Figure 3-5.

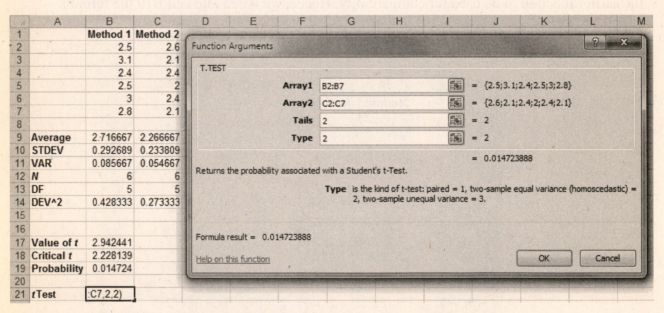

Figure 3-5 Worksheet showing T.TEST Function Arguments window.

With the cursor in the Array1 box, type B2:B7 or highlight these cells to insert B2:B7. With the cursor in the Array2 box, highlight cells C2:C7. Type 2 in the tails box, and 2 in the type box. When you click OK, the probability is returned in cell B21. Note that this is the same

value as calculated from the TDIST function. The advantage of the TTEST function is that we do

not have to calculate the value of *t* prior to doing the test.

Now we will use the Analysis ToolPak for the *t*-test. On the Data ribbon, select Data

Analysis…. From the Data Analysis window that opens select t-test: Two-Sample Assuming

Equal Variances as shown in Figure 3-6.

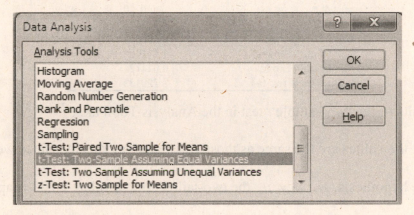

Figure 3-6 Data analysis window showing selection of two-sample t-test.

Click OK. In the new window that opens, place B2:B7 in the Variable 1 Range: box and C2:C7

in the Variable 2 Range: box as shown in Figure 3-7. Insert 0 in the box labeled Hypothesized

Mean Difference: and 0.05 in the Alpha: box. Select New Worksheet Ply:. Click the OK button.

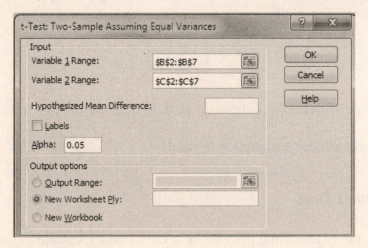

Figure 3-7 Two-sample *t* test window.

The results of the two-sample t test assuming equal variances should be as shown in Figure 3-8 after adjusting the column A width to accommodate the entries.

	A	B	C
1	t-Test: Two-Sample Assuming Equal Variances		
2			
3		Variable 1	Variable 2
4	Mean	2.716667	2.266667
5	Variance	0.085667	0.054667
6	Observations	6	6
7	Pooled Variance	0.070167	
8	Hypothesized Mean Difference	0	
9	df	10	
10	t Stat	2.942441	
11	P(T<=t) one-tail	0.007362	
12	t Critical one-tail	1.812461	
13	P(T<=t) two-tail	0.014724	
14	t Critical two-tail	2.228139	

Figure 3-8 Results of the two-sample t test in the Analysis ToolPak.

Note that the values are the same as those calculated previously. Since we are interested in the alternative hypothesis, H_a: $\mu_1 \neq \mu_2$, the two-tailed critical value, 2.228 is appropriate. Note that the two-tailed critical value of t corresponds to the 95% confidence level, while the one-tailed value corresponds to the 90% confidence level.

Excel can also perform a t test without assuming equal variances. Here instead of using the pooled variance, Excel uses the separate sample variances and calculates t from

$$t = \frac{\bar{x}_1 - \bar{x}_2}{\sqrt{\dfrac{s_1^2}{N_1} + \dfrac{s_2^2}{N_2}}}$$

The number of degrees of freedom is approximated. In our case the two tests give identical results since the equal variance assumption is valid.

Paired Two-Sample t Test

We will now perform the t test using pairs of data. Let us take Example 7-7 of FAC9 in which we compare two automated methods for determining glucose in serum. Both methods are done on the blood serum from six different patients. The test is to determine whether the methods are

different at the 95% confidence level. Hence the null hypothesis is H_0: $\mu_d = 0$ and the alternative

hypothesis, H_a: $\mu_d \neq 0$, where μ_d is the mean difference between pairs of data.

Enter the data into a blank worksheet as shown in Figure 3-9

	A	B	C
1		Method A	Method B
2	Patient 1	1044	1028
3	Patient 2	720	711
4	Patient 3	845	820
5	Patient 4	800	795
6	Patient 5	957	935
7	Patient 6	650	639

Figure 3-9 Worksheet for paired t test.

We will first manually calculate the t value for the mean difference using the equation

$$t = \frac{\bar{d} - 0}{s_d / \sqrt{N}}$$

where $\bar{d}$ is the average difference and s_d is the standard deviation of the difference. Type the

label **Difference** into cell D1. In cell D2 type

=B2-C2 [↵]

Use the fill handle to copy this formula into cells D3 through D7. In cell D8, type

=COUNT(D2:D7) [↵]

=AVERAGE(D2:D7) [↵]

=STDEV(D2:D7) [↵]

Label cells D8, D9 and D10 by typing **N**[↵], **Mean**[↵] and **SD**[↵] into cells E8 through

E10. Type the label **Value of t** in cell E11. In cell D11 type

$$=D9/(D10/SQRT(D8))[↵]$$

Your worksheet should look as shown in Figure 3-10

	A	B	C	D	E
1		Method A	Method B	Difference	
2	Patient 1	1044	1028	16	
3	Patient 2	720	711	9	
4	Patient 3	845	820	25	
5	Patient 4	800	795	5	
6	Patient 5	957	935	22	
7	Patient 6	650	639	11	
8				6	*N*
9				14.66667	Mean
10				7.763161	SD
11				4.627735	Value of *t*
12					

Figure 3-10 Worksheet after calculating the mean, the standard deviation and the *t* value.

We can now get Excel to perform the paired *t* test automatically. Select Data Analysis on the Data ribbon. Select t-Test: Paired Two Sample for Means. Fill in the boxes in the paired *t*-test window as shown in Figure 3-11 and select OK.

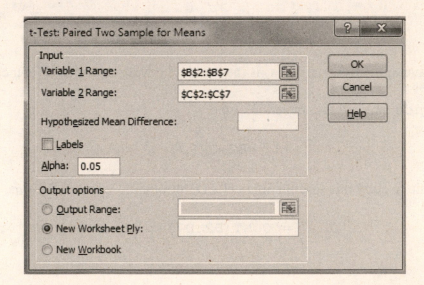

Figure 3-11 Paired *t*-test window.

A new worksheet should open with the results shown in Figure 3-12 after adjusting the column widths.

	A	B	C
1	t-Test: Paired Two Sample for Means		
2			
3		Variable 1	Variable 2
4	Mean	836	821.3333333
5	Variance	21466.8	20349.06667
6	Observations	6	6
7	Pearson Correlation	0.9989157	
8	Hypothesized Mean Difference	0	
9	df	5	
10	t Stat	4.6277348	
11	P(T<=t) one-tail	0.0028477	
12	t Critical one-tail	2.0150484	
13	P(T<=t) two-tail	0.0056954	
14	t Critical two-tail	2.5705818	

Figure 3-12 Results of paired t test.

Note here that Excel returns the same value of t (t Stat) as calculated manually. Since we are doing a two-tailed test, the appropriate critical value of t is 2.57 at the 95% confidence level and 5 degrees of freedom. Since $t > t_{crit}$, we reject the null hypothesis and conclude that the two methods give different results.

We can compare the paired t test to the unpaired test by choosing t-Test: Two-Sample Assuming Equal Variances from the Data Analysis window. This returns the new worksheet shown in Figure 3-13. Note here that merely averaging the results for the two methods gives a value of t of only 0.176, which is much smaller than the critical value of 2.228. Hence, if we did not pair the data, the large patient-patient variation would mask the method differences and cause us to erroneously conclude that the methods gave the same results.

	A	B	C
1	t-Test: Two-Sample Assuming Equal Variances		
2			
3		Variable 1	Variable 2
4	Mean	836	821.3333333
5	Variance	21466.8	20349.06667
6	Observations	6	6
7	Pooled Variance	20907.93333	
8	Hypothesized Mean Difference	0	
9	df	10	
10	t Stat	0.175685733	
11	P(T<=t) one-tail	0.432023724	
12	t Critical one-tail	1.812461102	
13	P(T<=t) two-tail	0.864047448	
14	t Critical two-tail	2.228138842	

Figure 3-13 Results of the *t* test without pairing the data.

F Test for Comparison of Variances

The *F* test is based on the null hypothesis that two population variances are equal, H_0: $\sigma_1^2 = \sigma_2^2$.

The test statistic is the ratio of two sample variances, $F = s_1^2 / s_2^2$. We calculate this statistic and

compare it with the critical value of *F* at the desired confidence level. Excel has two procedures

for performing the *F* test. One is the Excel 2010 worksheet function F.TEST(array1,array2)

which returns the two-tailed probability that the variances in the two arrays are not significantly

different (this function is FTEST(array1,array2) in Excel 2007. The other *F* test is a more

complete test in the Analysis ToolPak. We shall examine both of these methods here.

Let's use the data shown in Figure 3-1 that we previously used for the comparison of

means by the *t* test. We shall ask whether the variances are equal at the 95% confidence level.

Our alternative hypothesis will be the variance of method 1 exceeds that of method 2. Hence we

will use a one-tailed test and put the variance of method 1 in the numerator. Enter the data as in

Figure 3-1 or use the worksheet previously used for the *t* test. In a blank cell in column A below the data, type the label **F test**. Select the cell next to it in column B and click on the Insert

Function icon. Select Statistical and F.TEST or FTEST. This opens the F.TEST Function

arguments window shown in Figure 3-14.

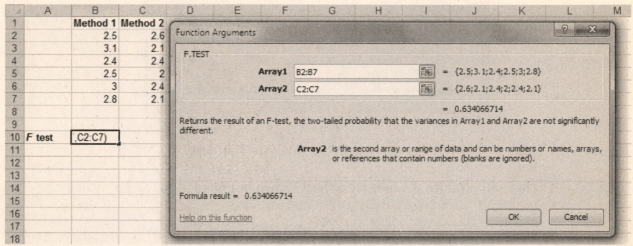

Figure 3-14 Data for *F* test with F.TEST Function Arguments window open.

Select cells B2:B7 for Array1 and C2:C7 for Array2 as shown. Click on OK. The probability

returned by the *F* test is 0.6341 or a 63.41% probability that the variances in the two arrays are

not different. Note again that the 2007 Excel function FTEST also works in Excel 2010.

Now we will use the Analysis ToolPak to perform the *F*-test. Select Data Analysis… on

the Data ribbon. Select F-Test Two-Sample for Variances from the Analysis Tools list. Click on

OK. This opens the F-Test Two-Sample for Variances window as shown in Figure 3-15. Select

cells B2:B7 for the Variable 1 Range: and C2:C7 for the Variable 2 Range: as shown. Enter 0.05

in the alpha box and check New Worksheet Ply:

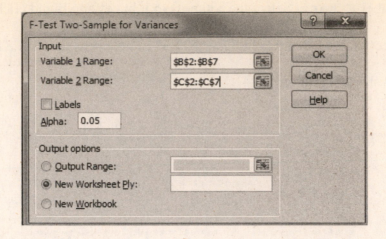

Figure 3-15 Window for performing *F* test in the Analysis ToolPak.

Click on OK. This will produce the new worksheet shown in Figure 3-16.

	A	B	C
1	F-Test Two-Sample for Variances		
2			
3		Variable 1	Variable 2
4	Mean	2.71666667	2.266666667
5	Variance	0.08566667	0.054666667
6	Observations	6	6
7	df	5	5
8	F	1.56707317	
9	P(F<=f) one-tail	0.31703336	
10	F Critical one-tail	5.05032906	

Figure 3-16 Results of the *F* test for comparison of variances.

Note that the calculated value of *F* is 1.567 and the critical value is 5.050. Since $F < F_{crit}$ we accept the null hypothesis and conclude that the variances are identical at the 95% confidence level. Note that the Analysis ToolPak *F*-test returns the probability that $F_{crit} \leq F$, a one-tailed probability. In this case it is 31.7% probable that the means are different. In any case we are led to conclude that there is no significant difference in the variances of the two methods.

ANOVA

There are several ways to do ANOVA in Excel. First, we can enter the formulas manually and have Excel do the calculations. A second way is to take advantage of the Analysis ToolPak. We illustrate both of these procedures on the data from Example 7-9 of FAC9. Here, results from the determination of calcium by 5 analysts are compared to see if the means differ significantly.

Entering ANOVA Formulas Manually

Enter the data from the table in Example 7-9 of FAC9 so that it appears as shown in Figure 3-17.

	A	B	C	D	E	F
1						
2						
3	Trial No.	Analyst 1	Analyst 2	Analyst 3	Analyst 4	Analyst 5
4	1	10.3	9.5	12.1	9.6	11.6
5	2	9.8	8.6	13.0	8.3	12.5
6	3	11.4	8.9	12.4	8.2	11.4
7	Means					
8	SDs					

Figure 3-17 Results from the determination of Ca (in mmol) by 5 analysts.

Now click on cell B7, and type

$$\text{=AVERAGE(B4:B6) [↵]}$$

Type in cell B8

$$\text{=STDEV(B4:B6) [↵]}$$

Select cell B7 and extend the selection to B8 by holding the mouse button and dragging. With both cells selected, click on the fill handle, and drag to the right to copy these formulas into cells C7 through F8. The spreadsheet should now appear as shown in Figure 3-18.

◢	A	B	C	D	E	F
1						
2						
3	Trial No.	Analyst 1	Analyst 2	Analyst 3	Analyst 4	Analyst 5
4	1	10.3	9.5	12.1	9.6	11.6
5	2	9.8	8.6	13.0	8.3	12.5
6	3	11.4	8.9	12.4	8.2	11.4
7	Means	10.5	9	12.5	8.7	11.83333
8	SDs	0.818535	0.458258	0.458258	0.781025	0.585947

Figure 3-18 ANOVA worksheet after calculating the means and standard deviations of the analysts results.

We will now have Excel calculate the grand mean $\bar{\bar{x}}$ in cell B9 by using the following equation.

$$\bar{\bar{x}} = \left(\frac{N_1}{N}\right)\bar{x}_1 + \left(\frac{N_2}{N}\right)\bar{x}_2 + \left(\frac{N_3}{N}\right)\bar{x}_3 + \cdots + \left(\frac{N_I}{N}\right)\bar{x}_I$$

Since each analyst has 3 results and there are 15 total, the grand mean can be obtained by multiplying the sum of the means by 3/15. Alternatively, since the number of results is the same for each analyst, the grand mean is the simple average of the 5 means. Type in cell A9

Grand[↵]

and in cell B9

=AVERAGE(B7:F7) [↵]

The grand mean of 10.50667 should now appear in cell B9. Now type the labels for calculating the remaining quantities in columns A, C, and E as shown in Figure 3-19.

	A	B	C	D	E	F
1						
2						
3	Trial No.	Analyst 1	Analyst 2	Analyst 3	Analyst 4	Analyst 5
4	1	10.3	9.5	12.1	9.6	11.6
5	2	9.8	8.6	13.0	8.3	12.5
6	3	11.4	8.9	12.4	8.2	11.4
7	Means	10.5	9	12.5	8.7	11.83333
8	SDs	0.818535	0.458258	0.458258	0.781025	0.585947
9	Grand	10.50667				
10	SSF		df SSF		MSF	
11	SSE		df SSE		MSE	
12	SST		df SST		F	

Figure 3-19 ANOVA worksheet after calculating the grand mean and adding labels.

To calculate the sum of the squares due to the factor, SSF, we use the equation:

$$\text{SSF} = N_1(\bar{x}_1 - \bar{\bar{x}})^2 + N_2(\bar{x}_2 - \bar{\bar{x}})^2 + N_3(\bar{x}_3 - \bar{\bar{x}})^2 + \cdots + N_I(\bar{x}_I - \bar{\bar{x}})^2$$

Enter into cell B10 the formula corresponding to this equation:

=3*(B7−B9) ^2+3* (C7−B9) ^2+3* (D7−B9) ^2+3* (E7−B9) ^2+3* (F7−B9) ^2[↵]

The result 33.80267 should appear in B10. To find the sum of the squares due to error, SSE, we use the equation

$$\text{SSE} = \sum_{j=1}^{N_1}(x_{1j} - \bar{x}_1)^2 + \sum_{j=1}^{N_2}(x_{2j} - \bar{x}_2)^2 + \sum_{j=1}^{N_3}(x_{3j} - \bar{x}_3)^2 + \cdots + \sum_{j=1}^{N_I}(x_{ij} - \bar{x}_I)^2$$

Enter into cell B11 the formula corresponding to this equation

=2*B8^2+2*C8^2+2*D8^2+2*E8^2+2*F8^2 [↵]

The total sum of squares SST can be found in cell B12 by adding the results of B10 and B11

=B10+B11 [↵]

Your worksheet should now appear as shown in Figure 3-20.

	A	B	C	D	E	F
1						
2						
3	Trial No.	Analyst 1	Analyst 2	Analyst 3	Analyst 4	Analyst 5
4	1	10.3	9.5	12.1	9.6	11.6
5	2	9.8	8.6	13.0	8.3	12.5
6	3	11.4	8.9	12.4	8.2	11.4
7	Means	10.5	9	12.5	8.7	11.83333
8	SDs	0.818535	0.458258	0.458258	0.781025	0.585947
9	Grand	10.50667				
10	SSF	33.80267	df SSF		MSF	
11	SSE	4.086667	df SSE		MSE	
12	SST	37.88933	df SST		F	

Figure 3-20 Worksheet after calculating the sums of squares.

The number of degrees of freedom (df) can be found in several ways. We could use the COUNT function as described in previously or enter these manually since they are easy to obtain. The df value for the factor levels (df SSF) is just the number of groups (5) minus one, or 4. The df value for error (df SSE) is the total number of data points (15) minus the number of groups (5), or 10. The total number of degrees of freedom (df SST) is the total number of measurements minus 1 or 14. We can thus enter these values into cells D10, D11 and D12, respectively.

Now we are ready to obtain the mean squares due to the factor levels and due to error. Enter into cell F10 for the mean square due to factor levels (MSF)

$$=B10/D10\,[\dashv]$$

and in cell F11 for the mean square due to error (MSE)

$$=B11/D11\,[\dashv]$$

The *F* value can now be calculated in cell F12 by entering

$$=F10/F11\,[\dashv]$$

Your worksheet should now be similar to that shown in Figure 3-21.

	A	B	C	D	E	F
1						
2						
3	Trial No.	Analyst 1	Analyst 2	Analyst 3	Analyst 4	Analyst 5
4	1	10.3	9.5	12.1	9.6	11.6
5	2	9.8	8.6	13.0	8.3	12.5
6	3	11.4	8.9	12.4	8.2	11.4
7	Means	10.5	9	12.5	8.7	11.83333
8	SDs	0.818535	0.458258	0.458258	0.781025	0.585947
9	Grand	10.50667				
10	SSF	33.80267	df SSF	4	MSF	8.450667
11	SSE	4.086667	df SSE	10	MSE	0.408667
12	SST	37.88933	df SST	14	F	20.67863

Figure 3-21 Final worksheet for manual entry of ANOVA formulas.

From the calculations above, an ANOVA table can be filled out as shown below:

Source of Variation	Sum of Squares (SS)	Degrees of Freedom (df)	Mean Square (MS)	F
Between groups	33.80267	4	8.450667	20.68
Within groups	4.086667	10	0.408667	
Total	37.88933	14		

The critical value of F for 4 degrees of freedom in the numerator and 10 in the denominator is 3.48 at the 5% probability level (see Table 7-4 of FAC9). Since our F value of 20.68 greatly exceeds the critical value, we reject H_0 and conclude that there is a significant difference among the analysts in determining calcium. You can document your worksheet as described previously.

The Analysis ToolPak ANOVA Tool

There is another way to perform ANOVA that takes advantage of the built-in procedure in Excel's Analysis ToolPak. The advantage of this method is that the formulas and calculations are done automatically. In addition to producing the F value, Excel gives the probability (P-value) of obtaining thecalculated F value . This allows the user to choose the rejection region for the null hypothesis based on the P-value. If the probability is smaller than the significance level α, we

reject H_0.

We will use the same data as the previous example taken from Example 7-9 of FAC9.

With the ANOVA worksheet of Figure 3-21 on your screen, select Data Analysis from the Data

ribbon. In the Data Analysis window, select Anova: Single Factor from the Analysis Tools list.

Click OK. To choose the Input Range: in the window that appears, highlight cells B4 through F6.

The remaining options in the window should be chosen as shown in Figure 3-22.

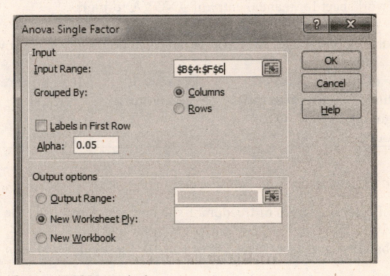

Figure 3-22 Single-factor ANOVA window.

Click OK. The ANOVA results should appear in a new worksheet that opens. Expand column A

so that the label is fully visible. The worksheet should appear as in Figure 3-23.

	A	B	C	D	E	F	G
1	Anova: Single Factor						
2							
3	SUMMARY						
4	Groups	Count	Sum	Average	Variance		
5	Column 1	3	31.5	10.5	0.67		
6	Column 2	3	27	9	0.21		
7	Column 3	3	37.5	12.5	0.21		
8	Column 4	3	26.1	8.7	0.61		
9	Column 5	3	35.5	11.83333	0.343333		
10							
11							
12	ANOVA						
13	Source of Variation	SS	df	MS	F	P-value	F crit
14	Between Groups	33.80267	4	8.450667	20.67863	7.97E-05	3.47805
15	Within Groups	4.086667	10	0.408667			
16							
17	Total	37.88933	14				

Figure 3-23 New worksheet with ANOVA results.

Note that Excel fills out the ANOVA table automatically. The *P-value*, is displayed in the next to last column of the ANOVA table. Note here the extremely low probability of obtaining an F value of 20.678 ($P = 7.97 \times 10^{-5}$). Excel also displays the critical value of F corresponding to the significance level chosen in options window. Because of the extremely low probability and the large value of F, we are quite confident that rejection of the null hypothesis is the appropriate decision. Hence, we conclude that there is a significant difference among the analysts in the calcium determination.

Distribution-Free Methods

The various statistical tests and the ANOVA methods just discussed require that samples be drawn from a normal population. For populations that are not normal, there are procedures termed **distribution-free** or **non-parametric** that offer improved performance over the standard tests. In addition, these procedures work almost as well when the underlying populations are normally distributed. There are distribution-free procedures for testing means, for testing

83

populations, and for ANOVA. We discuss here only the alternative to the t test, the Wilcoxon

Signed-Rank test for single samples and for paired data.

Wilcoxon Signed-Rank Test for Single Samples

In the Wilcoxon signed-rank test, the only assumptions made are that the samples are drawn

from a continuous, symmetric population with a mean and median μ. We hypothesize that μ is

equal to some known value μ_0.

The basic idea of the signed-rank test is to compute the differences between our result

and the known value $(x_i - \mu_0)$ and to rank them from smallest to largest disregarding signs at this

point. The ranks will be 1 for the smallest difference, 2 for the next smallest, and so on. We then

apply the sign of each result to the corresponding rank. In most cases, some of the differences

will be positive and some negative. We then calculate a test statistic s_{calc} that is the sum of the

positively signed ranks. If the mean μ is actually larger than μ_0, most of the deviations with a

large magnitude should be positive. We would then reject H_0 if s_{calc} is larger than a critical value,

c.

The null hypothesis and the various alternative hypotheses are:

$H_0: \mu = \mu_0$

$H_a: \mu > \mu_0$ reject if $s_{calc} \geq c_1$

$H_a: \mu < \mu_0$ reject if $s_{calc} \leq c_2$ where $c_2 = n(n+1)/2 - c_1$

$H_a: \mu \neq \mu_0$ reject is $s_{calc} \geq c$ or $s_{calc} \leq n(n+1)/2 - c$

To illustrate, we'll take an example of the determination of dissolved oxygen in a pond

into which sewage had been dumped. The fish in the pond need a dissolved O_2 level of at least

5.00 ppm to survive. The results shown in Figure 3-24 were collected over a several week

period.

	A	B	C
1			
2			
3		Dissolved O2, ppm	
4		4.92	
5		5.03	
6		5.60	
7		4.35	
8		4.70	
9		4.63	
10		4.58	
11		5.20	
12		4.26	
13		4.75	
14			
15	Avg.	4.80	

Figure 3-24 Dissolved oxygen levels in a pond.

We suspected that the data are not normally distributed, and so, we choose the Wilcoxon

signed-rank test to test the null hypothesis that the mean value is 5.00 ppm. The alternative

hypothesis is that the mean is less than 5.00 ppm so it is a one-tailed test. Enter the values shown

in Figure 3-24. Enter the μ_0 value of 5.00 ppm in cell B2. Compute the differences between the

data values and μ_0 in column C. In order to rank the values, we need the differences without

regard to sign. We can use Excel's absolute value function ABS to do this calculation. So, in cell

D4 type **=ABS(C4) [↵]**. Copy this formula in the cells D5:D13.

Now we want to rank these 10 values from smallest (1) to largest (10). For such a small

data set, these are easy to rank manually, but we will let Excel rank them. With the cursor on cell

E4, click on the Insert Function icon and choose Statistical. Select the function RANK.AVG

from the list. In the Function Arguments, enter cell D4 as the **Number** to be ranked. In the **Ref**

box enter $D$4:$D$13 to select the reference list for the ranking. Enter 1 in the Order box.

Selecting 0 or leaving the box blank, ranks the numbers in column D from largest to smallest.

Selecting 1 will rank them from smallest to largest. Copy the ranking into cells E5:E13. Your

spreadsheet should then look like Figure 3-25 after putting some titles on the columns.

	A	B	C	D	E
1					
2	μ_0	5.00			
3		Dissolved O2, ppm	$x_i - \mu_0$	Abs Value	Ranked
4		4.92	-0.08	0.08	2
5		5.03	0.03	0.03	1
6		5.60	0.60	0.6	8
7		4.35	-0.65	0.65	9
8		4.70	-0.30	0.3	5
9		4.63	-0.37	0.37	6
10		4.58	-0.42	0.42	7
11		5.20	0.20	0.2	3
12		4.26	-0.74	0.74	10
13		4.75	-0.25	0.25	4
14					
15	Avg.	4.80			

Figure 3-25 Worksheet after ranking the deviations.

Now we will have Excel put signs to the rankings. To do this, we will use Excel's logical

IF function. In cell F4, type **IF(C4>0,1,-1)[↵]** . This command tells Excel to insert 1 if the

number in C4 is positive and –1 if it is negative. Copy this command to cells F5:F13. In cell G4,

type **IF(F4>0,E4,0)[↵]**. This statement tells Excel to insert the rank in E4 if F4 is positive

and to insert 0 if F4 is negative. We should now have the ranks with positive deviations listed in

column G. All that is left is to sum these values to form our test statistic s_{calc}. In cell G15,

calculate the sum of the values in column G. This should return the value 12 for our test statistic.

We now need to find a table of the Critical Values for the Wilcoxon Signed-Rank test.

We can find such tables on-line or in a statistics book such as Devore's text.[1] The critical value

for a one-tailed test at the 95% confidence level is 11. Since our calculated value is greater than

---

[1] J. L. Devore, *Probability and Statistics for Engineering and the Sciences 7th ed.*, pp 332, 340. Boston, MA:
Brooks/Cole Publishing Co. 2012, Table A.13.

11 (barely), we must accept H_0 and conclude that the mean value is 5.00 ppm. The final

worksheet for the Wilcoxon test is shown in Figure 3-26 along with the documentation.

	A	B	C	D	E	F	G
1	Wilcoxon Signed-Rank Test						
2	μ_0	5.00					
3		Dissolved O2, ppm	$x_i - \mu_0$	Abs Value	Ranked	Sign	Positive ranks
4		4.92	-0.08	0.08	2	-1	0
5		5.03	0.03	0.03	1	1	1
6		5.60	0.60	0.6	8	1	8
7		4.35	-0.65	0.65	9	-1	0
8		4.70	-0.30	0.3	5	-1	0
9		4.63	-0.37	0.37	6	-1	0
10		4.58	-0.42	0.42	7	-1	0
11		5.20	0.20	0.2	3	1	3
12		4.26	-0.74	0.74	10	-1	0
13		4.75	-0.25	0.25	4	-1	0
14							
15	Avg.	4.80				s_{calc}	12
16	Documentation						
17	Cell C4=B4-$B$2						
18	Cell D4=ABS(C4)						
19	Cell E4=RANK.AVG(D4,$D$4:$D$13,1)						
20	Cell F4=IF(C4>0,1,-1)						
21	Cell G4=IF(F4>0,E4,0)						
22	Cell B15=AVERAGE(B4:B13)						
23	Cell G15=SUM(G4:G13)						

Figure 3-26 Final worksheet for signed-ranks test for a single sample

Since Excel does not have a built in signed-ranks test, entering and calculating the values

is a rather tedious, time-consuming task. For those who do non-parametric tests often, there are

Excel plug-ins or separate statistics packages, such as Minitab, SPSS, Origin, STATISTICA, and

many others to do non-parametric testing.

Wilcoxon Signed-Rank Test for Paired Data

A distribution-free alternative to the paired t-test is the signed-rank test for paired data. In this

test, we make a hypothesis about the differences d_i between the pairs using the signed-rank test

on the differences. Again a continuous, symmetric distribution is assumed. To illustrate, we will

take the same data used previously in this chapter from Example 7-7 of FAC9 for comparing two different methods for determining glucose in serum. The results are shown in Figure 3-27. Note

in this case that all differences are positive, so we can rank and sum them directly. Our test

statistic value is 21. In the tables the critical value for being 95% confident is also 21. Since s_{calc}

is greater than or equal to 21, we must reject the null hypothesis that there is no difference

between the methods and accept the two-tailed alternative that the methods do not give

equivalent results. This is then the same conclusion reached by the t test.

	A	B	C	D	E
1	Wilcoxon Signed-Rank Test for Paired Data				
2				Difference	Rank
3	Patient 1	1044	1028	16	4
4	Patient 2	720	711	9	2
5	Patient 3	845	820	25	6
6	Patient 4	800	795	5	1
7	Patient 5	957	935	22	5
8	Patient 6	650	639	11	3
9					
10				s_{calc}	21
11	Documentation				
12	Cell D3=B3-C3				
13	Cell E3=RANK.AVG(D3,$D$3:$D$8,1)				
14	Cell E10=SUM(E3:E8)				

Figure 3-27 Wilcoxon signed-rank tests for paired data.

Summary

In this chapter, we have used Excel to perform several important statistical tests, including t tests,

F tests, analysis of variance, and signed-rank tests. In most cases, Excel has built-in analysis

tools that allow us to carry out these tests without typing in the formulas manually. Most of the

results of these tools were verified by manual calculations. Using the built-in functions in Excel

greatly decreases the chances of making typographical errors and provides formatted tables

containing relevant statistics. The results for the Wilcoxon signed-rank tests were done manually

since Excel has no built-in procedures for these non-parametric tests.

Problems

1.   Sewage and industrial pollutants dumped into a body of water can reduce the dissolved

oxygen concentration and adversely affect aquatic species. In one study, weekly readings

are taken from the same location in a river over a two-month period.

Week Number	Dissolved O_2, ppm
1	4.9
2	5.1
3	5.6
4	4.3
5	4.7
6	4.9
7	4.5
8	5.1

Some scientists think that 5.0 ppm is a dissolved O_2 level that is marginal for fish to live.

Conduct a statistical test to determine whether the mean dissolved O_2 concentration is

less than 5.0 ppm at the 95% confidence level. State clearly the null and alternative

hypotheses.

2.   The spreadsheet below shows the results of two different analytical methods used to

determine Ca in a standard sample. Method A is the accepted method, and Method B is a

newly developed procedure.

	A	B	C	D
1	Analytical results for Ca in mg/100 mL			
2				
3		Method A	Method B	
4		20.4	21.0	
5		25.4	23.6	
6		25.1	22.2	
7		22.5	17.8	
8		24.9	24.2	
9		26.2	21.6	

(a)   Find the means of the two methods, the standard deviations, and the standard deviations

of the means.

(b)   Give the 95% confidence interval for the means of the two methods.

(c)   Do the variances differ at the 95% confidence level?

(d)   Apply the t test by entering the formulas manually in a spreadsheet, and test the null

hypothesis that the two means are equal at the 95% confidence level.

(e)   Use the Analysis ToolPak t test and test the same null hypothesis.

3.   The homogeneity of the chloride level in a water sample from a lake was tested by

analyzing portions drawn from the top and from near the bottom of the lake , with the

following results in ppm Cl:

Top	Bottom
26.30	26.22
26.43	26.32
26.28	26.20
26.19	26.11
26.49	26.42

(a)   Apply the t test at the 95% confidence level to determine if the means are

different.

(b)   Now use the paired t test and determine whether there is a significant difference

between the top and bottom values at the 95% confidence level.

(c)   Is a different conclusion drawn from using the paired t test than from just pooling

the data and using the normal t test for differences in means? If so, why?

4.   Two atomic absorption instruments are being compared for the determination of lead in a

biological sample. Instrument A is suspected of giving high results compared to

instrument B, which is a newer background-corrected unit. Six standard samples containing 40 µg/mL Pb were analyzed with each instrument by an identical analytical

procedure. The results are given in the spreadsheet below.

	A	B	C	D
1	Trace Pb by atomic absorption spectrometry			
2				
3		Instrument A	Instrument B	
4		46.8	35.7	
5		41.5	39.8	
6		43.6	42.6	
7		40.4	36.2	
8		45.9	45.7	
9		43.2	39.6	

(a)   Use a statistical test to determine if Instrument A gives a result higher than 40.0

µg/mL at the 95% confidence level. Set up the appropriate hypotheses and use an

appropriate test (one- or two-tailed).

(b)   Use a statistical test and determine if Instrument A gives a result significantly

higher than Instrument B at the 95% confidence level.

(c)   Determine whether there is a significant difference in the variances of the two

instruments at the 95% confidence level.

(d)   It is suspected that the data obtained come from a non-normal population. Use the

Wilcoxon signed-ranks test for a single sample to test whether instrument A gives

a result higher than 40.0 µg/mL at the 95% confidence level.

5.   Two different analytical methods were used to determine residual chlorine in sewage

effluents. Both methods were used on the same samples, but each sample came from

various locations with differing amounts of contact time with the effluent. The

concentration of Cl in mg/L was determined by the two methods and the following results

obtained:

Sample	Method A	Method B
1	0.39	0.36
2	0.84	1.35
3	1.76	2.56
4	3.35	3.92
5	4.69	5.35
6	7.70	8.33
7	10.52	10.70
8	10.92	10.91

(a)   What type of t test should be used to compare the two methods and why?

(b)   Do the two methods give different results? State and test the appropriate

hypotheses.

(c)   Does the conclusion depend on whether the 90%, 95%, or 99% confidence levels

are used?

(d)   Use the Wilcoxon signed-ranks test for paired data at the same confidence levels

as in part (c). Are the results different from those of the t test?

6.   Five different laboratories participated in an interlaboratory study involving

determinations of the iron level in water samples. The results below are replicate

determinations of the Fe concentration in ppm for laboratories A-E

Result No.	Lab A	Lab B	Lab C	Lab D	Lab E
1	10.3	9.5	10.1	8.6	10.6
2	11.4	9.9	10.0	9.3	10.5
3	9.8	9.6	10.4	9.2	11.1

(a)   State the appropriate hypotheses.

(b)   Do the laboratories differ at the 95% confidence level? At the 99% confidence

level ($F_{crit} = 5.99$)? At the 99.9% confidence level ($F_{crit} = 11.28$)?

(c)   Which laboratories are different from each other at the 95% confidence level?

7.   Four analysts carry out replicate sets of determinations of mercury determinations on the

      same analytical sample. The results in ppb Hg are shown in the table below.

Determination	Analyst 1	Analyst 2	Analyst 3	Analyst 4
1	10.24	10.14	10.19	10.19
2	10.26	10.12	10.11	10.15
3	10.29	10.04	10.15	10.16
4	10.23	10.07	10.12	10.10

(a)   State the appropriate hypotheses.

(b)   Do the analysts differ at the 95% confidence level? At the 99% confidence level

       ($F_{crit} = 5.95$)? At the 99.9% confidence level ($F_{crit} = 10.80$)?

(c)   Which analysts, differ from each other at the 95% confidence level?

8.   Four different fluorescence flow cell designs were compared to see if they were

      significantly different. The following results represented relative fluorescence intensities

      for 4 replicate measurements.

Measurement No.	Design1	Design 2	Design 3	Design 4
1	72	93	96	100
2	93	88	95	84
3	76	97	79	91
4	90	74	82	94

(a)   State the appropriate hypotheses.

(b)   Do the flow cell designs differ at the 95% confidence level?

(c)   If a difference was detected in part (b), which designs differ from each other at the

       95% confidence level?

9.   Three different analytical methods are compared for determining Ca. We are interested in

      knowing whether the methods differ. The results shown below represent the

      concentration of Ca in ppm determined by colorimetry, EDTA titration and atomic

      absorption spectrometry.

Repetition No.	Colorimetry	EDTA Titration	Atomic Absorption
1	3.92	2.99	4.40
2	3.28	2.87	4.92
3	4.18	2.17	3.51
4	3.53	3.40	3.97
5	3.35	3.92	4.59

(a)   State the null and alternative hypotheses.

(b)   Determine whether there are differences in the 3 methods at the 95% and 99% confidence levels.

(c)   If a difference was found at the 95% confidence level, determine which methods differ from each other.

10.   The following results represent the concentration of Mo in ppb in 24 plants randomly allocated to 4 different fertilizers.

	A	B	C	D	E
1	Determination of Mo in plant samples				
2					
3			Fertilizers		
4		A	B	C	D
5		6.2	6.3	6.8	5.6
6	ppb Mo	6.0	6.7	6.6	6.2
7		6.3	7.1	7.1	6
8		5.9	6.4	6.7	6.1
9			6.5	6.8	6.3
10			6.6	6.8	6.4
11					'6.3
12					5.9

(a)   Do the results indicate a difference in the means at the 95% confidence level? State the null and alternative hypotheses.

(b)   A difference in means indicates a difference among the fertilizers with regard to molybdenum uptake. Which fertilizers differ from each other?

Chapter 4

Least-Squares and Calibration Methods

Many of the standardization and calibration methods in analytical chemistry are based on the assumption of a linear relationship between the measured quantity and the concentration of the analyte. In particular, the **external standard method** involves preparing a **calibration curve** or **working curve** from the responses of a set of standards containing the analyte at known concentrations. The parameters of the linear relationship, such as the slope and intercept, are then calculated by least-squares analysis, which is also widely used in other calibration procedures such as the internal standard method and the method of standard additions. This chapter describes the use of Excel for obtaining least-squares estimates of slopes and intercepts and for predicting unknown concentrations. Multiple linear regression and polynomial linear regression are also discussed. Nonlinear regression methods are described in Chapter 13.

Linear Least-Squares Analysis

Linear least-squares analysis is fairly straightforward with Excel. This type of analysis can be accomplished in several ways: by entering the equations for linear regression manually, by employing the built-in functions of Excel, or by utilizing the regression data analysis tool. Because the built-in functions are the easiest of these options, we'll explore them in detail and see how they may be used to evaluate analytical data. We'll also briefly describe the regression analysis tool.

The Slope and Intercept

As an example, we will take the external standard calibration data for the chromatographic

determination of isooctane as described in example 7-7 of AC7 or Example 8-4 of FAC9. We

enter the data from Table 7-4 of AC7 or 8-1 of FAC9 so that it appears as shown in Figure 4-1.

	A	B	C
1		x	y
2		0.352	1.09
3		0.803	1.78
4		1.08	2.6
5		1.38	3.03
6		1.75	4.01
7			
8	Slope		
9	Intercept		
10			

Figure 4-1 Worksheet after entering data.

Now, click on cell B8, and then on the Home tab. Click on the Insert Function icon so that the

window shown in Figure 4-2 appears. Then click on the Statistical category.

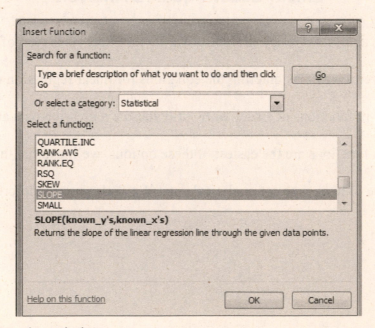

Figure 4-2 Insert Function window.

Use the mouse to scroll down the list of functions until you come to the SLOPE function, and the click on it. The function appears in bold under the Select a function: window, and a description of the function appears below it. Read the description of the slope function, and then click OK. The window shown in Figure 4-3 appears just below the formula bar.

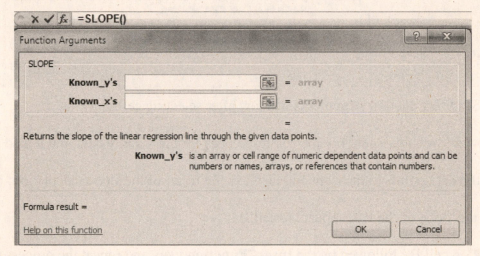

Figure 4-3 Function Arguments window for SLOPE function.

Look carefully at the information that is provided in the window and in the formula bar. The SLOPE() function appears in the formula bar with no arguments, so we must select the data that Excel will use to determine the slope of the line. Now click on the selection button at the right end of the Known_y's box, use the mouse to select cells C2:C6, and type **[↵]**. Similarly, click on the selection button for the Known_x's box, select cells B2:B6 followed by **[↵]**. The window should now appear as shown in Figure 4-4.

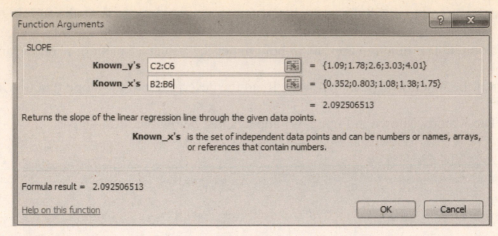

Figure 4-4 SLOPE Function Arguments window after selecting cells.

The window shows not only the cell references for the x and y data, but it also shows the first few of the data points to the right and displays the result of the slope calculation. Now click on OK, and the slope of the line appears in cell B8.

Click on cell B9 followed by the **Insert Function** icon, and repeat the process that we just carried out, except that now you should select the INTERCEPT function. When the intercept Function Arguments window appears, select the Known_y's and the Known_x's as before, and click OK. When you have finished, the worksheet should appear as shown in Figure 4-5.

	A	B	C
1		x	y
2		0.352	1.09
3		0.803	1.78
4		1.08	2.6
5		1.38	3.03
6		1.75	4.01
7			
8	**Slope**	2.092507	
9	**Intercept**	0.256741	

Figure 4-5 Worksheet after calculation of slope and intercept.

At this point you may wish to compare these results to those obtained for the slope and intercept in Example 8-4 of FAC9 or Example 7-7 of AC7. We should note at this point that Excel provides many digits that are not significant. We shall see how many figures are significant after we find the standard deviations of the slope and intercept.

Using LINEST

Another way to accomplish the least-squares analysis is by using the LINEST function. On the worksheet shown in Figure 4-5, use the mouse to select an array of cells two cells wide and five cells high, such as E2:F6. Then click on the Insert Function icon, select STATISTICAL and LINEST in the left and right windows, respectively, and click on OK. Select the Known_y's and Known_x's as before, then click on the box labeled Const and type **true**. Also type **true** in the box labeled Stats. When you click on each of the latter two boxes, notice that a description of the meaning of these logical variables appears below the box. In order to activate the LINEST function, you must now type the rather unusual keystroke combination `Ctrl+Shift+[ ↵]`. This keystroke combination must be used whenever you perform a function on an array of cells. The worksheet should now appear as in Figure 4-6.

▲	A	B	C	D	E	F	G
1		x	y				
2		0.352	1.09		2.092507	0.256741	
3		0.803	1.78		0.134749	0.158318	
4		1.08	2.6		0.987712	0.144211	
5		1.38	3.03		241.1465	3	
6		1.75	4.01		5.015089	0.062391	
7							
8	Slope	2.092507					
9	Intercept	0.256741					

Figure 4-6 Worksheet after implementing the LINEST function.

As you can see, cells E2 and F2 contain the slope and intercept of the least-squares line. Cells E3 and F3 are the respective standard deviations of the slope and intercept. Cell E4 contains the coefficient of determination (R^2). The standard deviation about regression (s_r, standard error of the estimate) is located in cell F4. The smaller the s_r value, the better the fit. The square of the standard error of the estimate is the mean square for the residuals (error). The value in cell E5 is the F statistic. Cell F5 contains the number of degrees of freedom associated with the error. Finally cells E6 and F6 contain the sum of the squares of the regression and the sum of the squares of the residuals, respectively. Note that the F value can be calculated from these latter quantities as described in Section 8D-2 of FAC9.

It is worth noting that the number of significant figures that we keep in a least-squares analysis depends on the use for which the data are intended. If the results are to be used to carry out further spreadsheet computations, wait until final results are computed before rounding to an appropriate number of significant figures. Excel provides 15 digits of numerical precision, and so, in general, spreadsheet computations will not contribute to the uncertainty in the final results. Final answers must be rounded to be consistent with the uncertainty in the original data, which is reflected in the standard deviations of the slope and intercept and the standard error of the estimate. The standard deviations of the slope and intercept in our example suggest that, at most, we should express both the slope and the intercept to only two decimal places. Thus, the least-squares results for the slope and intercept may be expressed as 2.09 ± 0.13 and 0.26 ± 0.16, respectively, or as 2.1 ± 0.1 and 0.3 ± 0.2.

The Analysis ToolPak Regression Tool

Yet a third way to do a linear least-squares analysis is to use the regression function in Excel's Analysis ToolPak. The advantage of this third mode is the production of a complete ANOVA table for the analysis. Select the **Data** tab and bring up the Data ribbon. At the far right, select **Data Analysis**. From the Data Analysis window, select **Regression** and then click on OK. Using the isooctane worksheet as before, select C2:C6 as the Input <u>Y</u> Range: and B2:B6 as the Input <u>X</u>

Range:. Select New Worksheet <u>Ply</u>: for the output. The results should appear as shown in Figure

4-7. Note you will have to expand some cell widths to see all the text.

In addition to the normal regression statistics, the output of Figure 4-7 shows a statistic

called the multiple correlation coefficient (multiple R) and the adjusted R square. The former is

beyond the scope of this discussion[1], while the latter is the R^2 value adjusted for the number of

parameters used in the fit (the adjusted R^2 includes a "price" to pay for adding more parameters

to improve the fit). In the ANOVA table the entries are the degrees of freedom (*df*), the sum of

squares (*SS*), the mean square values (*MS*), the *F* statistic, and the significance level of the *F*

value (probability of getting an *F* value this large by random chance alone). The bottom table

gives the slope and intercept, their standard deviations (standard errors), the *t* statistics for the

slope and intercept, the probabilities of getting these *t* values by random chance, and the upper

and lower values for the 95% confidence intervals for the slope and intercept. From this analysis,

we note that the linear model fits the data quite well. There is, however, significant uncertainty in

the intercept value.

---

[1] For further discussion of the multiple correlation coefficient, see J. L. Devore, *Probability and Statistics for Engineering and the Sciences,* 8th edition (Boston, MA: Duxbury, Brooks/Cole, 2012), p. 560.

	A	B	C	D	E	F	G
1	SUMMARY OUTPUT						
2							
3	*Regression Statistics*						
4	Multiple R	0.993837158					
5	R Square	0.987712297					
6	Adjusted R Square	0.983616396					
7	Standard Error	0.144211147					
8	Observations	5					
9							
10	ANOVA						
11		*df*	*SS*	*MS*	*F*	*Significance F*	
12	Regression	1	5.015089435	5.015089	241.1465	0.000580234	
13	Residual	3	0.062390565	0.020797			
14	Total	4	5.07748				
15							
16		*Coefficients*	*Standard Error*	*t Stat*	*P-value*	*Lower 95%*	*Upper 95%*
17	Intercept	0.256740511	0.158317598	1.62168	0.203322	-0.247096745	0.760578
18	X Variable 1	2.092506513	0.134749235	15.52889	0.00058	1.663674308	2.521339

Figure 4-7 Results using Analysis ToolPak Regression Tool.

Plotting a Graph of the Data and the Least-Squares Fit

It is customary and useful to plot a graph of the data and the least-squares fitted line. The built-in charting engine of Excel makes creating such plots relatively easy. There are several ways to display the data points and the predicted line simultaneously. One way is to plot the predicted values $\hat{y}$ and the experimental y values simultaneously. The predicted values for the isooctane data are given in Table 8-2 of FAC9. The easiest way is to have Excel add the line, called a trendline, itself.

To plot the points, select the xy data (Cells B2:C6) from the original isooctane worksheet. Click on the **Insert** tab to display the Insert ribbon. The Charts group on the Insert ribbon includes drop-down menus for seven charting types: column, line, pie, bar, area, scatter, and other charts. Since we want to plot xy data, display the **Scatter** drop-down shown in Figure 4-8 by clicking on the downward pointing arrow under the Scatter chart type.

Figure 4-8 Drop-down menu for Scatter chart type.

Choose the upper left Scatter type (Scatter with only markers). This will produce the

default graph shown in Figure 4-9.

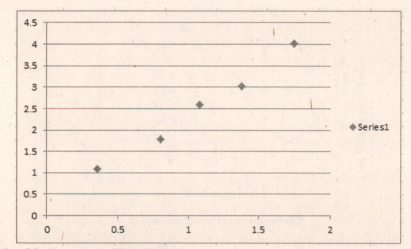

Figure 4-9 Graph of isooctane calibration data.

Next we will make the graph look more presentable for scientific data. Click on the chart

and note that the Chart Tools ribbon appears. We will first change the gridlines. Click on the

Layout tab on the Chart Tools ribbon to display the Layout gallery shown in Figure 4-10.

Figure 4-10 Chart layout tab.

Now click on **Gridlines** in the axes group and choose Major gridlines under the Primary

Vertical gridlines option. Note that the default graph has already chosen Major gridlines for the

Primary Horizontal gridlines. Now click on **Axis Titles** in the Labels group. Under the Primary

Horizontal Axis Title, choose Title Below Axis. Type Concentration of isooctane, mol % for the

horizontal axis title. Under the Primary Vertical Axis Title, choose **Rotated Title** and type Peak

area, arbitrary units. Note that the default font size is 10 point. You can change the font face and

font size by selecting the text to be changed, clicking on the Home tab to display the Home

ribbon and choosing a new font face and font size in the Font group. Change the Axis Titles to be

Arial font face and 11 point font size. Your chart should now appear as shown in Figure 4-11.

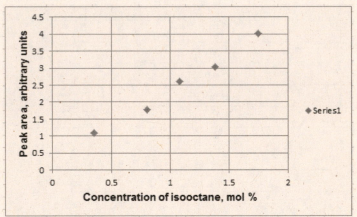

Figure 4-11 Plot of isooctane calibration data with gridlines and axes titles.

Next we will add the least-squares line. Now right-click on any data point and then click

on **Add Trendline….** Under Trendline Options, select **Linear**, and check both Display Equation

on chart and Display R-squared value on chart. Then click on Close. The weight of the line can

be adjusted by right-clicking on the line and selecting **Format Trendline….**Under Line Style,

select a line of 1 pt. Width. You can also move the equation and R^2 text to a more convenient

place as indicated in Figure 4-12.

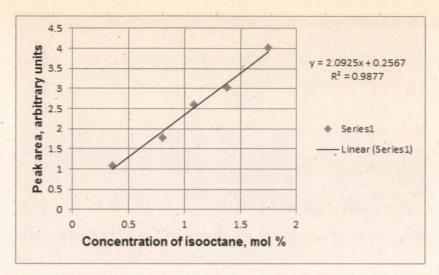

Figure 4-12 Plot of isooctane calibration curve with added trendline.

As an extension of this exercise, modify your spreadsheet to include a column of residuals. You can either enter them from Table 8-2 of FAC9 or include residuals when obtaining the ANOVA table in the Analysis ToolPak. You can even produce the chart of the fit and the residuals directly from the Regression window by checking the appropriate boxes. Be sure to save your spreadsheet in a file for reference and for future use.

Using Excel to Find Unknown Concentrations From a Calibration Curve

Now let's use the calibration curve from above to find the concentration of isooctane in a sample and the standard deviation in concentration. We'll use the data of Example 8-5 of FAC9 or Example 7-8 of AC7. In this example, a peak area of 2.65 was obtained for the unknown. Type the label **Unknown** in cell A7 of the original worksheet, and enter 2.65 in cell C7.

From the equation for a line, $y = mx + b$, we can rearrange to give the concentration x as

$$x = \frac{y-b}{m}$$

Add the label **Concentration of unknown** in cell A10 and enter in cell B10

=(C7-B9)/B8 [↵]

Your worksheet should now appear like that shown in Figure 4-13.

	A	B	C	D	E	F
1		x	y			
2		0.352	1.09		2.092507	0.256741
3		0.803	1.78		0.134749	0.158318
4		1.08	2.6		0.987712	0.144211
5		1.38	3.03		241.1465	3
6		1.75	4.01		5.015089	0.062391
7	Unknown		2.65			
8	Slope	2.092507				
9	Intercept	0.256741				
10	Concentration of unknown	1.143729				

Figure 4-13 Least-squares worksheet after entering unknown and calculating its concentration.

In order to find the standard deviation in concentration, we utilize Equation 8-18 of FAC9 or 7-18 of AC7. In this equation we need several additional quantities. First, we need s_r, the standard deviation about regression, also called the standard error in Y. We also need M, the number of replicate analyses of unknowns (1 in this case) and N, the number of points in the calibration curve (5 in this case). Finally we need S_{xx}, the sum of the squares of the deviations of x values from the mean x value and the mean y value. In column A, add the labels **standard error in Y, N, S_{xx}, y bar, M,** and **Standard deviation in c** in cells A11 through A16. Your worksheet should now appear as shown in Figure 4-14.

	A	B	C	D	E	F
1		x	y			
2		0.352	1.09		2.092507	0.256741
3		0.803	1.78		0.134749	0.158318
4		1.08	2.6		0.987712	0.144211
5		1.38	3.03		241.1465	3
6		1.75	4.01		5.015089	0.062391
7	Unknown		2.65			
8	Slope	2.092507				
9	Intercept	0.256741				
10	Conc. Unknown	1.143729				
11	Standard error in Y					
12	N					
13	S_{xx}					
14	y bar					
15	M					
16	Standard deviation in c					

Figure 4-14 Worksheet after entering labels for error analysis.

The standard error in Y could be found from Equation 8-15 of FAC9 or 7-15 of AC7. However, it is easier to make use of the built-in Excel function STEYX(), which returns the standard error in Y directly. Hence, select cell B11. Click on the Insert Function icon. In the Insert Function window, select the Statistical category and the STEYX function. Click OK. This opens the STEYX Function Arguments window shown in Figure 4-15.

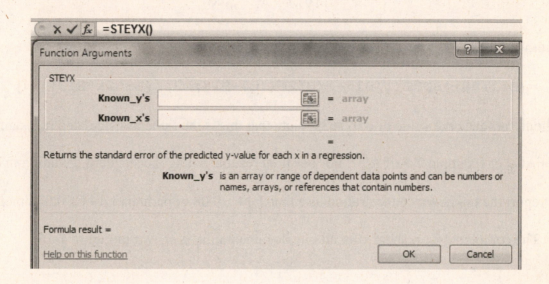

Figure 4-15 The STEYX Function Arguments window.

Select cells C2:C6 for the **Known_y's** and cells B2:B6 for the **Known_x's**. Click on OK. This should return the value of 0.144211 in cell B11. Note that this is the same value given by LINEST in cell F4 or by the Analysis ToolPak under Regression Statistics, Standard Error.

Now enter 5 for N in cell B12 or use the Count function as described earlier. In cell B13, we need to find the sum of the squares of the deviations of x from the mean x. This could be calculated from Equation 8-10 of FAC9 or 7-10 of AC7. The easiest way is to use Excel's built-in function **DEVSQ()** as discussed in Chapter 2. Thus, we enter in cell B13

$$\texttt{=DEVSQ(B2:B6) [↵]}$$

In cell B14, we need to find the average of the y values. To do this we type

=AVERAGE(C2:C6) [↵]

In cell B15, we type 1 or use the Count function for the number of repetitions of the unknown

results. Now, we are only left with finding the standard deviation in c. We use Equation 8-18 of

FAC8 or 7-18 of AC7, which is

$$s_c = \frac{s_r}{m}\sqrt{\frac{1}{M} + \frac{1}{N} + \frac{(\bar{y}_c - \bar{y})}{m^2 S_{xx}}}$$

Enter into cell B16, the formula corresponding to this equation

=(B11/B8)*SQRT(1/B15+1/B12+((C7-B14)^2)/(B8^2)*B13) [↵]

This should produce the value of 0.075633. Note that this is the same as calculated in Example

8-5 of FAC9 or Example 7-8 of AC7. For our final results, noting the standard deviation in c, we

would report the unknown concentration as either 1.14 ± 0.08 or perhaps 1.144 ± 0.076 mole %.

The completed worksheet after adding documentation is shown in Figure 4-16.

	A	B	C	D	E	F
1		x	y			
2		0.352	1.09		2.092507	0.256741
3		0.803	1.78		0.134749	0.158318
4		1.08	2.6		0.987712	0.144211
5		1.38	3.03		241.1465	3
6		1.75	4.01		5.015089	0.062391
7	Unknown		2.65			
8	Slope	2.092507				
9	Intercept	0.256741				
10	Concentration of unknown	1.143729				
11	Standard error in Y	0.144211				
12	N	5				
13	S_{xx}	1.145368				
14	y bar	2.502				
15	M	1				
16	Standard deviation in c	0.075676				
17						
18	Documentation					
19	Cell E2=LINEST(C2:C6,B2:B6,TRUE,TRUE)					
20	Cell B8=SLOPE(C2:C6,B2:B6)					
21	Cell B9=INTERCEPT(C2:C6,B2:B6)					
22	Cell B10=(C7-B9)/B8					
23	Cell B11=STEYX(C2:C6,B2:B6)					
24	Cell B12=COUNT(B2:B6)					
25	Cell B13=DEVSQ(B2:B6)					
26	Cell B14=AVERAGE(C2:C6)					
27	Cell B15=COUNT(C7)					
28	Cell B16=B11/B8*SQRT(1/B15+1/B12+((C7-B14)^2)/((B8^2)*B13))					

Figure 4-16 Completed least-squares worksheet after adding documentation.

The Internal Standard Method

In the *internal standard method*, a known amount of a reference species is added to all the samples, standards and blanks. The response signal is then not the analyte signal itself, but the *ratio* of the analyte signal to the reference species signal. A calibration curve is prepared where the y axis is the ratio of responses and the x axis is the analyte concentration in the standards as usual. The internal standard method is often used to compensate for errors if they influence both the analyte and the reference signals to the same proportional extent.

As an example, we'll use Example 8-7 of FAC9, which describes the determination of sodium by flame spectrometry. Lithium was added as an internal standard. The worksheet shown in Figure 4-17 can be constructed by entering the data from the table in Example 8-7 of FAC9.

The data are entered into columns A through C as shown. Cells D4 through D9 calculate the ratio of the sodium intensity to lithium intensity using the formula shown in documentation cell A22. Two plots are shown. In the upper plot, the normal calibration curve of sodium intensity versus sodium concentration is shown. Note that the linearity is poor. The lower plot shows the intensity ratio of sodium to lithium versus the sodium concentration. Note that the linearity of this plot is quite good. To determine the unknown concentration, we'll use the internal standard method since the linearity is much better. The concentration of the unknown is obtained from the linear regression equation by the method described in the previous section. The linear regression statistics are calculated in cells B11 through B20 using the approach described earlier. The statistics are found by the formulas in documentation cells A23 through A31. The sodium concentration in the unknown is found to be 3.55 ± 0.05 ppm.

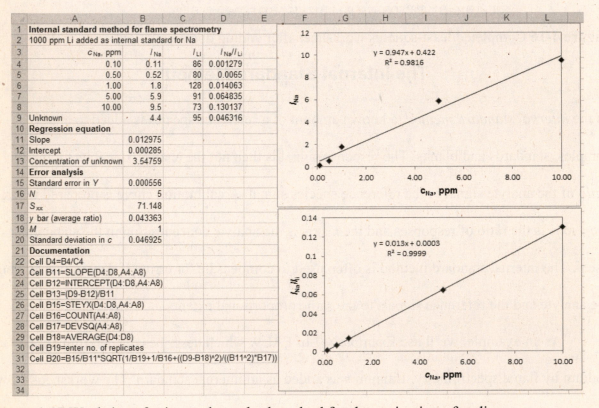

Figure 4-17 Worksheet for internal standard method for determination of sodium.

Method of Standard Additions

The method of standard additions is widely used when it is difficult or impossible to duplicate the sample matrix. The single point standard additions method is discussed in Section 8D-3 of FAC9 and Sections 19F-3 and 23A-2 of AC7. We'll discuss here the multiple additions method in which additions of known amounts of standard analyte solution are made to several portions of the sample. A calibration curve is then made of the analytical signal versus the volume of the standard solution added.

Principles of Multiple Standard Additions

Let's take an example in which strontium is determined in seawater by ICP emission spectrometry. There is a linear relationship between the emission signal I and the Sr concentration c. In the standard additions method, various volumes of a 3.00 g/L standard solution of Sr were added to 1.00 L of the sample solution and the emission signal was recorded after blank subtraction. The following results were obtained:

Volume of Sr added, mL	ICP emission signal, I
0.00	18.2
1.00	23.2
2.00	30.1
3.00	35.8
4.00	41.9
5.00	48.0

The concentration of Sr in the sample is to be determined by the multiple additions method.

We'll call the analytical signal for the sample without standard added, I_x and that after the addition, I_{x+s}. The concentration before any addition is c_x and that after an addition is c_{x+s}. If the analytical signal is proportional to the analyte concentration, we can write

$$I_x = kc_x$$

and

$$I_{x+s} = kc_{x+s}$$

where k is the proportionality constant. Now the concentration after the addition is the number of moles originally present plus the number of moles added divided by the total solution volume ($V_x + V_s$). Or,

$$c_{x+s} = \frac{c_x V_x + c_s V_s}{V_x + V_s}$$

Substituting into the equation for c_{x+s} and rearranging gives,

$$I_{x+s} = \frac{k V_x c_x + k V_s c_s}{V_x + V_s}$$

A plot of I_{x+s} as a function of V_s should give a straight line of the form

$$I_{x+s} = mV_s + b$$

where the slope m and intercept b are given by

$$m = \frac{kc_s}{V_x + V_s} \quad \text{and} \quad b = \frac{k V_x c_x}{V_x + V_s}$$

After solving both equations for k and setting them equal, we find,

$$k = \frac{m(V_x + V_s)}{c_s} = \frac{b(V_x + V_s)}{V_x c_x}$$

From this equation, the concentration of the unknown can be found as

$$c_x = \frac{bc_s}{mV_x}$$

It is useful in multiple standard additions to find the volume intercept when $I_{x+s} = 0$. The intercept on the volume axis, V_{int}, is given by

$$V_{int} = \frac{-b}{m}$$

From this quantity, the unknown concentration is readily obtained as

$$c_x = \frac{-V_{int}c_s}{V_x}$$

Finding the Unknown Concentration

The worksheet for this multiple additions problem is shown in Figure 4-18. The only differences in this multiple additions example from the normal least-squares examples described previously occur in the calculation of the concentration of the unknown. The concentration of the standard c_s is entered into cell B2, and the volume of the unknown used is entered into cell B3. The slope and intercept of the multiple additions curve are calculated in the usual manner in cells B12 and B13. The formulas corresponding to the equations for V_{int} and c_x from above are entered into cells B14 and B15 respectively. The units for the unknown concentration must now be obtained. The units for c_x are obtained as follows:

$$c_x = \frac{bc_s}{mV_x} \quad \text{in units of} \quad \frac{g/L}{L^{-1} \times mL} = g/mL \text{ or } mg/\mu L$$

Analyzing Errors

The error analysis shown in Figure 4-18 is similar to that for a normal least-squares analysis. The standard error in y is calculated in cell B17 with the STEYX() function. The sum of the squares S_{xx} is calculated in cell B19 with the DEVSQ() function. Because the volume added is the x independent variable, it is relatively straightforward to find the standard deviation in the volume. The formula is Equation 8-18 of FAC9 or 7-18 of AC7 with the $1/M$ term absent and the x value for the unknown entered as 0.00 since this is the emission signal at the x axis intercept.

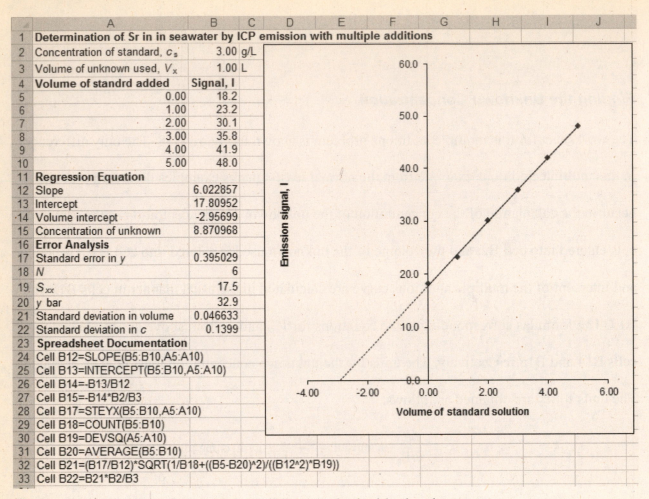

The table in the figure:

	A	B	C
1	**Determination of Sr in in seawater by ICP emission with multiple additions**		
2	Concentration of standard, c_s	3.00	g/L
3	Volume of unknown used, V_x	1.00	L
4	**Volume of standrd added**	**Signal, I**	
5	0.00	18.2	
6	1.00	23.2	
7	2.00	30.1	
8	3.00	35.8	
9	4.00	41.9	
10	5.00	48.0	
11	**Regression Equation**		
12	Slope	6.022857	
13	Intercept	17.80952	
14	Volume intercept	-2.95699	
15	Concentration of unknown	8.870968	
16	**Error Analysis**		
17	Standard error in y	0.395029	
18	N	6	
19	S_{xx}	17.5	
20	y bar	32.9	
21	Standard deviation in volume	0.046633	
22	Standard deviation in c	0.1399	
23	**Spreadsheet Documentation**		
24	Cell B12=SLOPE(B5:B10,A5:A10)		
25	Cell B13=INTERCEPT(B5:B10,A5:A10)		
26	Cell B14=-B13/B12		
27	Cell B15=-B14*B2/B3		
28	Cell B17=STEYX(B5:B10,A5:A10)		
29	Cell B18=COUNT(B5:B10)		
30	Cell B19=DEVSQ(A5:A10)		
31	Cell B20=AVERAGE(B5:B10)		
32	Cell B21=(B17/B12)*SQRT(1/B18+((B5-B20)^2)/((B12^2)*B19))		
33	Cell B22=B21*B2/B3		

Figure 4-18 Worksheet for multiple additions method with plot shown.

To convert the standard deviation in volume to the standard deviation in concentration, we use the following relationship obtained from the equation for c_x:

$$s_c = s_V \left(\frac{c_s}{V_x} \right)$$

where s_c is the standard deviation in c_x, and s_V is the standard deviation in volume. The formula corresponding to this equation is entered into cell B22.

Now that the standard deviation in c_x is known, we can report the concentration with the appropriate number of significant figures. We should report 8.9 ± 0.1 mg/μL or perhaps 8.87 ± 0.14 mg/μL as the concentration of Sr in the seawater sample.

Plotting the Curve

Also shown in Figure 4-18 is the standard additions plot. To construct this plot, select the x and y data as before and click on the Scatter drop-down in the Insert ribbon. Choose a Scatter with only markers plot. To get the x axis range to extend to negative values, right click on the x axis. Choose Format Axis…to display the Format Axis window. Under Axis Options, set the minimum to a fixed value -4.00. Use the right mouse button and click on a data point. Choose Add Trendline… and select Linear under Trend/Regression Type. Under Forecast select 3 units under Backward. This extends the plot backward 3 units before the first data point. You can try various formatting options to make the plot look better or to add a dashed line to the plot. Your final worksheet should look something like Figure 4-18. Save the worksheet with a file name like `multipleaddns.xls` for future use.

Multiple Linear Regression

Often in chemistry, we are interested in the influence of more than one variable on an outcome. For example, we might be interested in the effect of reagent concentration and pH on the concentration of an absorbing product or the influence of time and pressure on the yield of a gas-phase reaction. The Excel function LINEST and the Analysis ToolPak Regression tool can both be used in the case of multiple regression analysis. In the case of two variables x_1 and x_2 affecting result y, the linear equation treated by Excel is of the form:

$$y = m_1 x_1 + m_2 x_2 + b$$

This can be extended to any number (n) of variables influencing result y.

LINEST

To illustrate for two variables, we'll taken an example of a chemical reaction that produces an absorbing product. The yield of the product, and thus the absorbance, is affected by reaction time t (minutes) and reaction temperature T (°C). An experiment then determined A vs t and T with the results shown in Figure 4-19. Enter these results.

	A	B	C
1	x_1 (Time)	x_2 (Temp)	y (A)
2	16.7	30	1.05
3	17.4	42	0.55
4	18.4	47	0.515
5	16.8	47	0.515
6	18.9	43	0.455
7	17.1	41	0.38
8	17.3	48	0.365
9	18.2	44	0.35
10	21.3	43	0.34
11	21.2	50	0.265
12	20.7	56	0.225
13	18.5	60	0.155

Figure 4-19 Worksheet showing the influence of time and temperature on absorbance

We will first use LINEST to determine the regression equation and its statistics. Select an array of cells, three cells wide and five cells high, such as F14:H18. Then click on the Insert Function icon, select STATISTICAL and LINEST in the left and right windows, respectively, and click on OK. Select the cells C2:C13 as the Known_y's and the variables in cells A2:B13 as the Known_x's. Select TRUE for both the constant and the statistics.

Now type the unusual array activation command **Ctrl+Shift+[↵]** in order to activate LINEST. You should obtain the results shown in Figure 4-20. In columns D, E, and I, we have labeled the results shown in cells F14:H18. In cell F14 is the slope for variable x_2, m_2. The remainder of column F displays the standard deviation of m_2, the R^2 value, the F value, and the sum of the squares due to regression, just as LINEST shows for a single variable. Likewise, column G shows m_1, the standard deviation of m_1, the standard error of the estimate, s_r, the

number of degrees of freedom, df, and the sum of the squares of the residuals. Column H shows

the intercept, b and the standard deviation of b, with the remainder of the column being Non-

applicable (#NA).

	A	B	C	D	E	F	G	H	I
1	x_1 (Time)	x_2 (Temp)	y (A)						
2	16.7	30	1.05						
3	17.4	42	0.55						
4	18.4	47	0.515						
5	16.8	47	0.515						
6	18.9	43	0.455						
7	17.1	41	0.38						
8	17.3	48	0.365						
9	18.2	44	0.35						
10	21.3	43	0.34						
11	21.2	50	0.265						
12	20.7	56	0.225						
13	18.5	60	0.155						
14				m_2	m_1	-0.02252	-0.03296	2.075565	b
15				SD m_2	SD m_1	0.005356	0.024296	0.412587	SD b
16				R^2	s_r	0.767997	0.122273	#N/A	
17				F	df	14.89632	9	#N/A	
18				SSRegr	SSResid	0.445418	0.134555	#N/A	

Figure 4-20 LINEST results for multiple linear regression.

From these results, we can write the regression equation as:

$$A = -0.023T - 0.03t + 2.1$$

ANALYSIS TOOLPAK

We can accomplish the same multiple regression with the Analysis ToolPak Regression Tool.

With the same worksheet as used in Figure 4-20, Select Data Analysis on the Data ribbon. In the

Data Analysis window, choose Regression. Select cells C2:C13 as the Input Y range and cells

A2:B13 as the Input X range. Choose New Worksheet Ply: Click on OK. The new worksheet

shown in Figure 4-21 then opens. Note that there are now statistics for the two different slopes, but the rest of the statistics are the same as shown in Figure 4-7 for one variable. The ANOVA

table again gives valuable information regarding the regression and its significance.

	A	B	C	D	E	F	G
1	SUMMARY OUTPUT						
2							
3	Regression Statistics						
4	Multiple R	0.876354504					
5	R Square	0.767997216					
6	Adjusted R Square	0.716441042					
7	Standard Error	0.122272615					
8	Observations	12					
9							
10	ANOVA						
11		df	SS	MS	F	Significance F	
12	Regression	2	0.445417585	0.222709	14.89632	0.001395466	
13	Residual	9	0.134555331	0.014951			
14	Total	11	0.579972917				
15							
16		Coefficients	Standard Error	t Stat	P-value	Lower 95%	Upper 95%
17	Intercept	2.075564989	0.412587023	5.030611	0.000709	1.142228299	3.008902
18	X Variable 1	-0.032963892	0.024296273	-1.35675	0.207913	-0.08792588	0.021998
19	X Variable 2	-0.022517811	0.005355781	-4.20439	0.002292	-0.03463343	-0.0104

Figure 4-21 Multiple linear regression with the Analysis ToolPak.

Linear Regression With Polynomials

The techniques of LINEST and the Analysis ToolPak can be used to find the regression

coefficients for equations of higher order. To produce a curve for a higher-order function, a

power series is often used such as

$$y = a + bx + cx^2 + dx^3 + \cdots$$

	A	B	C	D
1	T, deg C			Sol. ppm
2	0			58.4
3	5			51.1
4	10			45.1
5	15			40.3
6	20			36.4
7	25			33.1
8	30			30.3
9	35			27.9
10	40			25.9
11	45			22.5
12	50			20.1
13	60			14.5
14	70			6.8

Figure 4-22 Data representing the solubility of oxygen in water.

Enter the data representing the solubility of oxygen in fresh water as a function of temperature as shown in Figure 4-22. We have purposely left columns B and C blank. It is good practice to use the equation with the lowest order first and then to add terms to improve the fit. Try a linear equation with the data in Figure 4-22. Plot a scatter plot and add a least-squares trendline to the plot. You will find a poor fit to a linear equation as shown in Figure 4-23a. Notice how the data points fall above, then below and then above the line. This indicates that a residual plot is sinusoidal as shown in Figure 4-23b. You can produce a residual plot by calculating the difference between the data value and that predicted by the linear equation. Such a sinusoidal residual plot indicates a higher order function may be more appropriate.

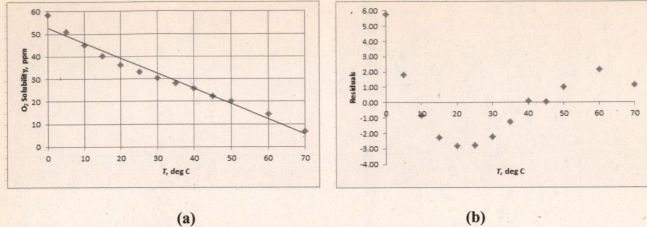

(a)   **(b)**

Figure 4-23 Solubility data with a linear trendline (a) and the residual plot (b).

Now, in cell B1 enter T^2 as the column title. Fill out the column with data representing

the square of the temperature shown in column A. Try LINEST with two x variables, T and T^2.

Try a scatter plot of Solubility versus Temperature, but use a polynomial Trendline of order 2.

You will find a better fit, but still not extremely good.

Now in cell C1, enter the title T^3 and compute the values in of T^3 in column C. Use

LINEST and produce the statistics shown in Figure 4-24. Notice LINEST has calculated the

coefficients which indicates the third-order equation

$$y = -0.0002x^3 + 0.0251x^2 - 1.504x + 58.11$$

represents the data. Plot a Scatter plot of Solubility versus Temperature and chose a polynomial

Trendline of order 3. The resulting plot is shown in Figure 4-25. Note that the equation displayed

on the chart is exactly the same as that given by LINEST. The advantage of LINEST is that it

produces the statistics of the coefficients. Now use the Analysis ToolPak Regression tool to

produce the statistic for a third-order polynomial including the ANOVA table.

	A	B	C	D	E
1	*T*, deg C	T^2	T^3	Sol. ppm	
2	0	0	0	58.4	
3	5	25	125	51.1	
4	10	100	1000	45.1	
5	15	225	3375	40.3	
6	20	400	8000	36.4	
7	25	625	15625	33.1	
8	30	900	27000	30.3	
9	35	1225	42875	27.9	
10	40	1600	64000	25.9	
11	45	2025	91125	22.5	
12	50	2500	125000	20.1	
13	60	3600	216000	14.5	
14	70	4900	343000	6.8	
15					
16		-0.0002	0.025135	-1.50373	58.10976
17		1.26E-05	0.001342	0.038925	0.296491
18		0.999545	0.359038	#N/A	#N/A
19		6595.75	9	#N/A	#N/A
20		2550.743	1.160176	#N/A	#N/A
21					
22	**Documentation**				
23	Cell B2=A2^2				
24	Cell C2=A2^3				
25	Cells B16:E20=LINEST results				

Figure 4-24 Worksheet showing LINEST results for a polynomial of third order.

You could also plot the residuals and note that they are much smaller than with the linear equation. The residuals are also more randomly distributed showing that the fit is much better.

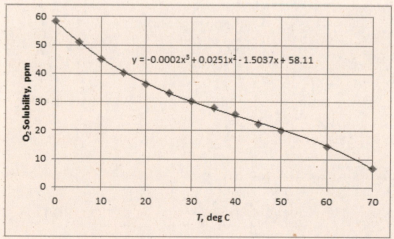

Figure 4-25 Scatter plot showing a third-order polynomial and the solubility data.

Summary

In this Chapter we explored several methods for calibration, standardization, and regression analysis. The method of linear least-squares was performed with several of Excel's functions. Unknown concentrations were calculated from least-squares analyses. The internal standard method and the method of multiple standard additions were investigated and used to obtain concentrations. Calibration curves were plotted with the aid of Excel's charting functions. Multiple linear regression and regression with polynomial functions were briefly treated. The tools introduced in this chapter are useful in many different situations in analytical chemistry.

Problems

1.   The sulfate ion concentration in natural water can be determined by using a turbidimeter

to measure the turbidity that results when an excess of $BaCl_2$ is added to a measured

quantity of the sample. A turbidimeter was calibrated with a series of standard Na_2SO_4

solutions. The following data were obtained in the calibration:

Conc. SO_4^{2-}, mg/L	Turbidimeter Reading, R
0.00	0.06
5.00	1.48
10.00	2.28
15.0	3.98
20.0	4.61

Assume that there is a linear relationship between the instrument reading and

concentration.

(a)   Determine the least-squares slope and intercept for the best straight line among

the points using Excel's built-in SLOPE() and INTERCEPT() functions.

(b)   Use LINEST to determine the regression statistics.

(c)   Use Excel to plot the data and add a linear regression trendline. Display the

equation and R^2 values on the chart.

(d)   Use the Analysis ToolPak to perform regression on the data, and find the R^2 value

and the adjusted R^2 value. Comment on the significance of the regression.

(e)   A sample containing an unknown concentration of sulfate gave a turbidimeter

reading of 2.84. Find the concentration of the sample, the absolute standard

deviation and the coefficient of variation.

(f)   Repeat the calculations in (d) assuming that the 2.84 was the mean of six

turbidimeter readings.

2.   The following data were obtained in calibrating a calcium ion-selective electrode for the

determination of pCa. A linear relationship between the potential E and pCa is known to

exist.

pCa = $-\log$ [Ca^{2+}]	E, mV
5.00	-53.8
4.00	-27.7
3.00	+2.7
2.00	+31.9
1.00	+65.1

(a)   Use Excel to find the least-squares expression for the best straight line through the

points. Plot the points and the least-squares line.

(b)   Use the Regression function from Excel's Analysis ToolPak, and report the

statistics given in the ANOVA table. Comment on the meaning of the ANOVA

statistics.

(c)   Calculate pCa for a serum solution in which the electrode potential was +10.7

mV. Find the absolute and relative standard deviations for pCa if the result was

from a single voltage measurement.

(d)   Find the absolute and relative standard deviations for pCa if the millivolt reading

in (c) was the mean of two replicate measurements. Repeat the calculation based

on the mean of eight measurements.

3.   The following are relative peak areas for chromatograms of standard solutions of methyl

vinyl ketone (MVK).

Concentration MVK, mmol/L	Relative Peak Area
0.500	3.76
1.50	9.16
2.50	15.03
3.50	20.42
4.50	25.33
5.50	31.97

(a)   Use Excel to find the equation of the least-squares line.

(b)   Find the Regression statistics using the Analysis ToolPak.

(c)   Plot the experimental points and superimpose the least-squares line.

(d)   A sample containing MVK yielded relative a peak area of 10.3. Find the

concentration of MVK in the solution.

(e)   Assume that the result in (d) represents a single measurement as well as the mean

of four measurements. Calculate the respective absolute and relative standard

deviations.

(f)   Repeat the calculations in (d) and (e) for a sample that gave a peak area of 22.8.

4.   The data in the table below were obtained during a colorimetric determination of glucose

in blood serum.

Glucose concentration, mM	Absorbance, A
0.0	0.002
2.0	0.150
4.0	0.294
6.0	0.434
8.0	0.570
10.0	0.704

(a)   Assuming that there is a linear relationship between the two variables, find the

least-squares estimates of the slope and intercept.

(b)   What are the standard deviations of the slope and intercept? What is the standard

error of the estimate?

(c)   Determine the 95% confidence intervals for the slope and intercept.

(d)   A serum sample gave an absorbance of 0.350. Find the 95% confidence interval

for glucose in the sample?

5.   The data in the table below represent electrode potential E vs. concentration c

E, mV	Concentration c in mol L^{-1}
106	0.20000
115	0.07940
121	0.06310
139	0.03160
153	0.02000
158	0.01260
174	0.00794
182	0.00631
187	0.00398
211	0.00200
220	0.00126
226	0.00100

(a)   Transform the data to E vs. $-\log c$ values.

(b)   Plot E vs. $-\log c$ and find the least-squares estimate of the slope and intercept.

Write the least-squares equation.

(c)   Find the 95 % confidence limits for the slope and intercept.

(d)   Use Excel's Analysis ToolPak to perform the F test, and comment on the

significance of regression.

(e)   Use LINEST to find the standard error of the estimate, the F statistic, and the

coefficient of determination.

6.   A study was made to determine the activation energy E_A for a chemical reaction. The rate

constant k was determined as a function of temperature T and the data in the table below

obtained.

Temperature, T, K	k, s^{-1}
599	0.00054
629	00025
647	0.0052
666	0.014
683	0.025
700	0.064

The data should fit a linear model of the form $\log k = \log A - E_A/(2.303RT)$, where A is a

preexponential factor, and R is the gas constant.

(a)   Fit the data to a straight line of the form $\log k = a - 1000b/T$.

(b)   Find the slope, intercept and standard error of the estimate.

(c)   Noting that $E_A = -b \times 2.303R \times 1000$, find the activation energy and its standard

deviation (Use $R = 1.987$ cal mol^{-1} K^{-1}).

(d)   A theoretical prediction gave $E_A = 41.00$ kcal mol^{-1} K^{-1}. Test the null hypothesis

that E_A is this value at the 95% confidence level.

7.   Water can be determined in solid samples by infrared spectroscopy. The water content of

calcium sulfate hydrates is to be measured using calcium carbonate as an internal

standard to compensate for some systematic errors in the procedure. A series of standard

solutions containing calcium sulfate dihydrate and a constant known amount of the

internal standard is prepared. The solution of unknown water content is also prepared

with the same amount of internal standard. The absorbance of the dihydrate is measured at one wavelength (A_{sample}) along with that of the internal standard at another wavelength

(A_{std}). The following results were obtained.

A_{sample}	A_{std}	% Water
0.15	0.75	4.0
0.23	0.60	8.0
0.19	0.31	12.0
0.57	0.70	16.0
0.43	0.45	20.0
0.37	0.47	Unknown

(a)   Plot the absorbance of the sample (A_{sample}) vs. the % water and use the regression

statistics to determine whether the plot is linear.

(b)   Plot the ratio A_{sample}/A_{std} vs % water and comment on whether use of the internal

standard improves the linearity from that in part (a). If it improves the linearity,

why?

(c)   Calculate the percentage of water in the unknown using the internal standard data.

8.   Potassium can be determined by flame emission spectrometry (flame photometry) using a

lithium internal standard. The following data were obtained for standard solutions of KCl

and an unknown containing a constant known amount of LiCl as the internal standard.

All the intensities were corrected for background by subtracting the intensity of a blank.

Concentration of K, ppm	Intensity of K Emission	Intensity of Li Emission
1.0	10.0	10.0
2.0	15.3	7.5
5.0	34.7	6.8
7.5	65.2	8.5
10.0	95.8	10.0
20.0	110.2	5.8
Unknown	47.3	9.1

(a)   Plot the K emission intensity vs. the concentration of K and determine the

        linearity from the regression statistics.

(b)   Plot the ratio of the K intensity to the Li intensity vs. the concentration of K and

        compare the resulting linearity to that in part (a). Why does the internal standard

        improve linearity?

(c)   Calculate the concentration of K in the unknown.

9.   Copper is to be determined in a vitamin sample by the method of multiple additions.

      Three aliquots of 5 mL each are taken from the same sample solution. Volumes of a 100

      ppm copper standard are added to all but the first aliquot. The volumes added and the

      responses of an atomic absorption instrument are given in the table below.

	A	B	C
1	Determination of Cu with multiple additions		
2	Conc standard	100	ppm
3	Volume unkn	5000	µL
4		Volume added, µL	Instrument response
5		0.0	0.20
6		5.0	0.31
7		10.0	0.45
8		15.0	0.57
9		20.0	0.67

(a)   Use linear regression analysis to find the least-squares straight line.

(b)   Find the volume intercept and use it to calculate the concentration of the

        unknown. Be sure to express the volume of the sample solution taken in the same

        units as the additions.

(c)   Find the standard deviation in the concentration of the unknown.

10.   Chromium was determined in a water sample by the multiple standard additions. Various

       volumes of a 500 ng/mL standard were added to 10.00 mL aliquots of the unknown chromium solution. Each solution was determined by spectrophotometry in triplicate with

the results shown below.

	A	B	C	D	E	F
1	Determination of Cr with multiple additions					
2	Conc. Of Standard	500	ng/mL			
3	Volume unknown	10000	µL			
4						
5		Volume added, µL	Response1	Response2	Response3	Mean
6		0	0.049	0.049	0.053	0.050333
7		10	0.078	0.092	0.086	0.085333
8		20	0.114	0.117	0.120	0.117
9		40	0.194	0.182	0.183	0.186333
10		80	0.300	0.301	0.313	0.304667

(a)   Use linear regression analysis to find the least-squares straight line.

(b)   Find the volume intercept and use that to obtain the concentration of the

unknown.

(c)   Find the standard deviation in the concentration of the unknown.

11.   Reagent R reacts with metal ion M to form a product P. The following data represent the

effect of the concentration of R on the absorbance A of the product at a constant

concentration of metal ion.

Concentration of R, mM	A
0	0.55
2	0.615
4	0.595
6	0.43
8	0.31

(a)   Try a linear fit of the data?

(b)   Try polynomial fits of orders 2, 3, and 4. Which of these arethe best model for the

data?

(c)   Examine the residuals for the various polynomial fits. Are there any trends in the

residual plots?

12.   A chemical to be used as a primary standard was tested for change in mass after exposure

to the air for different amounts of time. The relative humidity of the air was also measured

during exposure. The results are shown in the table below.

Mass change, mg (y)	Exposure time, min (x_1)	Rel. humidity, % (x_2)
4.3	4	20
5.5	5	20
6.8	6	20
8.0	7	20
4.0	4	30
5.2	5	30
6.6	6	30
7.5	7	30
2.0	4	40
4.0	5	40
5.7	6	40
6.6	7	40

(a)   Use a multiple regression model $y = m_1 x_1 + m_2 x_2 + b$ and find the LINEST

coefficients m_1 and m_2.

(b)   Find the standard deviations of m_1 and m_2 and the remaining statistics from

LINEST.

(c)   Use the Analysis ToolPak Regression tool for the same data.

(d)   Do the data fit the presumed model? Why or why not?

Chapter 5

Equilibrium, Activity and Solving Equations

In many chemical equilibrium problems in analytical chemistry, we find that we must solve equations of various degrees of complexity. Excel provides several tools for coping with these situations. In this chapter, we will use some of these tools to solve equations that arise in the study of equilibrium and activity.

A Quadratic Equation Solver

Many equilibrium problems involve solving a quadratic equation or making approximations so that a simple linear equation results. Here, we will use Excel to construct a general purpose quadratic equation solver that will be useful in many problems. To illustrate how to set up and use the solver, we will use Example 9-8 of FAC9 or Example 4-8 of AC7, which require us to find the hydronium ion concentration in a 2.0×10^{-4} M solution of anilinium hydrochloride. The quadratic equation has the form

$$x^2 + 2.51 \times 10^{-5}x - 5.02 \times 10^{-9} = 0$$

where x is the H_3O^+ concentration. You may recall that there are two roots to this equation given by the quadratic formula

$$x = \frac{-b \pm \sqrt{b^2 - 4ac}}{2a}$$

where a is the coefficient of x^2 (1 in this case), b is the coefficient of x (2.51×10^{-5}), and c is the constant (5.09×10^{-9}). In any given chemical problem, only one of the roots makes physical sense.

	A	B	C
1	**Quadratic Equation Solver**		
2	For the equation $ax^2 + bx + c = 0$		
3	a=Coeff of x^2	1	
4	b=Coeff of x	2.51E-05	
5	c=Constant	-5.02E-09	
6			
7	Positive root		
8	Negative root		

Figure 5-1 Labels for quadratic equation solver.

Let's begin with a blank worksheet and type the labels shown in Figure 5-1. Now enter

into cells B3, B4, and B5 the value of the coefficients a, b, and c. In cell B7, we will enter the

formula corresponding to the positive root of the equation. Type the following in cell B7:

$$=(-B4+SQRT(B4^2-4*B3*B5))/(2*B3)\ [\ \lrcorner\ ]$$

This should return the positive root **5.94049E-05**. Now type in cell B8,

$$=(-B4-SQRT(B4^2-4*B3*B5))/(2*B3)\ [\ \lrcorner\ ]$$

This should return the negative root **-8.45049E-05**. Since we are calculating the H_3O^+

concentration, a negative value makes no physical sense, so we retain the positive root, and the

solution is $[H_3O^+] = 5.94 \times 10^{-5}$ M.

	A	B	C
1	**Quadratic Equation Solver**		
2	For the equation $ax^2 + bx + c = 0$		
3	a=Coeff of x^2	1	
4	b=Coeff of x	2.51E-05	
5	c=Constant	-5.02E-09	
6			
7	Positive root	5.94049E-05	
8	Negative root	-8.45049E-05	
9			
10	**Documentation**		
11	Cells B3:B5=user-entered data		
12	Cell B7=(-B4+SQRT(B4^2-4*B3*B5))/(2*B3)		
13	Cell B8=(-B4-SQRT(B4^2-4*B3*B5))/(2*B3)		

Figure 5-2. Final worksheet for quadratic equation solver.

Add documentation to your worksheet and save it as **quadsolver.xls**. The final

worksheet should look similar to that shown in Figure 5-2. This worksheet should be useful for

solving any problem involving a quadratic equation. As practice, use the solver to find the

hydronium ion concentration in Example 9-7 of FAC9 or Example 4-7 of AC7. Compare the

results you obtain from the exact solution to the result of 9.2×10^{-3} M obtained by assuming the

H_3O^+ concentration is small compared to the 0.120 M initial concentration of HNO_2. You can

also practice by solving Equation 9-19 of FAC9 or Equation 4-18 of AC7 using the K_a values

and c_{HA} values given in Table 9-4 or 4-4.

Using Excel to Iterate

In some cases it is desirable to obtain a quick iterative solution to an equation like the quadratic

equation given in Example 9-8 of FAC9 or Example 4-8 of AC7. Excel's properties make it very

easy to find an iterative solution by the method of successive approximations (see Feature 9-4 of

FAC9 or Feature 4-5 of AC7). The quadratic equation can be rearranged to the form

$$x = \sqrt{5.02 \times 10^{-9} - 2.51 \times 10^{-5}x}$$

Start with a blank worksheet, click on cell A2, and type

=SQRT(5.02E-9-2.51E-5*A1) [↵]

The spreadsheet now appears as shown in in Figure 5-3, and the value in cell A2 is

precisely the answer obtained in the iteration of Feature 9-4 of FAC9 or Feature 4-5 of AC7.

Note that since cell A1 is blank, Excel interprets the relative cell reference to mean that the

content of A1 is zero. Thus, the second term inside the square-root function is zero, and cell A2

displays the square root of 5.02×10^{-9}.

Figure 5-3 Worksheet after first iteration.

Now, click on cell A2 and type **[Ctrl-c]**, which copies the contents of the cell into the clipboard. Then, highlight cells A3 to A10 by dragging with the mouse and type **[Ctrl-v]**. Your spreadsheet should now display the results shown in Figure 5-4.

Figure 5-4 Worksheet after several iterations.

Click on each of the cells A2 through A10 one at a time and observe how the formula changes in the formula bar. Each formula refers to the result in the cell directly above it, so for example, cell A3 contains the formula **=SQRT(0.00000000502-0.0000251*A2)**. The value of x calculated in A3 results from the previous value of x calculated in A2, the value in A4 is computed from the value in A3, and so on. Each new calculation is an *iteration*. Note that after four iterations, cell A6 contains the value 5.94×10^{-5}, which is sufficiently precise for virtually any purpose. Note that there is almost no change to 4 decimal places after 7 iterations. The value is said to have *converged* at this point. To see how many iterations are required to get the same result as the exact solution to five decimal places, increase the width of column A. Note that the

result in cell A10 is identical to that shown in Figure 5-2. Hence, 9 iterations are required.

Use this same approach to solve the problem of Example 9-7 of FAC9 or Example 4-7 of

AC7. You can also use the iterative approach to solve Equation 9-19 or 4-18 with the values of

K_a and c_{HA} given in Table 9-4 or 4-4.

Using Excel's Solver

Excel provides other tools to solve equations or to fit models to data. One of the most useful

tools is the Solver. Solver is not limited to finding the roots of quadratic equations. It can solve

polynomial equations of higher order such as cubic and quartic equations, nonlinear equations,

and even transcendental equations. We will use Solver here to solve equations of the type

encountered in equilibrium problems. Chapters 13-15 will consider the use of Solver in fitting

nonlinear models such as those encountered in kinetics, in chromatography and in other areas.

If Solver has not been installed on your computer, you must run the Excel Setup program

to install it. If it has been installed, but not enabled, you must enable Solver by clicking on Excel

Options in the File tab (Office button in Excel 2007). In the Excel options menu, click on **Add-**

Ins and select the Solver add-in. After installation and enabling, Solver should appear in the

Analysis group on the Data ribbon.

Solving a Quadratic Equation

We will first use Solver to obtain the H_3O^+ concentration in Example 9-8 of FAC8 or Example

4-8 of AC7, the same problem we solved previously with the quadratic formula and by iteration.

As before, the equation to be solved is $x^2 + 2.51 \times 10^{-5}x - 5.02 \times 10^{-9} = 0$. Begin with a

worksheet such as that shown in Figure 5-5.

	A	B	C
1	**Excel Solver Solution**		
2	For the equation $ax^2 + bx + c = 0$		
3	a=Coeff of x^2	1	
4	b=Coeff of x	2.51E-05	
5	c=Constant	-5.02E-09	
6			
7	Equation		
8	$x=$		

Figure 5-5 Worksheet for quadratic equation.

With Solver, we must set an Objective cell (called a "target cell" in Excel 2007), which contains the formula we want to maximize, minimize, or set to a certain value. In our case the target cell will contain our quadratic equation with unknown x. Thus, we type in cell B7

$$=B3*B8^2+B4*B8+B5 \ [\ \hookleftarrow \ ]$$

This formula in the target cell contains cell B8, which will be the value of x ($[H_3O^+]$) that we seek. Note that the number -5.02×10^{-9} appears in cell B7 since this is the result of the equation with $x = 0$.

Open Solver by clicking on its icon in the Analysis group. This should open the Solver Parameters dialog box shown in Figure 5-6. With the cursor in the **Se̲t Objective:** box (Set Target Cell in Excel 2007), click on cell B7. This inserts **$B$7** as the Objective (Target) cell and highlights the cell as shown. Next, in the line labeled **To:**, check **V̲alue of:**, and enter 0 in the box. This means that Solver will attempt to set the formula in cell B7 equal to 0.[1]

---

[1] For more information on Solver, see the website, http://www.solver.com.

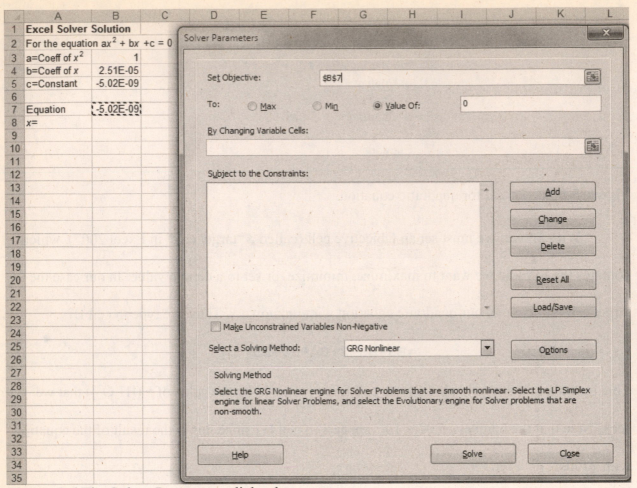

Figure 5-6 The Solver Parameters dialog box.

Now in the box labeled **By Changing Variable Cells:**, type B8. Solver will

systematically adjust the value in this cell until various constraints are satisfied, and the cell in

the Objective cell box (B7) reaches its target. Now we will add the constraints. In the cell label

Subject to the Constraints, click on Add. This displays the **Add Constraint** dialog box shown

in Figure 5-7. Type B8 in the **Cell Reference:** box, and select >= in the middle box. Type 0 in

the **Constraint:** box. The constraint forces the value of x found in cell B8 to be greater than zero,

since the negative root is physically meaningless. For the Solving Method, chose GRG nonlinear.

Simplex LP is used for linear problems and Evolutionary for non-smooth problems. Now click

on OK. This returns you to the **Solver Parameters** window.

Figure 5-7 The Add Constraint dialog box.

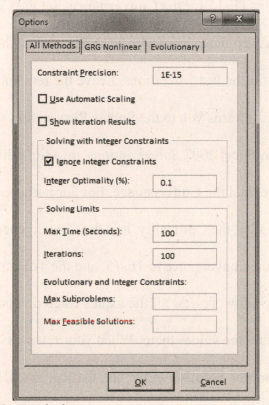

Figure 5-8 The **Solver Options** window.

We will now change the value of some options before completing the calculation. In the

Solver Parameters window, click on **Options**. This displays the **Solver Options** dialog box

shown in Figure 5-8. Under the **All Methods** tab, Leave the **Max Time:** box set at its default

value of 100 seconds. This option limits the amount of time that Solver will work on a problem.

The value can be set as high as 32,767 seconds. Leave the **Iterations:** box set at 100. This option

limits the number of intermediate calculations of the value in cell B7. Change the **Constraint**

Precision: box to 1E-15. Under **Integer Optimality(%),** enter 0.1. The Constraint precision controls how closely a constraint cell must be to the constraint value entered before Solver considers the constraint to be met. The Integer optimality (called tolerance in Excel 2007) represents the maximum difference that Solver should allow between the best solution and the optimal value. A higher optimality speeds up the process, but lowers the accuracy. Click on the **GRG Nonlinear** tab and enter 1E-15 for the **C**onvergence. Set the Convergence box to 1E-15. The Objective cell value is said to have reached *convergence* when the relative change is less than the number in this box for the last 5 iterations. Leave the remaining boxes set to default values, and click on OK. This returns you to the **Solver Parameters** window.

The Solver options in Excel 2007 Solver options are a little different. Here, you can set the **Precision**, the **Tolerance**, and the **Convergence**. For our purposes, these are essentially the same as the Constraint Precision, the Optimality, and the Convergence in Excel 2010. For this problem, set the Precision and Convergence to 1E-15, and the Tolerance to 0.1%.

In the **Solver Parameters** window, review all the values entered. You may want to go back to the **Solver Options** and review the values there. When your review is completed, click on the **S**olve button. Check **K**eep Solver Solution in the box that opens and click **OK**. The optimum value for *x* should then appear in cell B8 as shown in Figure 5-9. Note also that the final value for the equation when the search stopped is displayed in cell B7. In this case the final value is -3.25×10^{-16}, which is certainly close to zero and smaller than the convergence value. You may want to document the worksheet as shown. Also you can go back to the **Solver Options** and change the values of the precision and convergence and observe their effects on the results. Solutions always need to be checked to make sure that these values are set appropriately for the problem at hand. Note by clicking on Cell B8 that the value displayed in the formula bar is identical, to six figures, with that calculated with the quadratic equation or by iterative solution

(5.94049×10^{-5}). Save the worksheet for future use.

	A	B	C
1	**Excel Solver Solution**		
2	For the equation $ax^2 + bx + c = 0$		
3	a=Coeff of x^2	1	
4	b=Coeff of x	2.51E-05	
5	c=Constant	-5.02E-09	
6			
7	Equation	-3.25E-16	
8	$x=$	5.94E-05	
9			
10	**Documentation**		
11	Cells B3:B5=user entered data		
12	Cell B7=B3*B8^2+B4*B8+B5		
13	Cell B8=Solver calculated result		
14	**Constraints**		
15	Cel B8>=0		

Figure 5-9 Final results of Solver search for solution to quadratic equation.

Solving a Cubic Equation

We will now use Solver to obtain a solution for a cubic equation. The problem chosen is

Example 9-4 of FAC9 or Example 4-4 of AC7. In this example, the molar solubility of $Ba(IO_3)_2$

is to be calculated. The following cubic equation results:

$$0.5000x^3 + 0.0200x^2 - 1.57 \times 10^{-9} = 0$$

Set up a new worksheet as shown in Figure 5-10.

	A	B	C
1	**Cubic Equation**		
2			
3	Coeff of x^3	0.5	
4	Coeff of x^2	0.02	
5	Coeff of x	0	
6	Constant	-1.57E-09	
7			
8	Equation		
9	$x=$		
10			

Figure 5-10 Initial entries for cubic equation solution.

In cell B8, type the following formula corresponding to the cubic equation above.

$$=B3*B9\text{^}3+B4*B9\text{^}2+B5*B9+B6 \;[\;↵\;]$$

Note that we have included the x term, even though its coefficient is zero for this problem.

Including x makes our worksheet more general so it can be used for problems in which x is not

zero. Now invoke Solver, chose cell B8 as the target cell, and select it to be set equal to zero.

Choose B9 as the cell to be changed and add the constraint that B9 should be ≥ 0. In the **Solver**

Options dialog box, set the **Constraint Precision** and **Convergence** to be 1E-15, and the

Integer Optimality to be 0.1%. Your final worksheet should look like Figure 5-11 after adding

documentation. Note that the result, with the appropriate number of significant figures, is

identical to that obtained with approximations in the example in FAC9 or AC7. Also, the final

value for the equation is very small (8.60×10^{-16}) and less than the convergence value. The value

of x in this example represents the equilibrium concentration of iodate. The solubility is $\frac{1}{2}[IO_3^-]$

or 1.40×10^{-4} M with the appropriate number of significant figures.

	A	B	C	D
1	**Cubic Equation**			
2				
3	Coeff of x^3	0.5		
4	Coeff of x^2	0.02		
5	Coeff of x	0		
6	Constant	-1.57E-09		
7				
8	Equation	8.60E-16		
9	$x=$	0.000279		
10				
11	**Documentation**			
12	Cells B3:B6=user-entered data			
13	Cell B8=B3*B9^3+B4*B9^2+B5*B9+B6			
14	Cell B9=Solver-calculated result			
15	**Constraints**			
16	Cell B9>=0			

Figure 5-11 Final worksheet for cubic equation.

Solving a Quartic Equation

Now try solving the following equation

$$x^4 + 8x^3 + 19.25x^2 - 17.5x - 122 = 0$$

Find the positive and negative roots by first setting the constraint that $x \geq 0$ and later changing it

to $x \leq 0$. The two solutions should be $x = 2$ and $x = -4$.

Exploring the Ionic Strength Dilemma

As you read Chapter 10 of FAC9 or Chapter 9 of AC7, you probably noticed that most of the

cases treated in these chapters involve only weak electrolytes. Slightly soluble salts and weak

acids often contribute only slightly to the ionic strength of their solutions when an inert

electrolyte is present. What happens when a salt is soluble enough to substantially increase the

ionic strength of the solution? As an example, consider a solution of 0.01000 M NaCl with

excess solid TlCl ($K_{sp} = 1.7 \times 10^{-4}$) added. If we choose to neglect activity coefficients, we have

a common-ion solubility problem. First, we write the solubility product constant and the mass

balance expression.

$$K_{sp} = [Tl^+][Cl^-]$$
$$[Cl^-] = [Na^+] + [Tl^+]$$

If we substitute the second equation into the first, we obtain

$$K_{sp} = [Tl^+]([Tl^+] + [Na^+]) = [Tl^+]([Tl^+] + 0.01000) = 1.7 \times 10^{-4}$$

If we use our quadratic equation solver or Excel's Solver, we find that

$$[Tl^+] = 8.96 \times 10^{-3} \text{ M}$$

So in this case, the solubility of thallium chloride nearly doubles the ionic strength of the

solution. This result will have a significant effect on the activity coefficients of the ions in the solution, which in turn will affect the solubility. Because solubility influences ionic strength, and ionic strength affects solubility, this problem is made somewhat difficult.

An Iterative Solution

One way to solve for the solubility of TlCl using Solver is to find an iterative solution. Here, the Tl^+ concentration from our first calculation is used to calculate the ionic strength and then the activity coefficients. These in turn are used to obtain a new concentration. This iteration is continued until the results reach a steady value.

The ionic strength of the solution is given by

$$\mu = \frac{1}{2}\sum_i c_i z_i^2$$

$$= \frac{1}{2}\{[Na^+](+1)^2 + [Tl^+](+1)^2 + [Cl^-](-1)^2\}$$

$$= \frac{1}{2}([Na^+] + [Tl^+] + [Cl^-])$$

The Debye-Hückel equation gives the activity coefficients of the ions in terms of the ionic strength and the ion-size parameters. We can find the ion-size parameters from Table 10-2 of FAC9 or Table 9-1 of AC7,

$$\log\gamma_{Tl} = -\frac{0.51(+1)^2\sqrt{\mu}}{1+(0.33)(2.5)\sqrt{\mu}}$$

$$\log\gamma_{Cl} = -\frac{0.51(-1)^2\sqrt{\mu}}{1+(0.33)(3.0)\sqrt{\mu}}$$

Finally, we write the mass balance expression as

$$[Cl^-] = [Tl^+] + [Na^+] = [Tl^+] + 0.0100$$

These four equations plus the solubility product expression contain five unknowns: γ_{Tl}, γ_{Cl}, $[Tl^+]$, $[Cl^-]$, and μ. In principle, we can solve the system of equations for the unknown quantities. Instead, we will seek an iterative solution.

Let's begin by typing in labels and constants as shown in column A and cells B2 and B3 of Figure 5-12.

	A	B	C
1	**Thallium Chloride Solubility**		
2	K_{sp}	1.70E-04	
3	[Na$^+$]	0.01	
4			
5	**Assume Gammas = 1**		
6	Equation		
7	[Tl$^+$]		

Figure 5-12 Initial worksheet for thallium chloride solubility problem.

We first assume that activity coefficients are unity and use Solver to find the Tl^+ concentration. From the K_{sp} expression and the mass balance, we can write

$$[Tl^+]^2 + 0.01[Tl^+] - 1.7 \times 10^{-4} = 0$$

In cell B6 we type the formula corresponding to this quadratic equation for $[Tl^+]$

$$=B7\char94 2+B3*B7-B2\ [\ \dashv]$$

Set an initial value of 0.013 for the $[Tl^+]$. This value is $\sqrt{K_{sp}}$ and is just an initial guess assuming no NaCl is added. Set cell B6 as the target cell for Solver and have it find the solution by changing cell B7. Set the Solver options as you did for the quadratic and cubic equations previously with **Precision** and **Convergence** at 1E-15 and **Integer Optimality** at 0.1%. Add the constraint that cell B7 should be ≥ 0. Solver should return the value shown for the Tl^+ concentration in cell B7 in Figure 5-13.

	A	B	C
1	**Thallium Chloride Solubility**		
2	K_{sp}	1.70E-04	
3	[Na$^+$]	0.01	
4			
5	**Assume Gammas = 1**		
6	Equation	-9.00E-17	
7	[Tl$^+$]	0.008964	
8	[Cl]	0.018964	
9	μ	0.018964	
10	γ_{Tl}	0.864834	
11	γ_{Cl}	0.867349	

Figure 5-13 Worksheet after Solver solution for [Tl$^+$].

In cell B8 we find the Cl$^-$ concentration from the mass balance ([Cl$^-$] = [Tl$^+$] + [Na$^+$]).

$$=B7+\$B\$3 \ [\ \hookleftarrow]$$

In cell B9, we find the ionic strength from $\mu = \frac{1}{2}([Na^+] + [Tl^+] + [Cl^-])$.

$$=0.5*(\$B\$3+B7+B8) \ [\ \hookleftarrow]$$

In cells B10 and B11, we enter the formulas corresponding to the Debye-Hückel equation for the

activity coefficients of Tl$^+$ and Cl$^-$ as

$$=10\verb|^|((-0.51*SQRT(B9))/(1+0.33*2.5*SQRT(B9))) \ [\ \hookleftarrow]$$

$$=10\verb|^|((-0.51*SQRT(B9))/(1+0.33*3*SQRT(B9))) \ [\ \hookleftarrow]$$

We are now ready to use these activity coefficients to obtain a new value for [Tl$^+$]. We

use the full expression for the solubility product of TlCl as

$$K_{sp} = a_{Tl^+} a_{Cl^-} = [Tl^+][Cl^-]\gamma_{Tl}\gamma_{Cl} = 1.70 \times 10^{-4}$$

The new value of the Tl$^+$ concentration is then found from combining this equation and the mass

balance equation

$$K_{sp} = [Tl^+]([Tl^+] + 0.0100)\gamma_{Tl}\gamma_{Cl}$$

Rearranging this and substituting the values for the activity coefficients gives us the following:

$$\gamma_{Tl}\gamma_{Cl}[Tl^+]^2 + 0.0100\gamma_{Tl}\gamma_{Cl}[Tl^+] - 1.70 \times 10^{-4} = 0$$

We then enter the formula corresponding to this equation into cell E6 as

```
=B10*B11*E7^2+B10*B11*0.01*E7-$B$2[ ↵]
```

where cell E7 will contain the new value for the Tl^+.concentration, $[Tl^+]'$. Now invoke Solver

again, with cell E6 as the target cell. Have Solver change cell E7. As the initial value for $[[Tl^+]'$,

use the value found previously in cell B7. Use the same options as before, but set as a constraint

the condition that cell E7 should be ≥ 0. Solver should converge and return a value of 0.010863.

This new value allows us to find a second value for the Cl^- concentration, $[Cl^-]'$, since $[Cl^-]' =$

$[Tl^+]' + [Na^+]$ by mass balance. From these new concentrations of Tl^+ and Cl^-, we find a new

ionic strength and new values for the activity coefficients, and so on until we get an unchanging

value for $[Tl^+]$. The final worksheet for this problem is shown in Figure 5-14. For each new

value for $[Tl^+]$, we use the previous value as our initial guess. The options for Solver are the

same for each calculation.

Note here that convergence actually occurs after just three tries. The fourth try shows

only a very small improvement over the third. The solubility of TlCl is 0.011 M, which is 18%

higher than that calculate by assuming unity activity coefficients (0.00896 M).

You can try changing the value of the added electrolyte, NaCl, by changing the value in

cell B3 from 0.01 to 0.02, etc. When you do this you will have to use Solver again with the new

values. This procedure is somewhat tedious and time consuming. An easier solution is shown

next.

	A	B	C	D	E	F	G	H	I	J	K
1	Thallium Chloride Solubility										
2	K_{sp}	1.70E-04									
3	[Na$^+$]	0.01									
4											
5	Assume gammas = 1			Using Gammas			Next Try			Next One	
6	Equation	-9.00E-17		Equation	-1.26E-16		Equation	-8.63E-16		Equation	-2.71E-19
7	[Tl$^+$]	0.008964		[Tl$^+$]	0.010863		[Tl$^+$]	0.010952		[Tl$^+$]	0.010956
8	[Cl$^-$]	0.018964		[Cl$^-$]	0.020863		[Cl$^-$]	0.020952		[Cl$^-$]	0.020956
9	μ	0.018964		μ	0.020863		μ	0.020952		μ	0.020956
10	γ_{Tl}	0.864834		γ_{Tl}	0.859368		γ_{Tl}	0.85912		γ_{Tl}	0.859108
11	γ_{Cl}	0.867349		γ_{Cl}	0.862088		γ_{Cl}	0.861849		γ_{Cl}	0.861838
12											
13	Documentation							Constraints			
14	Cell B6=B7^2+B3*B7-B2							For each calculation, the [Tl$^+$] is set >=0			
15	Cell B7=initial guess or Solver result										
16	Cell B8=B7+$B$3										
17	Cell B9=0.5*($B$3+B7+B8)										
18	Cell B10=10^((-0.51*SQRT(B9))/(1+0.33*2.5*SQRT(B9)))										
19	Cell B11=10^((-0.51*SQRT(B9))/(1+0.33*3*SQRT(B9)))										
20	Cell E6=B10*B11*E7^2+B10*B11*0.01*E7-$B$2										
21	Cell E7=initial guess or Solver result										
22	Cell E8=E7+$B$3										
23	Cell E9=0.5*($B$3+E7+E8)										
24	Cell E10=10^((-0.51*SQRT(E9))/(1+0.33*2.5*SQRT(E9)))										
25	Cell E11=10^((-0.51*SQRT(E9))/(1+0.33*3*SQRT(E9)))										
26	Cell H6=E10*E11*H7^2+E10*E11*0.01*H7-$B$2										
27	Cell H7=initial guess or Solver result										
28	Cell K6=H10*H11*K7^2+H10*H11*0.01*K7-$B$2										
29	Cell K7=initial guess or Solver result										

Figure 5-14 Final worksheet for thallium chloride solubility problem.

Using Solver to Obtain the Solution Directly

Solver is also capable of finding the solution directly. In this case we need to let Solver vary not only the Tl$^+$ concentration, but also the Cl$^-$ concentration and the ionic strength. We write a large formula containing all the variables and let Solver find the best values subject to the mass balance constraint, the ionic strength definition, and the constraint that concentrations are positive values.

Let's start with a worksheet similar to that shown in Figure 5-15. We have three major variables, the two concentrations and the ionic strength. The activity coefficients are found from these variables. Type in the initial values shown for [Tl$^+$], [Cl$^-$] and μ. We could enter many initial values, but we have some idea from the K_{sp} what the concentrations are, or we could enter

the values from assuming that the activity coefficients are unity.

	A	B	C
1	**Thallium Chloride Solubility**		
2	K_{sp}	1.70E-04	
3	$[Na^+]$	0.01	
4	**Variables**		
5	$[Tl^+]$		
6	$[Cl^-]$		
7	μ		
8	**Calculated Values**		
9	γ_{Tl}		
10	γ_{Cl}		
11			
12	Equation		

Figure 5-15 Initial worksheet for TlCl solubility problem.

Now enter into cells B9 and B10 the formulas corresponding to the Debye-Hückel

equation as before. The final entry in cell B12 will be the full equation for calculating the Tl^+

concentration.

$$\gamma_{Tl}\gamma_{Cl}[Tl^+]^2 + 0.0100\gamma_{Tl}\gamma_{Cl}[Tl^+] - K_{sp} = 0$$

The appropriate Excel formula is

$$=10\text{^}((-0.51*\text{SQRT}(B7))/(1+0.33*2.5*\text{SQRT}(B7)))*$$

$$10\text{^}((-0.51*\text{SQRT}(B7))/(1+0.33*3*\text{SQRT}(B7)))*B5\text{^}2$$

$$+B3*10\text{^}((-0.51*\text{SQRT}(B7))/(1+0.33*2.5*\text{SQRT}(B7)))*$$

$$10\text{^}((-0.51*\text{SQRT}(B7))/(1+0.33*3*\text{SQRT}(B7)))*B5-B2$$

Set up Solver to adjust the result of this formula to zero by changing cells B5:B7. Add the

constraints **B6=B5+B3** and **B7=0.5*B3+0.5*B6+0.5*B7**. These are the formulas for

finding the Cl^- concentration from mass balance and the ionic strength. Add the additional

constraints that B5, B6, and B7 should be greater than 0. In the **Solver Options** window set the constraint precision and convergence values to 1E-15 and the optimality to 0.001%. When

Solver has converged, the final worksheet should look similar to that shown in Figure 5-16. Note

that these values are very close to those calculated by iteration. You can now easily change the

concentration of NaCl and see its influence on solubility. You will again have to invoke Solver

after each change to obtain the new value.

	A	B	C	D	E	F	G	H	I	J
1	Thallium Chloride Solubility									
2	K_{sp}	1.70E-04								
3	$[Na^+]$	0.01								
4	Variables									
5	$[Tl^+]$	0.010956								
6	$[Cl^-]$	0.020956								
7	μ	0.020956								
8	Calculated Values									
9	γ_{Tl}	0.859108								
10	γ_{Cl}	0.861838								
11										
12	Equation	1.7E-16								
13										
14	Documentation									
15	Cell B5:B7=Solver results									
16	Cell B9=10^((-0.51*SQRT(B7))/(1+0.33*2.5*SQRT(B7)))									
17	Cell B10=10^((-0.51*SQRT(B7))/(1+0.33*3*SQRT(B7)))									
18	Cell B12=10^((-0.51*SQRT(B7))/(1+0.33*2.5*SQRT(B7)))*10^((-0.51*SQRT(B7))/(1+0.33*3*SQRT(B7)))*B5^2									
19	+B3*10^((-0.51*SQRT(B7))/(1+0.33*2.5*SQRT(B7)))*10^((-0.51*SQRT(B7))/(1+0.33*3*SQRT(B7)))*B5-B2									
20										
21	Constraints									
22	Cells B5:B7>=0									
23	Cell B6=B3+B5									
24	Cell B7=0.5*B3+0.5*B5+0.5*B6									

Figure 5-16 Final worksheet for TlCl solubility using Solver to optimize 3 variables.

You may also wish to collect the solubility data as a function of ionic strength and make a plot

similar to that shown in Figure 5-17. If you like, compare the results obtained here to those

obtained by assuming the Debye-Hückel limiting law holds. Which ionic strength values produce

significant deviations from the limiting law?

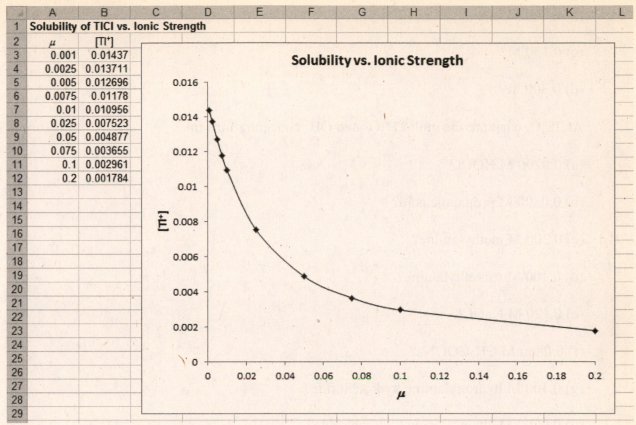

Figure 5-17 Worksheet collecting solubility data for TlCl as a function of ionic strength.

Summary

In this Chapter, we have learned how to use Excel to find numerical solutions to complex chemical equilibrium problems . A general purpose quadratic equation solver was developed and iterative solutions to quadratic equations were found. The use of Excel's Solver tool for solving a variety of equilibrium problems was explored. These tools and techniques should prove useful throughout the remainder of this book.

Problems

1.   The solubility-product constant for $Ce(IO_3)_3$ is 3.2×10^{-10}. What is the Ce^{3+} concentration in a solution prepared by mixing 50.0 mL of 0.0500 M Ce^{3+} with 50.00 mL of

        (a) water?

(b) 0.050 M?

(c) 0.150 M?

(d) 0.300 M?

2.   At 25°C, what are the molar H_3O^+ and OH^- concentrations in

(a) 0.0200 M HOCl?

(b) 0.0800 M propanoic acid?

(c) 0.200 M methylamine?

(d) 0.100 M trimethylamine?

(e) 0.120 M NaOCl?

(f) 0.0860 M CH_3COONa?

(g) 0.100 M hydroxylamine hydrochloride?

(h) 0.0500 M ethanolamine hydrochloride?

3.   For each of the solutions in Problem 2, find the hydronium ion concentration for 10%, 20%, 30%, 40%, 50%, 60%, 70%, 80%, 90%, 100%, 200%, 300%, 400%, and 500% of the listed molar analytical concentration. Use a separate worksheet for each species. Use a separate column to solve for the hydronium ion concentration using the iterative method. Note any trends in the results as the concentration of the indicated species changes.

4.   At 25°C, what is the hydronium ion concentration in

(a) 0.100 M chloroacetic acid?

(b) 0.100 M sodium chloroacetate?

(c) 0.0100 M methylamine?

(d) 0.0100 M methylamine hydrochloride?

(e) aniline hydrochloride?

(f) 0.200 M HIO_3?

5.   For each of the solutions in Problem 4, find the hydronium ion concentration for 10%, 20%, 30%, 40%, 50%, 60%, 70%, 80%, 90%, 100%, 200%, 300%, 400%, and 500% of the listed molar analytical concentration. Use a separate worksheet for each species. Note any trends in the results as the concentration of the indicated species changes.

6.   A 0.8720-g sample of a mixture consisting solely of sodium bromide and potassium bromide yields 1.505 g of silver bromide. What are the percentages of the two salts in the sample?

7.   A 0.6407-g sample containing chloride and iodide ions gave a silver halide precipitate weighing 0.4430 g. This precipitate was then strongly heated in a stream of Cl_2 gas to convert the AgI to AgCl; on completion of this treatment, the precipitate weighed 0.3181 g. Calculate the percentage of chloride and iodide in the sample.

8.   A 6.881-g sample containing magnesium chloride and sodium chloride was dissolved in sufficient water to give 500 mL of solution. Analysis for the chloride content of a 50.0-mL aliquot resulted in the formation of 0.5923 g of AgCl. The magnesium in a second 50.0-mL aliquot was precipitated as $MgNH_4PO_4$; on ignition 0.1796 g of $Mg_2P_2O_7$ was found. Calculate the percentage of $MgCl_2 \cdot 6H_2O$ and of NaCl in the sample.

9.   Calculate the solubilities of the following compounds in a 0.0167 M solution of $Ba(NO_3)_2$ using (1) activities and (2) molar concentrations:

(a) $AgIO_3$. (c) $BaSO_4$.

(b) $Mg(OH)_2$ (d) $La(IO_3)_2$.

10.   For the compounds in Problem 9,

(a)   vary the concentration of $Ba(NO_3)_2$ from 0.0001 M to 1.

(b)   plot ps vs. pc, where pc is the negative logarithm of the concentration of

        $Ba(NO_3)_2$.

11.   Design and construct a spreadsheet to calculate activity coefficients in a format similar to

        Table 10-2 of FAC9 or Table 9-1 of AC7. Enter values of α_X in column A beginning with

        cell A5, and enter ionic charges in column B, beginning with cell B5. Enter in row 4, the

        same set of values for ionic strength listed in Table 10-2 (FAC9) or 9-1 (AC7). Enter the

        formula for the activity coefficients in cells C5:G5. Be sure to use absolute cell

        references for ionic strength in your formulas for the activity coefficients. Finally, copy

        the formulas for the activity coefficients into the rows below row 5 by dragging the fill

        handle downward. Compare the activity coefficients that you calculate to those in Table

        Table 10-2 (FAC9) or 9-1 (AC7). Do you find any discrepancies? If so, explain how they

        arise?

Chapter 6

The Systematic Approach to Equilibria: Solving Many Equations

In Chapter 11 of FAC9 and Chapter 10 of AC7, we explore the systematic approach for solving chemical equilibrium problems. We learned that if we know the equilibria involved in a given system, we can write a corresponding system of equations that allows us to find the concentrations of all species in the system. Briefly, we write the chemical equations and the corresponding equilibrium constant expressions for these equilibria. Then we write the charge balance and mass balance equations for the system. After checking to make sure that the total number of independent equations is equal to the number of unknown concentrations, we then solve the system of equations for the unknown concentrations. With some practice, the equations may often be solved by making suitable assumptions about the relative concentrations of one or more species involved in the mass balance and charge balance equations.

There are several approaches we can use with Excel to solve the equations involved in equilibrium problems. We have already used Excel's Solver to provide solutions for some of these problems. In this Chapter we will expand our use of the Solver to find solutions to complex equilibrium situations. In addition, Excel has another slightly different way to solve complex equations, called Goal Seek that will be described in this Chapter. Finally, in some cases we can solve systems of simultaneous linear equations quite nicely using determinants and matrix operations. We briefly describe in this chapter how to use Excel for determinant and matrix manipulations.

Using Solver for Complex Equilibrium Problems

In Chapter 5, we used Solver to find the solutions to polynomial equations, such as quadratic and cubic equations, and to compute equilibrium concentrations in cases where activity corrections are needed. In this chapter, we'll use the systematic approach and Solver to calculate the equilibrium concentrations of all of the species in a chemical system by using a set of 3 equations and 3 unknowns. Let's use Example 11-5 from FAC9 or example 10-5 from AC7 in which we find the solubility of $Mg(OH)_2$ in water. The systematic approach yields three independent equations, one for K_{sp}, one for K_w, and the mass balance expression. The equilibrium constants can be written:

$$K_{sp} = [Mg^{2+}][OH^-]^2 = 7.1 \times 10^{-12}$$

$$K_w = [H_3O^+][OH^-] = 1.00 \times 10^{-14}$$

The charge balance and mass balance expressions are identical, and can be written

$$[OH^-] = 2[Mg^{2+}] + [H_3O^+]$$

There are two ways we can use Solver to find the three unknown concentrations. First, we can have Solver find the three concentrations directly by solving the three equations above. Second, we can write the mass balance equation in terms of only one variable and have Solver find the concentration.

Finding the Concentrations Simultaneously with Solver

To find the three concentrations simultaneously, we use the mass balance expression as the equation to be optimized, and the two equilibria as constraints. Start with the worksheet shown in Figure 6-1.

	A	B	C
1	The solubility of Mg(OH)$_2$		
2	K_{sp}	7.10E-12	
3	K_w	1.00E-14	
4			
5	Equation		
6	[OH$^-$]		
7	[Mg^{2+}]		
8	[H$_3$O$^+$]		

Figure 6-1 The initial worksheet for the solubility problem.

We rearrange the mass balance expression to $[OH^-] - 2[Mg^{2+}] - [H_3O^+] = 0$, and type in cell B5

$$=B6-2*B7-B8\ [\ \lrcorner\ ]$$

We now need initial estimates of the three variables [OH$^-$], [Mg^{2+}], and [H$_3$O$^+$]. These estimates should be fairly close to the final solutions since Excel has no knowledge of what is or is not chemically reasonable. As an easy guess, we could assume all equilibria are independent so that $[OH^-] = [H_3O^+] = 1.0 \times 10^{-7}$ M and $[Mg^{2+}] \approx \sqrt[3]{10^{-12}} = 10^{-4}$ M.

Put these values for the initial concentrations in cells B6 through B8. Notice that there is now a nonzero value in cell B5.

Activate Solver, choose cell B5 as the target cell and have Solver adjust the value to zero, by varying cells B5:B8. From the solubility product expression, $[Mg^{2+}] = K_{sp}/[OH^-]^2$. Hence, add a constraint that **B7=B2/B6^2**. From the ion product of water, $[H_3O^+] = K_w/[OH^-]$, so add another constraint that **B8=B3/B6**. Make certain the box Make Unconstrained Variables Non-Negative is not checked (this is called Assume Non-Negative in Excel 2007). In the Solver Options dialog box, choose a Constraint Precision and Convergence of 1E-13 and an Optimality of 0.1%. Have Solver find the solution. If Solver cannot find the solution in 100 iterations, you may have to adjust the Precision and Convergence to slightly smaller values (1E-14, for example) or increase the number of iterations. When done correctly, the results should be as

shown in Figure 6-2.

	A	B	C	D
1	The solubility of Mg(OH)$_2$			
2	K_{sp}	7.10E-12		
3	K_w	1.00E-14		
4				
5	Equation	3.22E-20		
6	[OH$^-$]	2.42E-04		
7	[Mg^{2+}]	1.21E-04		
8	[H$_3$O$^+$]	4.13E-11		
9				
10	Documentation			
11	Cell B5=B6-2*B7-B8			
12	Cell B6:B8=initial guess or Solver result			
13				
14	Constraints			
15	Cell B7=B2/B6^2			
16	Cell B8=B3/B6			

Figure 6-2 Final worksheet for Solver solution to solubility problem.

Finding a Single Concentration with Solver

The problem with using Solver to find the three concentrations simultaneously is that the user

must find the right combination of initial estimates, constraint precision, optimality (tolerance),

convergence and number of iterations. This happens because in this case the system of equations

is highly nonlinear. We can simplify the calculations by writing the mass balance equation in

terms of one variable.

Without making assumptions, we can substitute for Mg^{2+} and H$_3$O$^+$ in terms of [OH$^-$] as

follows

$$[OH^-] = \frac{2K_{sp}}{[OH^-]^2} + \frac{K_w}{[OH^-]}$$

Rearranging this equation, we obtain the following cubic equation in [OH$^-$].

$$[OH^-]^3 - K_w[OH^-] - 2K_{sp} = 0$$

Substituting the constants we find,

$$[OH^-]^3 - 1.00 \times 10^{-14}[OH^-] - 2 \times 7.1 \times 10^{-12} = 0$$

We have seen in the previous chapter how to solve such cubic equations using Solver. You can

retrieve the worksheet used for the cubic equation in Figure 5-11 or begin with a blank

worksheet. Enter the coefficients of x^3, x^2, and x and the constant as before. Enter the cubic

equation above as the equation to be optimized. Since x is $[OH^-]$, use 1.00×10^{-7} M as the initial

estimate as before. Bring up Solver and set the equation to zero by changing the value of x. Set

the constraint that $x \geq 0$. In the Solver options, use a Precision and Convergence of 1E-−15 and a

Tolerance of 0.1%. Your final worksheet should appear as shown in Figure 6-3. Note that the

OH^- concentration is the same as we found in the three variable solution.

	A	B	C	D
1	The solubility of Mg(OH)$_2$			
2	K_{sp}	7.10E-12		
3	K_w	1.00E-14		
4				
5	Coeff of x^3	1		
6	Coeff of x^2	0		
7	Coeff of x	-1.00E-14		
8	Constant	-1.42E-11		
9				
10	Equation	-7.63E-19		
11	x	2.42E-04		
12				
13	Documentation			
14	Cell B5	1		
15	Cell B6	0		
16	Cell B7	-1.00E-14		
17	Cell B8	1.42E-11		
18	Cell B10=B5*B11^3+B6*B11^2+B7*B11+B8			
19	Cell B11=initial guess or Solver result			
20				
21	Constraints			
22	Cell B11>=0			

Figure 6-3 Final worksheet for cubic equation solution to Mg(OH)$_2$ solubility problem.

Now to find the other concentrations, we invoke the solubility product to find $[Mg^{2+}]$ and

the ion product of water to find $[H_3O^+]$. The advantage of this single variable approach is that the initial estimates, the constraint precision, the convergence and the optimality (tolerance) are not as critical as in the previous example.

Goal Seek

Another approach to solving multiple equilibrium problems is to use an Excel feature called Goal Seek. The Goal Seek function allows the cubic equation that we solved above using Solver to be solved numerically.

What is Goal Seek and How Does it Work?

Goal Seek is a tool for finding a value of a cell referenced by a formula that will give a specific result for the formula. To accomplish this task, Excel uses an iterative process similar to that used by Solver. A common example of the use of Goal Seek might be to find out what final exam score you need in order to earn an A in one of your courses. You would enter a formula into a spreadsheet for calculating your grade from the individual scores on exams, homework, problem sets, and other items. You would then use Goal Seek to find the value for the final exam that would yield a final score that would earn you an A. The grade problem can be solved algebraically, but for complex problems, there may be no straightforward algebraic solution, or the algebraic solution may require more time and effort than it is worth. Hence, Goal Seek can be quite a valuable tool in your computation arsenal.

Finding the Solubility of Mg(OH)$_2$ with Goal Seek

Let's now use Goal Seek to find the solubility of Mg(OH)$_2$, which we just found using Solver. To begin, open Excel, click on the File tab (Office Button), and choose **Excel Options**. Then click on the Formulas. Under Calculation options, select **Automatic** and **Enable iterative**

calculation. Enter 500 in the Ma<u>x</u>imum iterations: box and 0.0001 in the Maximum <u>C</u>hange:

box. Finally click on OK. Goal Seek will attain the most accurate results when it is allowed to

iterate a large number of times and when the maximum change in consecutive solutions is set to

a very small number. The value of 0.0001 for the maximum change is the absolute amount by

which the results of consecutive iterations change before Excel concludes that it has found the

solution to the equation. The values selected for the number of iterations and the maximum

change depend on the type of problem at hand and the magnitude of the goal that is sought. In

this case our Goal will be 1.000, so a maximum change of 0.0001 seems appropriate. Some

experimentation may be required to achieve an acceptable solution for a given equation. Large

numbers of iterations and small values for the maximum change generally produce longer

computation times.

	A	B	C	D	E
1	The solubility of Mg(OH)$_2$				
2	K_{sp}	7.10E-12			
3	K_w	1.00E-14			
4					
5	Left Side	Right Side		Right Side/Left Side	
6	1.00E+00	1.42E-11		7.04E+10	
7					
8	[OH⁻]	[H$_3$O⁺]	[Mg^{2+}]		
9	1.00E+00	7.04E-04	7.10E-12		
10					
11	Documentation				
12	Cell A6=initial estimate of Goal Seek solution				
13	Cell B6=2*B2/(A6^2)+B3/A6				
14	Cell D6=A6/B6				
15	Cell A9=A6				
16	Cell B9=B3/B6				
17	Cell C9=B2/(A6^2)				

Figure 6-4 Worksheet for Goal Seek solution to solubility problem.

Start with a blank worksheet and type the headings and values for K_{sp} and K_w in cells B2

and B3 as shown in Figure 6-4. Give the Goal Seek function a starting value by typing 1 in cell

A6. Now type the formulas that you see in the documentation section in cells B6, D6, A9, B9, and C9. Note that the formula you typed in cell B6 is identical to the mass balance expression arranged in terms of [OH⁻]. That is,

$$[OH^-] = \frac{2K_{sp}}{[OH^-]^2} + \frac{K_w}{[OH^-]}$$

Now, we are ready to solve the equation. Our goal is to find a value of the hydroxide concentration for which the right hand side of the equation is equal to the left hand side. When the two sides are equal, their ratio located in cell D6 will be one. Goal Seek uses a systematic search method to vary the contents of cell A6 until this goal is met. Now make cell D6 the active cell by clicking on it. Goal Seek is located in the Data Tools group under What If Analysis on the Data ribbon. Click on this group and choose **Goal Seek** to bring up the Goal Seek dialog box shown in Figure 6-5.

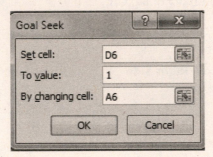

Figure 6-5 Goal Seek dialog box.

Note that D6 appears in the **Set cell**: box. Click on the **To value:** box, and enter 1, and enter A6 in the **By changing cell:** box. Now click on OK to start Goal Seek and observe the results. If everything works properly, your spreadsheet should appear as shown in Figure 6-6.

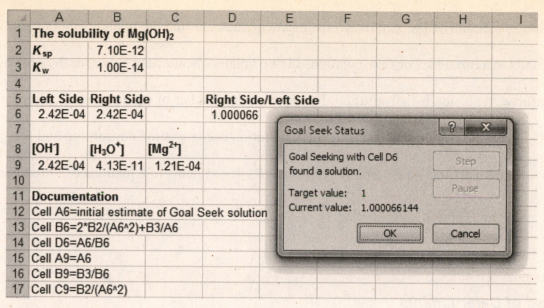

Figure 6-6 Goal Seek solution to solubility problem.

The Goal Seek Status window shows us that Goal Seek found a value in cell A6 that caused cell D6 to become 1 within the tolerance that we set for calculation options at the beginning of this exercise. The left-hand side of the equation (cell A6) is [OH⁻], which is displayed in cell A9. From this value, we find the concentrations of H_3O^+ and Mg^{2+} shown in cells B9 and C9, respectively. Now, you may click on OK. As an additional check on the validity of the solution, you may compare these results to those found by Solver, and you could enter a formula to calculate the solubility product for magnesium hydroxide from $[Mg^{2+}]$ and $[OH^-]$. This result can be compared to the value in cell B2.

Goal Seek can be used to solve problems of considerably higher complexity than shown here. It is a worthy addition to your Excel tools.

Solving Simultaneous Linear Equations

In some chemical equilibrium problems two or more simultaneous linear equations must be solved to obtain the desired result. For instance, Example 12-3 of FAC9 or Example 8-3 of AC7 involves the solution of two simultaneous linear equations to find the relative amounts of two

compounds in a mixture. For two equations in two unknowns, we may solve for the unknowns in

straightforward fashion by substitution. However, if the number of equations and unknowns

exceeds three, solution of the equations by substitution becomes time consuming and fraught

with error. Excel affords an extremely easy way to solve systems of equations of virtually any

size using determinants or matrix inversion. We will consider both of these methods in this

section.

Method of Determinants

Let us begin by considering the following pair of equations.

$$3x - 7y = 5$$

$$2x + y = 9$$

The determinant of the coefficients can be written

$$D = \begin{vmatrix} 3 & -7 \\ 2 & 1 \end{vmatrix}$$

By Cramer's rule, the solution to this system of equations is written

$$x = \frac{D_1}{D}, \qquad\qquad y = \frac{D_2}{D}$$

where

$$D_1 = \begin{vmatrix} 5 & -7 \\ 9 & 1 \end{vmatrix} \qquad \text{and} \qquad D_2 = \begin{vmatrix} 3 & 5 \\ 2 & 9 \end{vmatrix}$$

Note that D_1 is obtained by replacing the first column of D (x coefficients) with the constants 5

and 9, while D_2 is obtained by replacing the last column of D (y coefficients) with the same

constants.

Let's begin by constructing a worksheet similar to that shown in Figure 6-7. Here we have reproduced the coefficients and constants in cells B6:D7. In cells B11:C12 we have entered the elements of determinant D_1, while in cells B15:C16 we have entered the elements of D_2.

	A	B	C	D
1	Method of Determinants			
2	Equations	3x - 7y =5		
3		2x + y =9		
4				
5		Coefficients		Constants
6		3	-7	5
7		2	1	9
8	D			
9				
10		Coefficients		Constants
11		5	-7	
12		9	1	
13	D_1			
14		Coefficients		Constants
15		3	5	
16		2	9	
17	D_2			
18				
19	x			
20	y			

Figure 6-7. Initial worksheet for determinant solution to simultaneous equations.

To calculate D, we enter the Excel formula **=MDETERM(B6:C7)** in cell B8. Likewise to obtain D_1 and D_2, we enter **=MDETERM(B11:C12)** in cell B13 and **=MDETERM(B15:C16)** in cell B17. These formulas can be entered manually or chosen by Excel's Insert Function button in the formula bar. The solutions for x and y will then be calculated in cells B19 and B20. Note that we have also included the equation in rows 2 and 3 for reference purposes. The final results are shown in Figure 6-8 along with the documentation.

	A	B	C	D
1	**Method of Determinants**			
2	Equations	3x - 7y =5		
3		2x + y =9		
4				
5		**Coefficients**		**Constants**
6		3	-7	5
7		2	1	9
8	D	17		
9				
10		**Coefficients**		**Constants**
11		5	-7	
12		9	1	
13	D_1	68		
14		**Coefficients**		**Constants**
15		3	5	
16		2	9	
17	D_2	17		
18				
19	x	4		
20	y	1		
21				
22	**Documentation**			
23	Cell B8=MDETERM(B6:C7)			
24	Cell B13=MDETERM(B11:C12)			
25	Cell B17=MDETERM(B15:C16)			
26	Cell B19=B13/B8			
27	Cell B20=B17/B8			

Figure 6-8 Final worksheet for solving simultaneous equations by the method of determinants.

In applying this method to solving simultaneous equations, the determinant D must be nonzero. The method can be extended to more equations and more unknowns. For three simultaneous equations, Cramer's rule gives the solution for z as D_3/D, where D_3 is the determinant obtained by replacing the z coefficient column with the constant column. You may want to extend this method to the following equations:

$$x + 5y +2z = 6$$
$$2x + 2y - z = -4$$
$$3x - y - 2z = 2$$

Matrix Operations

The details of matrix algebra are beyond the scope of this presentation, but we'll demonstrate here the Excel matrix functions that are necessary for solving simultaneous equations. Let's begin by considering the same pair of equations as before.

$$3x - 7y = 5$$

$$2x + y = 9$$

From the two equations, we form two arrays of numbers: the *coefficient matrix* and the *constant matrix*. The coefficient matrix in matrix notation is

$$A = \begin{bmatrix} 3 & -7 \\ 2 & 1 \end{bmatrix}$$

and the constant matrix is

$$C = \begin{bmatrix} 5 \\ 9 \end{bmatrix}$$

The system of two equations may be written as

$$\mathbf{AX = C}$$

That is, the product of the coefficient matrix and the solution matrix is the constant matrix, where the solution matrix for the system of equations is written

$$\mathbf{X = A^{-1}C}$$

This equation says that if we find the inverse of the coefficient matrix and multiply it by the constant matrix, we obtain the solution matrix. Thus, we require two matrix operations to find the solution matrix: the matrix inverse function *minverse()* and the matrix multiplication function *mmult()*. Now let's try out these Excel functions using the example cited above.

Beginning with a clean spreadsheet, type the labels and arrays of coefficients and constants as shown in Figure 6-9.

	A	B	C	D	E
1	**Matrix Operations**				
2	Coefficient Matrix			Constant Matrix	
3	3	-7		5	
4	2	1		9	
5					
6	Inverse of Coefficient Matrix			Solution Matrix	
7					

Figure 6-9 Initial worksheet for matrix operations.

Then highlight cells A7:B8 and type

=minverse(a3:b4)[Ctrl+Shift+↵]

Recall that you must simultaneously depress the Control key, the Shift key, and the Enter key to accomplish array operations. Your spreadsheet should now appear as in Figure 6-10.

	A	B	C	D	E
1	**Matrix Operations**				
2	Coefficient Matrix			Constant Matrix	
3	3	-7		5	
4	2	1		9	
5					
6	Inverse of Coefficient Matrix			Solution Matrix	
7	0.058824	0.411765			
8	-0.11765	0.176471			

Figure 6-10 Matrix operations worksheet after finding the matrix inverse.

Now to determine the solution matrix, highlight cells D7:D8, and type

=mmult(a7:b8,d3:d4)[Ctrl+Shift+↵]

and your spreadsheet will appear as shown in Figure 6-11.

	A	B	C	D	E
1	**Matrix Operations**				
2	Coefficient Matrix			Constant Matrix	
3	3	-7		5	
4	2	1		9	
5					
6	Inverse of Coefficient Matrix			Solution Matrix	
7	0.058824	0.411765		4	
8	-0.11765	0.176471		1	

Figure 6-11 Matrix operations worksheet after finding the solution matrix.

Excel has given the same results as found by the determinant method, $x = 4$ and $y = 1$.

Now, to further test the *minverse()* and *mmult()* functions, confirm the results of Example 12-3 of

FAC9 or Example 8-3 of AC7. You can simply enter the x and y coefficients in cells A3:B4 and

the constants in cells D3:D4, and Excel will recalculate the solution matrix for you. You don't

need to reenter the matrix functions as long as you have the same number of equations and

unknowns.

Systems with Large Numbers of Equations and Unknowns

Consider the following system of four equations in four unknowns.

$$2w + 5x - y + 4z = 13$$

$$-3w - 2y + 6z = 12$$

$$2x + 3y + 10z = 0$$

$$5w + 3x + 2y = -5$$

Use the matrix functions of Excel to solve these four equations for x, y, z, and w as shown in the

spreadsheet in Figure 6-12. Note that if a coefficient does not appear in an equation, it must be

entered as a zero in the coefficient matrix.

169

	A	B	C	D	E	F	G
1	**Matrix Operations**						
2	Coefficient Matrix					Constant Matrix	
3	2	5	-1	4		13	
4	-3	0	-2	6		12	
5	0	2	3	10		0	
6	5	3	2	0		-5	
7							
8	Inverse Matrix					Solution Matrix	
9	-0.3	0.5	-0.18	0.62		-1	
10	0.433333	-0.5	0.126667	-0.47333		2	
11	0.1	-0.5	0.26	-0.34		-3	
12	-0.11667	0.25	-0.00333	0.196667		0.5	
13							
14	**Documentation**						
15	Cell B9:D12=MINVERSE(A3:D6)						
16	Cell F9:F12=MMULT(A9:D12,F3:F6)						

Figure 6-12 Solution of four equations in four unknowns using Excel's matrix functions.

The solution matrix gives $w = -1$, $x = 2$, $y = -3$, and $z = 0.5$. It is for systems of three or more

equations and unknowns that the Excel matrix functions are most useful.

Some Additional Examples of Equilibrium Problems

We'll consider here a few additional examples of using Excel to find solutions to chemical

equilibrium problems. Many other examples are found in later Chapters.

Solubility at a Fixed pH

In Example 11-7 of FAC9 and Example 10-6 of AC7, we calculate the solubility of calcium

oxalate at pH 4.00. The complexity of this example arises because oxalate ion is the fully

deprotonated form of the diprotic acid oxalic acid. By using Solver, we will calculate the

solubility of calcium oxalate not only at pH 4.00, but also at other values in the pH range 1-5.

We will also find the concentrations of the other species in the solution and plot the solubility as

a function of pH.

For calcium oxalate, we can write the solubility product as

$$K_{sp} = [Ca^{2+}][C_2O_4^{2-}] = 1.70 \times 10^{-9}$$

We can also write the two deprotonation equilibrium constants as

$$K_1 = \frac{[HC_2O_4^-][H_3O^+]}{[H_2C_2O_4]} = 5.60 \times 10^{-2}$$

$$K_2 = \frac{[C_2O_4^{2-}][H_3O^+]}{[HC_2O_4^-]} = 5.42 \times 10^{-5}$$

	A	B	C	D
1	Calcium Oxalate Solubility at Fixed pH			
2	K_{sp}	1.70E-09		
3	K_1	5.60E-02		
4	K_2	5.42E-05		
5	$[H_3O^+]$	1.00E-04		
6				
7	Equation			
8	$[C_2O_4^{2-}]$			
9	$[HC_2O_4^-]$			
10	$[H_2C_2O_4]$			
11	$[Ca^{2+}]$			

Figure 6-13 Initial worksheet for calcium oxalate solubility problem.

Set up a worksheet as shown in Figure 6-13. Use as initial estimates 1×10^{-5} M as the concentrations of the species in cells B8:B11. This is approximately the solubility if oxalate had no acid/base properties. Since all the oxalate comes from dissolution of calcium oxalate, we can write for the mass balance equation

$$[Ca^{2+}] = [C_2O_4^{2-}] + [HC_2O_4^-] + [H_2C_2O_4]$$

We'll use this expression as the equation to be solved. Enter the corresponding formula in cell B7 and have Solver set it equal to zero. Your constraints should include the solubility product constant, and the two acid dissociation constants. Finally, another constraint should be that cells B8:B11 must be equal to or exceed zero (no negative concentrations, please). This later constraint can either be entered explicitly in the Solver constraints box or in the Solver Options dialog box you can check the Make Unconstrained Variables Non-Negative box (Assume Non-Negative in Excel 2007). Enter a Constraint Precision and Convergence value of 1E-12 and a Optimality (Tolerance) value of 0.0001. It is also a good idea to check Use Automatic Scaling in a problem in which the values may vary over several orders of magnitude. Your final worksheet should look similar to Figure 6-14. It is a good idea to test whether Solver has found a correct solution. You can multiply the Ca^{2+} concentration and the $C_2O_4^{2-}$ concentration and see if they equal the solubility product. Also you could see if the various concentrations yield K_1 and K_2. This is always good practice with complex problems because of the frequent dependence ofanswers on initial estimates, and the values used for Constraint Precision, Optimality, and Convergence.

	A	B	C	D
1	Calcium Oxalate Solubility at Fixed pH			
2	K_{sp}	1.70E-09		
3	K_1	5.60E-02		
4	K_2	5.42E-05		
5	$[H_3O^+]$	1.00E-04		
6				
7	Equation	-8.23E-21		
8	$[C_2O_4^{2-}]$	2.44E-05		
9	$[HC_2O_4^-]$	4.51E-05		
10	$[H_2C_2O_4]$	8.05E-08		
11	$[Ca^{2+}]$	6.96E-05		
12				
13	Documentation			
14	Cell B7=B11-B8-B9-B10			
15	Cell B8:B11=Solver results			
16				
17	Constraints			
18	Cells B8:B11>=0			
19	Cell B8=B2/B11			
20	Cell B8=B4*B9/B5			
21	Cell B9=B3*B10/B5			

Figure 6-14 Final worksheet for finding solubility of CaC_2O_4 at pH 4.0

Now we'll enter different values for pH and have Solver find the new concentrations.

Calculate the solubility for every 0.5 pH units over the range of pH 1.0 to 5.0. Use Excel's charting feature to plot solubility vs. pH as shown in Figure 6-15. Note here that we have

manually entered the $[Ca^{2+}]$ values after Solver found the concentrations. In our chart, we have

specified a logarithmic y axis because the concentrations vary over several orders of magnitude.

You can also plot the concentrations of the other species on this same chart if desired.

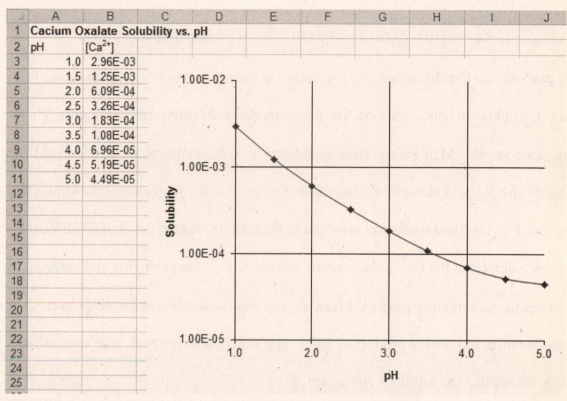

Figure 6-15 Worksheet for plotting calcium oxalate solubility vs. pH.

Solubility When the pH is Unknown.

As our final exercise in this chapter, we will find the solubility of calcium oxalate when the pH is

unknown. Again, we will use Solver to carry out the calculations. The six equations needed are

those described in Feature 11-1 of FAC8 and Feature 10-1 of AC7 and repeated here.

$$[Ca^{2+}] = [C_2O_4^{2-}] + [HC_2O_4^-] + [H_2C_2O_4] \qquad \text{mass balance}$$

$$2[Ca^{2+}] + [H_3O^+] = 2[C_2O_4^{2-}] + [HC_2O_4^-] + [OH^-] \qquad \text{charge balance}$$

$$K_{sp} = [Ca^{2+}][C_2O_4^{2-}] = 1.70 \times 10^{-9} \qquad \text{solubility product}$$

$$K_1 = \frac{[HC_2O_4^-][H_3O^+]}{[H_2C_2O_4]} = 5.60 \times 10^{-2} \qquad \text{acid dissociation 1}$$

$$K_2 = \frac{[C_2O_4^{2-}][H_3O^+]}{[HC_2O_4^-]} = 5.42 \times 10^{-5} \qquad \text{acid dissociation 2}$$

$$K_w = [H_3O^+][OH^-] = 1.00 \times 10^{-14} \qquad \text{ion product of water}$$

In this case, we use the charge balance expression as the formula to be optimized since it contains the unknown concentrations. The other equations are entered as constraints. The initial estimates are 1×10^{-5} M as before. Because the various concentrations are so widely different, you should check Use Automatic Scaling in the Solver Options. Also change the Optimality setting to 0.1%. The final worksheet is shown in Figure 6-16. Again, you should check the answer by multiplying the Ca^{2+} concentration and the $C_2O_4^{2-}$ concentration and ensuring that the product equals the solubility product. Likewise, you can check the acid dissociation constants, the mass balance, and the ion product of water. You should always check your spreadsheet results to ensure that the solutions are correct.

	A	B	C	D
1	**Calcium Oxalate Solubility at Unknown pH**			
2	K_{sp}	1.70E-09		
3	K_1	5.60E-02		
4	K_2	5.42E-05		
5	K_w	1.00E-14		
6				
7	Equation	-5.40E-21		
8	$[C_2O_4^{2-}]$	4.12E-05		
9	$[HC_2O_4^-]$	5.73E-08		
10	$[H_2C_2O_4]$	7.71E-14		
11	$[Ca^{2+}]$	4.13E-05		
12	$[H_3O^+]$	7.54E-08		
13	$[OH^-]$	1.33E-07		
14				
15	**Documentation**			
16	Cell B7=2*B11+B12-2*B8-B9-B13			
17	Cells B8:B13=Solver results			
18				
19				
20	**Constraints**			
21	Cell B8=B4*B9/B12			
22	Cell B9=B10*B3/B12			
23	Cell B11=B2/B8			
24	Cell B11=B8+B9+B10			
25	Cell B13=B5/B12			
26	Assume Non-Negative values			

Figure 6-16 Worksheet for finding calcium oxalate solubility when pH is unknown.

Summary

In this Chapter we have seen how to use Solver, Goal Seek, and Matrix Methods to find

solutions to the complex equations that arise in many applications of the systematic approach to

equilibrium problems. These tools are extremely useful, but must be used with care with highly

nonlinear problems. Solutions should always be checked to ensure that constraints are met and

that solutions to equations make sense chemically.

Problems

1.   Calculate the molar solubility of Ag_2CO_3 in a solution with a fixed H_3O^+ concentration of

      (a) 1.0×10^{-6} M   (c) 1.0×10^{-9} M

      (b) 1.0×10^{-7} M   (d) 1.0×10^{-11} M

2.   Calculate the molar solubility of $BaSO_4$ in solutions where the $[H_3O^+]$ is

      (a)   2.0 M   (c)   0.50 M

      (b)   1.0 M   (d)   0.10 M

3.   Calculate the molar solubility of CuS in the following solutions:

      (a)   pH fixed at 1.0

      (b)   pH fixed at 4.0

      (c)   water

4.   Find the molar solubility of CdS in the following solutions:

      (a)   pH fixed at 1.0

      (b)   pH fixed at 4.0

      (c)   water

5.   Find the molar solubility of $PbCO_3$ in the following solutions:

      (a)   pH fixed at 7.0

      (b)   pH fixed at 5.0

      (c)   pH fixed at 9.0

      (d)   water

6.   The equilibrium constant for the formation of $CuCl_2^-$ is given by

$$Cu^+ + 2Cl^- \rightleftharpoons CuCl_2^-$$

$$\beta_2 = \frac{[CuCl_2^-]}{[Cu^+][Cl^-]^2} = 7.9 \times 10^4$$

What is the solubility of CuCl in a solution having a molar analytical NaCl concentration

of

(a)   1.0 M?   (d) 1.0×10^{-3} M?

(b) 1.0×10^{-1} M?   (e) 1.0×10^{-4} M?

(c) 1.0×10^{-2} M?

7.   In contrast to many salts, calcium sulfate is only partially dissociated in aqueous solution:

$$CaSO_4(aq) \rightleftharpoons Ca^{2+} + SO_4^{2-}$$

$$K_d = 5.2 \times 10^{-3}$$

The solubility product of $CaSO_4$ is 2.6×10^{-5}. Calculate the solubility of $CaSO_4$ in (a)

water and (b) 0.0100 M Na_2SO_4. In addition find the percent of undissociated $CaSO_4$ in

each solution.

8.   Find the solubility of $Ca(OH_2)$ in water using (a) Solver and (b) Goal Seek.

9.   Find the solubility of lead oxalate in water using Solver.

10.   Calculate the molar solubility of CdS as a function of pH from pH 10 to pH 1. Find

values for every 0.5 pH unit and use the charting function to plot solubility vs. pH.

11.   Calculate the molar solubility of TlS as a function of pH over the range of pH 10 to pH 1.

Find values every 0.5 pH unit and use the charting function to plot solubility vs. pH.

12.   Use Goal Seek instead of Solver to find the solubility of calcium oxalate at pH 4.0.

        Discuss the ease of using Goal Seek compared to Solver. Can Goal Seek be used to find

        the solubility when the pH is unknown.

13.   The solubility of calcium oxalate at pH 3.0 is 1.83×10^{-4} M. Use Solver and Goal Seek

        to find the solubility product at this pH value.

14.   The solubility of barium oxalate in water at 25°C is 0.010 g/100 mL. Use this information

        and Solver to find the pH of the solution.

Chapter 7

Neutralization Titrations and Graphical Representations

One of Excel's major features is the ability to plot data simply and rapidly. We explored some graphical methods in Chapter 4 when dealing with least-squares and other calibration methods. In this chapter, we show how Excel can help us calculate and display the data from acid-base titrations. In addition, we'll explore some other methods for plotting titration and equilibrium data. We begin here with a strong acid/strong base titration and then extend our discussion to titrations of weak acids.

Calculating and Plotting a Strong Acid/Strong Base Titration Curve

In Example 14-1 of FAC9 and Example 12-1 of AC7 we constructed a curve for the titration of 50.00 mL of 0.0500 M HCl with 0.1000 M NaOH. To calculate pH as a function of volume of titrant added, we split the titration curve into three parts: the pre-equivalence point region after adding NaOH, the equivalence point, and points beyond the equivalence point. In each of these regions of the curve, we made certain assumptions regarding relative concentrations of solution species to simplify the calculations. In this application, we construct titration curves for this same example by two different methods. We first use a stoichiometric method, identical to that used in Example 14-1 (FAC9) or 12-1 (AC7). In the second method, we use the charge-balance equation to generate the titration curves (see Feature 14-1 of FAC9 or 12-1 of AC7).

Stoichiometric Method

In this method, we divide the titration curve into the pre-equivalence point region, the post-equivalence point region and the equivalence point itself. The titration reaction is

$$HCl + NaOH \rightarrow NaCl + H_2O$$

Let's begin by labeling a blank worksheet and entering the known quantities as shown in Figure

7-1. The initial concentrations of acid and base are c_{HCl}^0 and c_{NaOH}^0. We'll obtain the pH values for

each volume of NaOH shown in cells A8 through A32.

	A	B	C	D	E
1	Strong Acid/Strong Base Titration, Stoichiometric Method				
2	K_w	1.00E-14			
3	c_{HCl}^0	0.0500			
4	V_{HCl}	50.00			
5	c_{NaOH}^0	0.1000			
6					
7	V_{NaOH}, mL	[H_3O^+]	[OH]	pOH	pH
8	0.00				
9	5.00				
10	10.00				
11	15.00				
12	20.00				
13	24.00				
14	24.20				
15	14.40				
16	14.60				
17	14.80				
18	14.90				
19	24.99				
20	25.00				
21	25.01				
22	25.20				
23	25.20				
24	25.40				
25	25.60				
26	25.80				
27	26.00				
28	30.00				
29	35.00				
30	40.00				
31	45.00				
32	50.00				

Figure 7-1 Spreadsheet for strong acid/strong base titration.

The initial pH is just the pH of a 0.0500 M solution of HCl with [H_3O^+] = 0.0500 M. In

the pre-equivalence point region, it is relatively easy to find the hydronium ion concentration

from the stoichiometry. In this region, we haven't added enough NaOH to completely neutralize

the HCl, and hence unreacted HCl remains. The [H_3O^+] is equal to the unreacted HCl divided by

the total solution volume since adding NaOH changes the volume.

$$[H_3O^+] \approx c_{HCl} = \frac{\text{amount HCl (mmol) remaining after adding NaOH}}{\text{total solution volume}}$$

$$= \frac{\text{original no. mmol HCl} - \text{no. mmol NaOH added}}{\text{total solution volume}}$$

For the first volume of titrant (0.00 mL), we can enter into cell B8 the formula corresponding to this equation:

$$= (\$B\$3*\$B\$4-\$B\$5*A8)/(\$B\$3+A8)\,[\dashv]$$

Note here that the product of cells B3 and B4 is the original number of millimoles of HCl present. Since this is constant, we make the cell references absolute references by adding a dollar sign ($) before the letter and number. The product of cells B5 and A8 is the number of millimoles of titrant added, while the denominator is the total volume of solution after the addition. Note that since the volume of NaOH is variable, we make this cell reference a relative reference by not adding the dollar sign. As we fill in the entries for other volumes, this reference will change.

In cell E8, we calculate the pH from

$$pH = -\log[H_3O^+]$$

Hence, we enter the formula =-log10(B8)[⏎] in cell E8. We express the value in E8 to 2 decimal places. We could also calculate [OH$^-$] and pOH from the ion product constant of water, but this isn't necessary. Your worksheet should appear as shown in Figure 7-2.

Now, we can fill in the remaining calculations for volumes prior to the equivalence point. Select cells B8:E8 and use the fill handle to complete the entries through row 19. Your worksheet should then appear as shown in Figure 7-3.

	A	B	C	D	E
1	Strong Acid/Strong Base Titration, Stoichiometric Method				
2	K_w	1.00E-14			
3	c^0_{HCl}	0.0500			
4	V_{HCl}	50.00			
5	c^0_{NaOH}	0.1000			
6					
7	V_{NaOH}, mL	[H_3O^+]	[OH]	pOH	pH
8	0.00	0.05			1.30
9	5.00				
10	10.00				
11	15.00				

Figure 7-2 Worksheet showing calculation of first volume entry.

	A	B	C	D	E
1	Strong Acid/Strong Base Titration, Stoichiometric Method				
2	K_w	1.00E-14			
3	c^0_{HCl}	0.0500			
4	V_{HCl}	50.00			
5	c^0_{NaOH}	0.1000			
6					
7	V_{NaOH}, mL	[H_3O^+]	[OH]	pOH	pH
8	0.00	0.05			1.30
9	5.00	0.0363636			1.44
10	10.00	0.025			1.60
11	15.00	0.0153846			1.81
12	20.00	0.0071429			2.15
13	24.00	0.0013514			2.87
14	24.20	0.0010782			2.97
15	24.40	0.0008065			3.09
16	24.60	0.0005362			3.27
17	24.80	0.0002674			3.57
18	24.90	0.0001335			3.87
19	24.99	1.334E-05			4.88

Figure 7-3 Worksheet after using the fill handle to complete the pre-equivalance point calculations.

At the equivalence point, we've added enough NaOH to react completely with the HCl initially present. The only [H_3O^+] comes from the dissociation of water. Hence, at the equivalence point, [H_3O^+] = [OH^-] and

$$[H_3O^+] = \sqrt{K_w}$$

In cell B20, we thus enter the formula **=SQRT (B2) [↵]**. We then calculate pH inCell E20.

In the post-equivalence point region, there is excess NaOH. We calculate the concentration of NaOH from the excess added. Thus,

$$[OH^-] \approx c_{NaOH} = \frac{\text{no. of mmol NaOH added} - \text{original no. of mmol HCl present}}{\text{total solution volume}}$$

In cell C21, we enter the formula shown in the documentation section of Figure 7-4. Use the fill handle to complete the entries for column C. In cell D21, we calculate pOH from pOH = $-$log10([OH$^-$]). In cell E21, we calculate pH from pH = pK_w $-$ pOH. After extending columns D and E through the remaining entries, you worksheet should look like Figure 7-4 after completing the documentation.

Now we're ready to plot the data. Select cells A8:A32 and D8:D32. You can select non-adjacent ranges by selecting the first range and then holding down the **CTRL** key while selecting the second range. Click on the Insert tab and select **Scatter with only markers** as the type of plot. Format the chart. With the chart selected, chose **Design** on the **Chart Tools** menu and select **Move Chart Location**. Locate the chart as a new sheet called Chart 1. Add a smooth line so that the titration curve appears as shown in Figure 7-5.

We can also perform calculations for the strong acid/strong base system at lower concentrations as indicated in Table 14-2 and Figure 14-3 of FAC9 and Table 12-2 and Figure 12-3 of AC7. Figure 7-6 shows the spreadsheet and combined titration curves for the different concentrations.

	A	B	C	D	E
1	Strong Acid/Strong Base Titration, Stoichiometric Method				
2	K_w	1.00E-14			
3	c^0_{HCl}	0.0500			
4	V_{HCl}	50.00			
5	c^0_{NaOH}	0.1000			
6					
7	V_{NaOH}, mL	$[H_3O^+]$	[OH]	pOH	pH
8	0.00	0.05			1.30
9	5.00	0.0363636			1.44
10	10.00	0.025			1.60
11	15.00	0.0153846			1.81
12	20.00	0.0071429			2.15
13	24.00	0.0013514			2.87
14	24.20	0.0010782			2.97
15	24.40	0.0008065			3.09
16	24.60	0.0005362			3.27
17	24.80	0.0002674			3.57
18	24.90	0.0001335			3.87
19	24.99	1.334E-05			4.88
20	25.00	1.00E-07			7.00
21	25.01		1.33316E-05	4.88	9.12
22	25.20		0.000265957	3.58	10.42
23	25.20		0.000265957	3.58	10.42
24	25.40		0.000530504	3.28	10.72
25	25.60		0.000793651	3.10	10.90
26	25.80		0.001055409	2.98	11.02
27	26.00		0.001315789	2.88	11.12
28	30.00		0.00625	2.20	11.80
29	35.00		0.011764706	1.93	12.07
30	40.00		0.016666667	1.78	12.22
31	45.00		0.021052632	1.68	12.32
32	50.00		0.025	1.60	12.40
33	Documentaion				
34	Cell B8=($B$3*$B$4-$B$5*A8)/(A8+$B$4)				
35	Cell E8=-LOG10(B8)				
36	Cell B20=SQRT(B2)				
37	Cell C21=($B$5*A21-$B$3*$B$4)/(A21+$B$4)				
38	Cell D21=-LOG10(C21)				
39	Cell E21=-LOG10($B$2)-D21				

Figure 7-4 Final worksheet for calculating pH for various volumes of NaOH added.

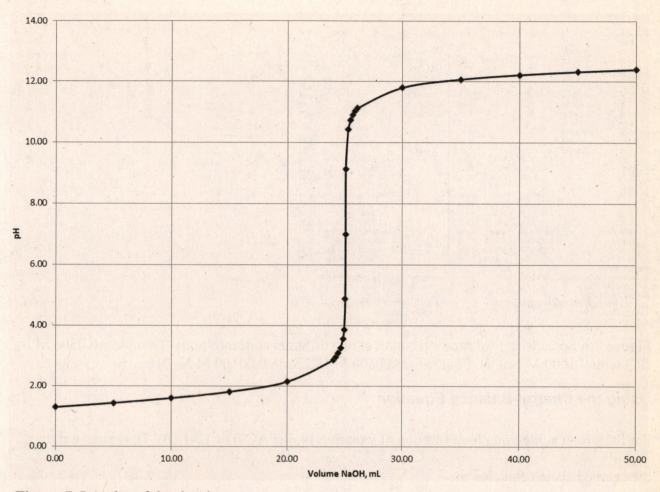

Figure 7-5 A plot of the titration curve.

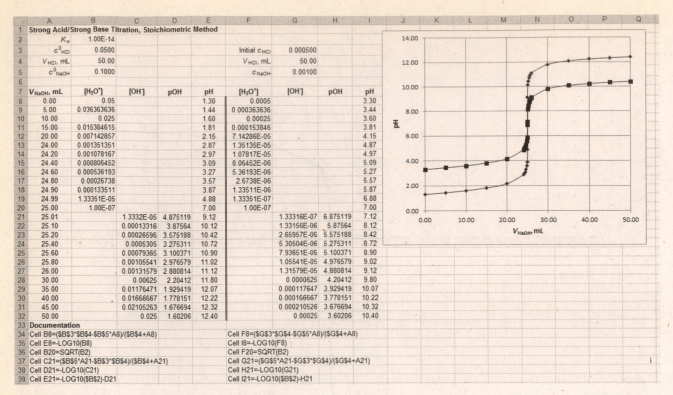

Strong Acid/Strong Base Titration, Stoichiometric Method

K_w	1.00E-14							
c^0_{HCl}	0.0500				Initial c_{HCl}	0.000500		
V_{HCl}, mL	50.00				V_{HCl}, mL	50.00		
c^0_{NaOH}	0.1000				c_{NaOH}	0.00100		

V_{NaOH}, mL	$[H_3O^+]$	$[OH^-]$	pOH	pH	$[H_3O^+]$	$[OH^-]$	pOH	pH
0.00	0.05			1.30	0.0005			3.30
5.00	0.036363636			1.44	0.00036363636			3.44
10.00	0.025			1.60	0.00025			3.60
15.00	0.015384615			1.81	0.000153846			3.81
20.00	0.007142857			2.15	7.14286E-05			4.15
24.00	0.001351351			2.87	1.35135E-05			4.87
24.20	0.001078167			2.97	1.07817E-05			4.97
24.40	0.000806452			3.09	8.06452E-06			5.09
24.60	0.000536193			3.27	5.36193E-06			5.27
24.80	0.00026738			3.57	2.6738E-06			5.57
24.90	0.000133511			3.87	1.33511E-06			5.87
24.99	1.33351E-05			4.88	1.33351E-07			6.88
25.00	1.00E-07			7.00	1.00E-07			7.00
25.01		1.3332E-05	4.875119	9.12		1.33316E-07	6.875119	7.12
25.10		0.00013316	3.87564	10.12		1.33156E-06	5.87564	8.12
25.20		0.00026596	3.575188	10.42		2.65957E-06	5.575188	8.42
25.40		0.0005305	3.275311	10.72		5.30504E-06	5.275311	8.72
25.60		0.00079365	3.100371	10.90		7.93651E-06	5.100371	8.90
25.80		0.00105541	2.976579	11.02		1.05541E-05	4.976579	9.02
26.00		0.00131579	2.880814	11.12		1.31579E-05	4.880814	9.12
30.00		0.00625	2.20412	11.80		0.0000625	4.20412	9.80
35.00		0.01176471	1.929419	12.07		0.000117647	3.929419	10.07
40.00		0.01666667	1.778151	12.22		0.000166667	3.778151	10.22
45.00		0.02105263	1.676694	12.32		0.000210526	3.676694	10.32
50.00		0.025	1.60206	12.40		0.00025	3.60206	10.40

Documentation

Cell B8=($B$3*$B$4-$B$5*A8)/($B$4+A8)
Cell E8=-LOG10(B8)
Cell B20=SQRT(B2)
Cell C21=($B$5*A21-$B$3*$B$4)/($B$4+A21)
Cell D21=-LOG10(C21)
Cell E21=-LOG10($B$2)-D21

Cell F8=($G$3*$G$4-$G$5*A8)/($G$4+A8)
Cell I8=-LOG10(F8)
Cell F20=SQRT(B2)
Cell G21=($G$5*A21-$G$3*$G$4)/($G$4+A21)
Cell H21=-LOG10(G21)
Cell I21=-LOG10($B$2)-H21

Figure 7-6 Spreadsheet for titration curves at two different concentrations. Triangles=0.0500 M HCl with 0.1000 M NaOH, Squares=0.000500 M HCl with 0.00100 M NaOH.

Using the Charge-Balance Equation

For the strong acid/strong base titration of Example 14-1 (FAC9) or 12-1 (AC7), we can write

the charge balance equation as:

$$[H_3O^+] + [Na^+] = [OH^-] + [Cl^-]$$

where

$$[Na^+] = \frac{V_{NaOH}\,c^0_{NaOH}}{V_{NaOH} + V_{HCl}} \qquad \text{and} \qquad [Cl^-] = \frac{V_{HCl}\,c^0_{HCl}}{V_{NaOH} + V_{HCl}}$$

For volumes of NaOH in the pre-equivalence point region, we have unreacted $[H_3O^+]$, so

$[OH^-] \ll [Cl^-]$. Thus

$$[H_3O^+] = [Cl^-] - [Na^+] = \frac{V_{HCl}\,c^0_{HCl}}{V_{NaOH} + V_{HCl}} - \frac{V_{NaOH}\,c^0_{NaOH}}{V_{NaOH} + V_{HCl}} = \frac{V_{HCl}\,c^0_{HCl} - V_{NaOH}\,c^0_{NaOH}}{V_{NaOH} + V_{HCl}}$$

The hydronium ion concentration and the pH in the pre-equivalence point region are shown in cells B8:B19 and D8:D19 in Figure 7-7.

At the equivalence point, $[Na^+] = [Cl^-]$ and $[H_3O^+] = [OH^-]$, so

$$[H_3O^+] = \sqrt{K_w}$$

The hydronium ion concentration and the pH are shown in cells B20 and D20 of Figure 7-7, respectively.

In the post equivalence point region, there is excess base so that $[H_3O^+] \ll [Na^+]$. The charge balance equation then becomes

$$[OH^-] = [Na^+] - [Cl^-] = \frac{V_{NaOH}c^0_{NaOH} - V_{HCl}c^0_{HCl}}{V_{NaOH} + V_{HCl}}$$

The $[OH^-]$ and the pH are computed in cells C21:C32 and D21:D32 of Figure 7-7.

	A	B	C	D	E	F
1	**Strong Acid/Strong Base Titration, Charge Balance Method**					
2	K_w	1.00E-14				
3	c^0_{HCl}	0.0500				
4	V_{HCl}, mL	50.00				
5	c^0_{NaOH}	0.1000				
6						
7	V_{NaOH}, ml	[H$_3$O$^+$]	[OH]	pH		
8	0.00	0.05		1.30		
9	5.00	0.036363636		1.44		
10	10.00	0.025		1.60		
11	15.00	0.015384615		1.81		
12	20.00	0.007142857		2.15		
13	24.00	0.001351351		2.87		
14	24.20	0.001078167		2.97		
15	24.40	0.000806452		3.09		
16	24.60	0.000536193		3.27		
17	24.80	0.00026738		3.57		
18	24.90	0.000133511		3.87		
19	24.99	1.33351E-05		4.88		
20	25.00	0.0000001		7.00		
21	25.01		1.33E-05	9.12		
22	25.10		0.000133	10.12		
23	25.20		0.000266	10.42		
24	25.40		0.000531	10.72		
25	25.60		0.000794	10.90		
26	25.80		0.001055	11.02		
27	26.00		0.001316	11.12		
28	30.00		0.00625	11.80		
29	35.00		0.011765	12.07		
30	40.00		0.016667	12.22		
31	45.00		0.021053	12.32		
32	50.00		0.025	12.40		
33	**Documentation**					
34	Cell B8=($B$3*$B$4-$B$5*A8)/($B$4+A8)					
35	Cell D8=-LOG10(B8)					
36	Cell B20=SQRT($B$2)					
37	Cell C21=($B$5*A21-$B$3*$B$4)/($B$4+A21)					
38	Cell D21=-LOG10($B$2)-(-LOG10(C21))					

Figure 7-7 Spreadsheet for titration curve using the charge-balance equation.

Weak Acid-Strong Base Titrations

In this application, we will compute the titration curve for a weak acid/strong base system by the stoichiometric method presented above and by a master equation approach. The latter allows us to obtain a single equation for the entire titration.

Stoichiometric Method

We will illustrate the computations of the stoichiometric approach with Example 14-3 (FAC9) or

12-6 (AC7) which considers the titration of 50.00 mL of 0.1000 M acetic acid with 0.1000 M

sodium hydroxide. The titration curve can be divided into four regions: (1) the initial region

before any base has been added where we have a solution of the weak acid; (2) the pre-

equivalence point region where we have a series of buffer solutions; (3) the equivalence point

where we have only the conjugate base of the weak acid; and (4) the post-equivalence point

region where we have excess strong base.

Let's begin with a spreadsheet such as that shown in Figure 7-8 entering the appropriate

values for the constants, initial concentrations and volumes. For the initial volume (0.00 mL), we

must find the pH of a 0.1000 M HOAc solution. The hydronium ion concentration is found using

Equation 9-22 of FAC9 or Equation 4-21 of AC7.

$$[H_3O^+] = \sqrt{K_a c_{HOAc}}$$

The pH is computed in the usual way.

	A	B	C	D	E	F	G	H	I
1	Weak Acid/Strong Base Titration, Stoichiometric Method								
2	K_a	1.75E-05							
3	K_w	1.00E-14							
4	Vol. HOAc	50.00							
5	c^0_{HOAc}	0.1000							
6	c^0_{NaOH}	0.1000							
7	Vol NaOH, ml	c_{HOAc}	c_{OAc^-}	c_{NaOH}	System	[H₃O⁺]	[OH⁻]	pOH	pH
8	0.00								
9	10.00								
10	20.00								
11	30.00								
12	40.00								
13	45.00								
14	49.00								
15	49.90								
16	50.00								
17	50.10								
18	51.00								
19	55.00								
20	60.00								
21	70.00								
22	80.00								
23	90.00								

Figure 7-8 Initial entries for weak acid-strong base titration

As soon as we add the first increment of base, we have a buffer solution. In fact, a series of buffers is formed for each addition of base prior to the equivalence point. The analytical concentrations of the two constituents of the buffer are given by:

$$c_{HOAc} = \frac{\text{initial no. of mmoles HOAc} - \text{no. of mmoles of base added}}{\text{total solution volume}}$$

$$c_{OAc^-} = \frac{\text{no. of mmoles of base added}}{\text{total solution volume}}$$

Hence we enter into cell B9 the formula =($B$4*$B$5–$B$6*A9)/($B$4+A9) [↵] and into cell C9 =$B$6*A9/($B$4+A9) [↵]. These formulas can be copied into cells B10:B15. The [H₃O⁺] can be found from the dissociation constant as

$$[H_3O^+] = K_a \frac{[HOAc]}{[OAc^-]} \approx K_a \frac{c_{HOAc}}{c_{OAc^-}}$$

We then enter into cell F9, the formula corresponding to the approximation and find the pH in cell I9.

At the equivalence point, all the acetic acid has been converted to sodium acetate. The solution is thus that of the conjugate base of the weak acid. A solution of sodium acetate is somewhat basic due to the reaction

$$OAc^- + H_2O \rightleftharpoons HOAc + OH^-$$

From the stoichiometry, we can write

$$[HOAc] = [OH^-]$$

$$[OAc^-] = c_{OAc^-} - [OH^-] \approx c_{OAc^-}$$

Substituting into the base dissociation constant of OAc^-, we have

$$K_b = \frac{K_w}{K_a} = \frac{[HOAc][OH^-]}{[OAc^-]} \approx \frac{[OH^-]^2}{c_{OAc^-}}$$

or

$$[OH^-] = \sqrt{K_b c_{HOAc}}$$

We enter the formula corresponding to this equation in cell G16.

After the equivalence point, we have excess base. Both the excess base and the acetate ion are sources of OH^-. Since the excess base suppresses the reaction of OAc^- with H_2O, we can safely assume that the $[OH^-]$ is equal to the analytical concentration of NaOH, or

$$[OH^-] \approx c_{NaOH} = \frac{\text{no. mmoles NaOH added} - \text{no. mmoles HOAc initially present}}{\text{total solution volume}}$$

We thus enter the formula corresponding to this equation into cell G17. We can calculate pOH in cell H17 and pH from $pK_w - pOH$ in cell I17. Since these same calculations hold through the rest of the titration, we copy the formulas in row 17 through row 23. The final spreadsheet and

documentation along with a titration curve are shown in Figure 7-9.

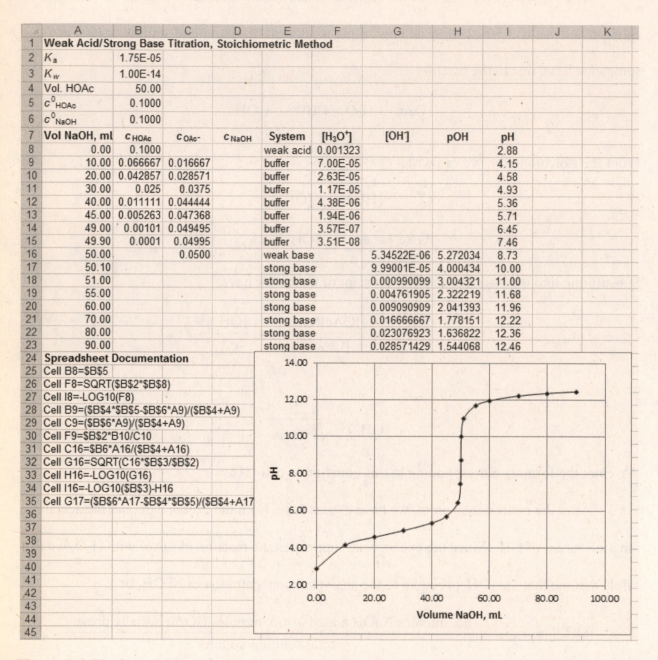

	A	B	C	D	E	F	G	H	I	J	K
1	Weak Acid/Strong Base Titration, Stoichiometric Method										
2	K_a	1.75E-05									
3	K_w	1.00E-14									
4	Vol. HOAc	50.00									
5	c^0_{HOAc}	0.1000									
6	c^0_{NaOH}	0.1000									
7	Vol NaOH, ml	c_{HOAc}	c_{OAc^-}	c_{NaOH}	System	$[H_3O^+]$	[OH⁻]	pOH	pH		
8	0.00	0.1000			weak acid	0.001323			2.88		
9	10.00	0.066667	0.016667		buffer	7.00E-05			4.15		
10	20.00	0.042857	0.028571		buffer	2.63E-05			4.58		
11	30.00	0.025	0.0375		buffer	1.17E-05			4.93		
12	40.00	0.011111	0.044444		buffer	4.38E-06			5.36		
13	45.00	0.005263	0.047368		buffer	1.94E-06			5.71		
14	49.00	0.00101	0.049495		buffer	3.57E-07			6.45		
15	49.90	0.0001	0.04995		buffer	3.51E-08			7.46		
16	50.00		0.0500		weak base		5.34522E-06	5.272034	8.73		
17	50.10				stong base		9.99001E-05	4.000434	10.00		
18	51.00				stong base		0.000990099	3.004321	11.00		
19	55.00				stong base		0.004761905	2.322219	11.68		
20	60.00				stong base		0.009090909	2.041393	11.96		
21	70.00				stong base		0.016666667	1.778151	12.22		
22	80.00				stong base		0.023076923	1.636822	12.36		
23	90.00				stong base		0.028571429	1.544068	12.46		
24	Spreadsheet Documentation										
25	Cell B8=$B$5										
26	Cell F8=SQRT($B$2*$B$8)										
27	Cell I8=-LOG10(F8)										
28	Cell B9=($B$4*$B$5-$B$6*A9)/($B$4+A9)										
29	Cell C9=($B$6*A9)/($B$4+A9)										
30	Cell F9=$B$2*B10/C10										
31	Cell C16=$B6*A16/($B$4+A16)										
32	Cell G16=SQRT(C16*$B$3/$B$2)										
33	Cell H16=-LOG10(G16)										
34	Cell I16=-LOG10($B$3)-H16										
35	Cell G17=($B$6*A17-$B$4*$B$5)/($B$4+A17										
36											
37											
38											
39											
40											
41											
42											
43											
44											
45											

Figure 7-9 Final spreadsheet for weak acid/strong base titration with the stoichiometric method

Master Equation Approach

For the master equation approach, we'll consider a hypothetical acid HA. The system of interest is the titration of 50.00 mL of 0.1000 M HA with 0.1000 M NaOH. The K_a of the acid is 1.00×10^{-5}. At any point in the titration, the charge-balance equation is given by

$$[Na^+] + [H_3O^+] = [A^-] + [OH^-]$$

We'll now substitute and rearrange this equation to obtain an equation in $[H_3O^+]$ as a function of the volume of NaOH added, V_{NaOH}. The sodium ion concentration is just the number of millimoles of NaOH added divided by the total volume of solution in milliliters, or

$$[Na^+] = \frac{c^0_{NaOH} V_{NaOH}}{V_{NaOH} + V_{HA}}$$

Mass balance gives the total concentration of species containing A, c_T, as

$$c_T = [HA] + [A^-] = \frac{[A^-][H_3O^+]}{K_a} + [A^-]$$

Solving for the concentration of A^- yields

$$[A^-] = \left(\frac{K_a}{[H_3O^+] + K_a} \right) c_T$$

The concentration of OH^- can be found from the ion product of water

$$[OH^-] = \frac{K_w}{[H_3O^+]}$$

If we substitute these latter two equations into the charge balance equation, we get

$$[Na^+] + [H_3O^+] = \frac{c_T K_a}{[H_3O^+] + K_a} + \frac{K_w}{[H_3O^+]}$$

By rearranging this equation into the standard form, we obtain the master system equation for the entire titration

$$[H_3O^+]^3 + \left(K_a + [Na^+]\right)[H_3O^+]^2 + \left(K_a[Na^+] - c_T K_a - K_w\right)[H_3O^+] - K_w K_a = 0$$

This direct, master equation can be solved using Excel's Solver as we did for a cubic equation in Chapter 5. For every volume of NaOH added, we must solve the cubic equation above for $[H_3O^+]$. Because we must make fairly decent initial estimates and set values for the Constraint Precision, Optimality (Tolerance) and Convergence appropriately, this becomes a somewhat time-consuming task. To avoid this, let's use Excel's Goal Seek to solve this equation.

Rearranging the charge balance expression from before, we can write

$$[H_3O^+] = \frac{c_T K_a}{[H_3O^+] + K_a} + \frac{K_w}{[H_3O^+]} - [Na^+]$$

In this form, we can find the value of $[H_3O^+]$ that equalizes the left and right sides of this equation. Let us begin with a worksheet set-up as shown in Figure 7-10.

	A	B	C	D	E	F	G
1	Weak Acid/Strong Base Titration, Master Equation Approach						
2	K_a	1.00E-05					
3	K_w	1.00E-14					
4	V_{HA}	50.00					
5	c^0_{HA}	0.1000					
6	c^0_{NaOH}	0.1000					
7							
8	V_{NaOH}	$[Na^+]$	c_T	$[H_3O^+]$	pH	Right side	Right/Left

Figure 7-10 Initial worksheet for acid-base titration.

Before invoking Goal Seek, enter the Excel options menu by clicking the File tab (Office button in Excel 2007). Make sure that **Automatic** and **Enable iterative calculations** are checked under **Formulas, Calculation options**. Also set **Maximum Iterations:** to 500 and **Maximum Change:** to 0.0001. Enter the following volumes of NaOH into column A beginning with cell A9: 0.00, 5.00, 10.00, 20.00, 25.00, 30.00, 40.00, 45.00, 49.00, 49.90, 49.95, 49.99, 50.00, 50.01, 50.05, 50.10, 51.00, 55.00, 60.00, and 70.00. For each volume of titrant added, we will calculate the Na^+ concentration, the total concentration of acid after dilution, the H_3O^+ concentration, and the pH. We would not have to calculate $[Na^+]$ and c_T for each volume, but doing so makes it less likely to make mistakes in the right side formula. Enter appropriate formulas for $[Na^+]$ and c_T in cells B9 and C9, respectively. The H_3O^+ concentration in cell D9 will be the left side of the equation, while the right side of the equation in cell F9 will be the formula corresponding to the rearranged charge balance equation, `=C9*$B$2/(D9+$B$2)+$B$3/D9-B9`. In cell G9, type the formula for the ratio of cell F9 to D9. Also enter into cell E9 the formula for calculating pH from $[H_3O^+]$. Since the K_a for HA is 1.00×10^{-5}, and we know that the solution will be acidic prior to the equivalence point, enter 1.00E-06 for the initial value of $[H_3O^+]$ in cell D9. Copy this value into cells D10:D28.

We will begin by doing the calculation for the first volume before proceeding to subsequent volumes. Select cell G9 and then choose **Goal Seek…** from the **Data Tools** group on the Data tab. In the dialog box, have Excel set cell G9 to a value of 1 by changing cell D9. This should produce the result shown in Figure 7-11.

	A	B	C	D	E	F	G
1	Weak Acid/Strong Base Titration, Master Equation Approach						
2	K_a	1.00E-05					
3	K_w	1.00E-14					
4	V_{HA}	50.00					
5	c^0_{HA}	0.1000					
6	c^0_{NaOH}	0.1000					
7							
8	V_{NaOH}	$[Na^+]$	c_T	$[H_3O^+]$	pH	Right side	Right/Left
9	0.00	0	0.1	9.95E-04	3.00	9.95E-04	1.00E+00

Figure 7-11 Worksheet after first calculation.

Note that you may have to format some of the numbers in scientific notation to see the same values.

Now we will do the remaining calculations. Select cells B9:C9, and use the fill handle to copy the formulas entered for [Na$^+$] and c_T into the remaining column B and column C cells (through row 28). Select cells E9:G9 and copy the formulas into cells E10:G28. Use Goal Seek to find the [H$_3$O$^+$] for each volume of titrant. Note that Goal Seek fails to find the correct value for the 49.90 mL volume. Hence, we will have to change the initial value of [H$_3$O$^+$] at this point. Note that [H$_3$O$^+$] became progressively smaller and reached a value of 2.04×10^{-7} M at a volume of 49.00 mL. So, enter a value of 1.00E-09 for the [H$_3$O$^+$] in cell D18 and copy this value into cells D19:D28. Continue using Goal Seek for each volume until it again fails to find an appropriate value (row 23). Note again that the [H$_3$O$^+$] has fallen substantially below the initial estimate for the previous calculation. Change the [H$_3$O$^+$] estimate to 1.00E-13 for the remaining calculations (cells D23:D28). Prepare a plot of pH vs. Volume of NaOH. When you have completed the calculations, the plot, and added documentation, your worksheet should appear as shown in Figure 7-12.

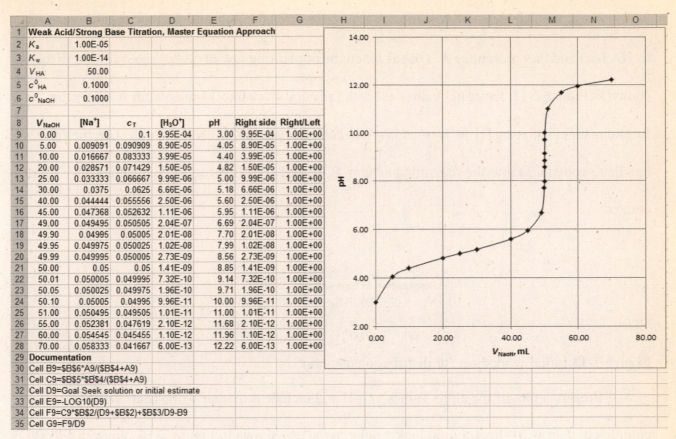

Figure 7-12 Final worksheet for acid-base titration curve using Goal Seek.

Try other values of K_a and of c_{HA} or c_{NaOH}. You can also try using Solver instead of Goal

Seek to obtain the pH values.

Distribution of Species: α Plots

One of the most useful methods of visualizing how various species vary in concentration as a

function of a variable like pH is the so-called *distribution diagram*, or α-plot. In this type of

diagram, the fraction α of a given species is plotted as a function of a logarithmic variable such

as pH. For a monoprotic acid,

$$\alpha_0 = \frac{[HA]}{c_T} = \frac{[H_3O^+]}{[H_3O^+]+K_a} \; ; \text{ and } \quad \alpha_1 = \frac{[A^-]}{c_T} = \frac{K_a}{[H_3O^+]+K_a}$$

and $\alpha_0 + \alpha_1 = 1$.

197

Here, we'll select a weak acid with a K_a value of 1.00×10^{-5} and find the fraction present

as HA (α_0) and that present as A^- (α_1) as functions of pH. Begin with a worksheet similar to that

shown in Figure 7-13. Enter pH values every 0.5 pH unit from 0 to 14 in column A. Enter the

formulas for calculating $[H_3O^+]$, α_0 and α_1 into cells B5, C5, and D5 respectively. Use the fill

handle to copy these into the remaining column B, C, and D cells for which pH values have been

entered.

	A	B	C	D
1	Distribution of Species			
2	K_a	1.00E-05		
3				
4	pH	$[H_3O^+]$	α_0	α_1

Figure 7-13 Initial worksheet for distribution diagram

Prepare a plot of the alpha values vs. pH and include it in your worksheet. When

completed, your worksheet should look similar to the one shown in Figure 7-14. You can add a

legend from the **Chart Tools, Layout** menu to identify the α values, or use text boxes and

arrows as shown. The text box can be added from the **Insert** ribbon.

	A	B	C	D	E	F	G	H	I	J	K	L
1	**Distribution of Species**											
2	K_a	1.00E-05										
3												
4	**pH**	**[H₃O⁺]**	**α_0**	**α_1**								
5	0.00	1.0000	1.0000	0.0000								
6	0.50	0.3162	1.0000	0.0000								
7	1.00	0.1000	0.9999	0.0001								
8	1.50	0.0316	0.9997	0.0003								
9	2.00	0.0100	0.9990	0.0010								
10	2.50	3.16E-03	0.9968	0.0032								
11	3.00	1.00E-03	0.9901	0.0099								
12	3.50	3.16E-04	0.9693	0.0307								
13	4.00	1.00E-04	0.9091	0.0909								
14	4.50	3.16E-05	0.7597	0.2403								
15	5.00	1.00E-05	0.5000	0.5000								
16	5.50	3.16E-06	0.2403	0.7597								
17	6.00	1.00E-06	0.0909	0.9091								
18	6.50	3.16E-07	0.0307	0.9693								
19	7.00	1.00E-07	0.0099	0.9901								
20	7.50	3.16E-08	0.0032	0.9968								
21	8.00	1.00E-08	0.0010	0.9990								
22	8.50	3.16E-09	0.0003	0.9997								
23	9.00	1.00E-09	0.0001	0.9999								
24	9.50	3.16E-10	0.0000	1.0000								
25	10.00	1.00E-10	0.0000	1.0000								
26	10.50	3.16E-11	0.0000	1.0000								
27	11.00	1.00E-11	0.0000	1.0000								
28	11.50	3.16E-12	0.0000	1.0000								
29	12.00	1.00E-12	0.0000	1.0000								
30	12.50	3.16E-13	0.0000	1.0000								
31	13.00	1.00E-13	0.0000	1.0000								
32	13.50	3.16E-14	0.0000	1.0000								
33	14.00	1.00E-14	0.0000	1.0000								
34	**Documentation**											
35	Cell B5=10^-A5											
36	Cell C5=B5/(B5+$B$2)											
37	Cell D5=$B$2/(B5+$B$2)											

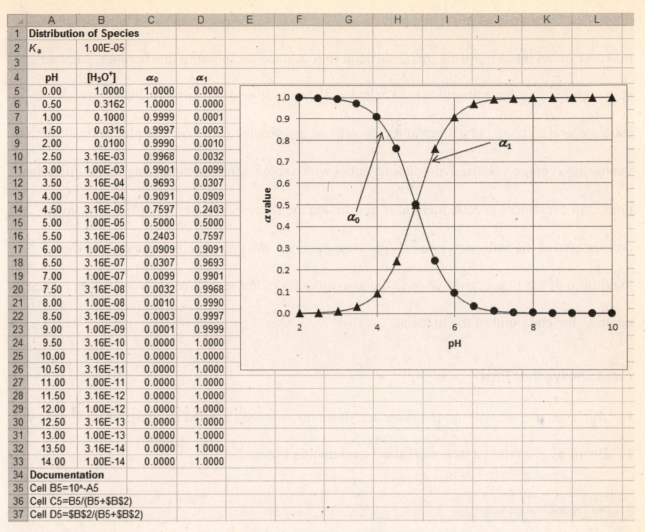

Figure 7-14 Final worksheet showing alpha values and distribution diagram.

After finishing your worksheet, try different values of K_a. Determine the value of pH where $\alpha_0 = \alpha_1$ for different K_a values. How is that pH value related to pK_a? In Chapter 8, we will make more extensive use of distribution diagrams in our coverage of polyfunctional acids and bases.

Derivative Plots

Locating the precise end point in a titration can be difficult especially with a sigmoid-shaped curve such as that illustrated in Figure 7-12. The traditional method was to estimate the inflection point visually in the steeply rising portion of the curve. As the pK_a of the acid becomes larger,

however, it becomes more and more difficult to locate the end point in this manner. Another

approach for acid-base titrations is to calculate the derivative of the titration curve, which we

estimate as the change in pH per unit change in titrant volume ($\Delta pH/\Delta V$). A plot of this function

versus the average volume V produces a curve with a maximum corresponding to the inflection

point. An even more precise method is to estimate the second derivative ($\Delta^2 pH/\Delta V^2$) which

crosses zero at the inflection point. With these methods, the volume corresponding to the

maximum of the first derivative or to the zero crossing of the second derivative crosses is then

taken as the end point of the titration.

First Derivative Plot

For this exercise, we will use the same acid-base titration data as used in constructing Figure 7-

12. Begin by preparing a new worksheet as shown in Figure 7-15.

	A	B	C	D	E	F	G	H	I	J
1	Derivative Titration Curve									
2	K_a	1.00E-05								
3	K_w	1.00E-14								
4	V_{HA}	50.00								
5	c_{HA}	0.1000								
6	c_{NaOH}	0.1000								
7				First Derivative				Second Derivative		
8	V_{NaOH}	pH	Midpoint Vol	ΔpH	ΔV	$\Delta pH/\Delta V$	Midpoint Vol	$\Delta(\Delta pH/\Delta V)$	ΔV	$\Delta^2 pH/\Delta V^2$

Figure 7-15 Initial worksheet for derivative titration curves.

Copy the volumes of NaOH from column A of the worksheet used to prepare Figure 7-12. Copy

the pH values from column E of this same worksheet. Since the values in Column E of Figure 7-

12 are calculated, you can't paste them directly into a new worksheet. One way to obtain only

the values is to right click the mouse on cell B9 of the new worksheet after copying the data, and

selecting **Paste Special…** from the resulting menu. Under **Paste** in the **Paste Special** window, click on the **Values** button, and click OK. This will paste only the pH values into column B, not the formulas. Format this column of data to contain only 2 figures after the decimal point.

Now we'll compute the first derivative of the titration curve. We'll first compute in column C, the average or midpoint volume to be associated with the calculated first derivative. Enter into cell C10, the formula **=(A9+A10)/2,** to find the average of the volumes in cells A9 and A10. Copy this formula into cells C11:C28. In column D, we'll find the difference in pH corresponding to this average volume. In cell D10, enter the formula **=B10−B9**, and copy it into cells D11:D28. In column E, we calculate the change in volume, ΔV, corresponding to this change in pH. In cell E10, enter the formula **=A10−A9**. Copy this into cells E11:E28. The first derivative is then found in column F, by dividing the ΔpH values in column D by the ΔV values in column E. Enter the appropriate formula in cell F10, and copy it into cells F11:F28. Your worksheet should now appear as shown in Figure 7-16.

We'll next construct a first derivative plot of ΔpH/ΔV vs. the midpoint volume. Select cells C10:C28 and F10:F28 for the plot. Choose a Scatter plot as the chart type. Your plot should appear as show in Figure 7-17 when completed. Note that the spike at about 50.00 mL makes it much easier to locate the end point than the normal titration curve. You can investigate how changing the scale on the *x* axis allows you to better estimate the volume at which the maximum occurs. Right click on the *x* axis and choose Format Axis from the menu that appears. Under Axis Options, you can manually choose the maximum and minimum volumes for the plot. Expansion of the scale shows that more data points are probably needed in the equivalence point region. Also, with real data any small differences or uncertainties in pH are amplified by taking the derivative. Any noise in the data can cause an error because the rate of change of the noise can be higher than the rate of change of pH. Computer averaging and smoothing of the derivative

data is often used in automatic titrators to reduce this problem.

	A	B	C	D	E	F
1	Derivative Titration Curve					
2	K_a	1.00E-05				
3	K_w	1.00E-14				
4	V_{HA}	50.00				
5	c_{HA}	0.1000				
6	c_{NaOH}	0.1000				
7				First Derivative		
8	V_{NaOH}	pH	Midpoint Vol	ΔpH	ΔV	ΔpH/ΔV
9	0.00	3.00				
10	5.00	4.05	2.50	1.05	5.00	0.2097
11	10.00	4.40	7.50	0.35	5.00	0.0698
12	20.00	4.82	15.00	0.43	10.00	0.0425
13	25.00	5.00	22.50	0.18	5.00	0.0352
14	30.00	5.18	27.50	0.18	5.00	0.0352
15	40.00	5.60	35.00	0.43	10.00	0.0426
16	45.00	5.95	42.50	0.35	5.00	0.0704
17	49.00	6.69	47.00	0.74	4.00	0.1840
18	49.90	7.70	49.45	1.01	0.90	1.1175
19	49.95	7.99	49.93	0.30	0.05	5.9037
20	49.99	8.56	49.97	0.57	0.04	14.3069
21	50.00	8.85	50.00	0.29	0.01	28.6002
22	50.01	9.14	50.01	0.29	0.01	28.5938
23	50.05	9.71	50.03	0.57	0.04	14.2940
24	50.10	10.00	50.08	0.29	0.05	5.8906
25	51.00	11.00	50.55	0.99	0.90	1.1044
26	55.00	11.68	53.00	0.68	4.00	0.1705
27	60.00	11.96	57.50	0.28	5.00	0.0562
28	70.00	12.22	65.00	0.26	10.00	0.0263

Figure 7-16 Worksheet for first derivative calculation.

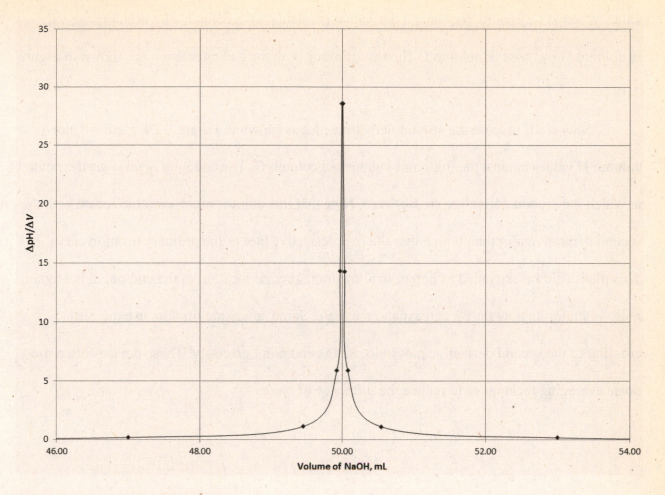

Figure 7-17 First derivative plot.

Second Derivative Plot

Now we'll compute the second derivative of the titration curve. The first step is to calculate

another average or midpoint volume to associate with each value of the second derivative.

Column G in Figure 7-18 shows this volume which is obtained from the average midpoint

volume in column C by the formula `=(C11+C10)/2` for the first value in cell C11. Column H

contains the numerators for the second derivative. Each of these values is the change in the first

derivative and is computed by the formula `=F11-F10` for the first value in cell H11. The

volume changes to be used in the denominator of the second derivative are calculated in column

I from the differences in the midpoint volumes of column C. The formula is `=C11-C10` for the value in cell I11.. Finally, the second derivative is calculated in column J by dividing the values in column H by those in column I. The results along with the documentation are shown in Figure 7-18.

Now we'll prepare the second derivative plot as shown in Figure 7-19. Here we plot column H values against the midpoint volumes in column G. The endpoint is taken as the point at which the second derivative crosses zero. Note that this point is much easier to locate with the second derivative plot than with either the first derivative plot or the ordinary titration curve. This plot could be expanded as before to allow more precise location of the end point, but again, noise often limits how far we can expand the scale. Some automatic titrators use the zero crossing of the second derivative plot to locate the end point precisely. These devices often use noise averaging techniques to reduce the influence of noise.

	A	B	C	D	E	F	G	H	I	J
1	Derivative Titration Curve									
2	K_a	1.00E-05								
3	K_w	1.00E-14								
4	V_{HA}	50.00								
5	c_{HA}	0.1000								
6	c_{NaOH}	0.1000								
7				First Derivative				Second Derivative		
8	V_{NaOH}	pH	Midpoint Vol	ΔpH	ΔV	ΔpH/ΔV	Midpoint Vol	Δ(ΔpH/ΔV)	ΔV	$\Delta^2pH/\Delta V^2$
9	0.00	3.00								
10	5.00	4.05	2.50	1.05	5.00	0.2097				
11	10.00	4.40	7.50	0.35	5.00	0.0698	5.00	-0.1399	5.00	-0.02798
12	20.00	4.82	15.00	0.43	10.00	0.0425	11.25	-0.0272	7.50	-0.00363
13	25.00	5.00	22.50	0.18	5.00	0.0352	18.75	-0.0073	7.50	-0.00097
14	30.00	5.18	27.50	0.18	5.00	0.0352	25.00	0.0000	5.00	2.07E-06
15	40.00	5.60	35.00	0.43	10.00	0.0426	31.25	0.0074	7.50	0.000985
16	45.00	5.95	42.50	0.35	5.00	0.0704	38.75	0.0278	7.50	0.003712
17	49.00	6.69	47.00	0.74	4.00	0.1840	44.75	0.1135	4.50	0.025233
18	49.90	7.70	49.45	1.01	0.90	1.1175	48.23	0.9335	2.45	0.381038
19	49.95	7.99	49.93	0.30	0.05	5.9037	49.69	4.7862	0.47	10.07614
20	49.99	8.56	49.97	0.57	0.04	14.3069	49.95	8.4032	0.05	186.7385
21	50.00	8.85	50.00	0.29	0.01	28.6002	49.98	14.2933	0.03	571.7326
22	50.01	9.14	50.01	0.29	0.01	28.5938	50.00	-0.0064	0.01	-0.63943
23	50.05	9.71	50.03	0.57	0.04	14.2940	50.02	-14.2999	0.03	-571.995
24	50.10	10.00	50.08	0.29	0.05	5.8906	50.05	-8.4034	0.05	-186.742
25	51.00	11.00	50.55	0.99	0.90	1.1044	50.31	-4.7862	0.47	-10.0761
26	55.00	11.68	53.00	0.68	4.00	0.1705	51.78	-0.9339	2.45	-0.38118
27	60.00	11.96	57.50	0.28	5.00	0.0562	55.25	-0.1144	4.50	-0.02541
28	70.00	12.22	65.00	0.26	10.00	0.0263	61.25	-0.0298	7.50	-0.00398
29	Documentation									
30	Cell C10=(A9+A10)/2			Cell G11=(C11+C10)/2						
31	Cell D10=B10-B9			Cell H11=F11-F10						
32	Cell E10=A10-A9			Cell I11=C11-C10						
33	Cell F10=D10/E10			Cell J11=H11/I11						

Figure 7-18. Worksheet for calculating first and second derivatives.

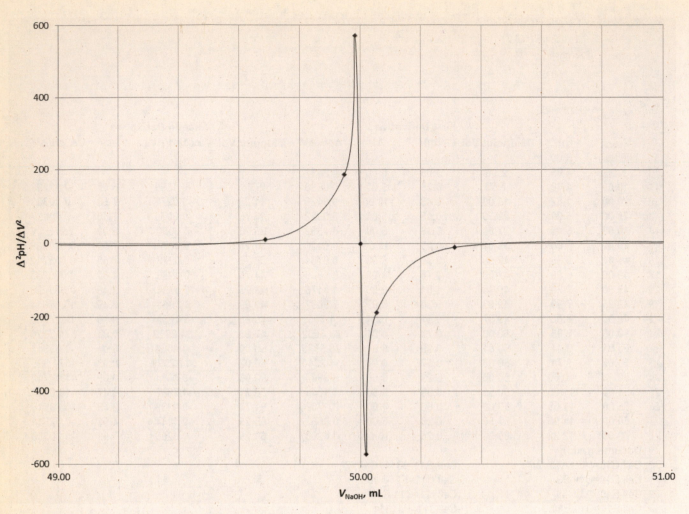

Figure 7-19 A plot of the second derivative data vs. midpoint volume.

Producing a Combination Plot

In cases like the second derivative plot, it would be desirable to plot simultaneously the titration

curve and the second derivative curve. This is called in Excel, a *combination chart*. Combination

charts allow us to plot sets of data with different *x*- or *y*-axis scales. For most scientific data, a

combination Scatter plot is appropriate.

Let's consider making a combination plot with the same *x*-axis, but two different *y*-axes.

First, select the primary data series to be plotted. For example, cells A9:A28 and B9:B28 in

Figure 7-18. Create a Scatter plot as before. Right click on a data point in the plot and choose

Select Data… This will bring up the **Select Data Source** window as shown in Figure 7-20.

Click the **Add** button to add a second series. In the Edit Series window that appears name the

series Series 2. Position the cursor in the **Series X Values**: window. Select cells A11:A28.

Position the cursor in the **Series Y Values**: window, erase anything Excel has put there, and

select cells F11:F28. The Edit Series windows should appear as shown in Figure 7-21.

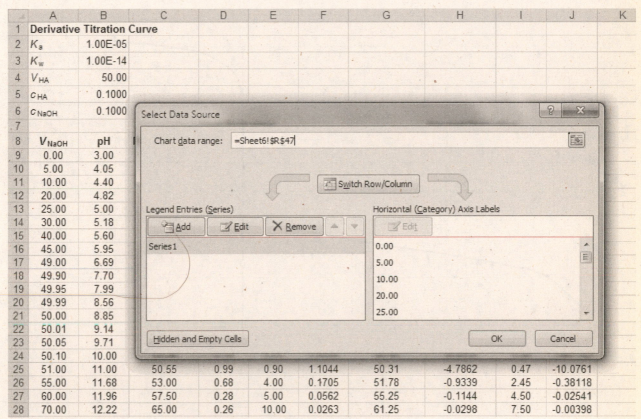

Figure 7-20. Worksheet showing Select Data Source window.

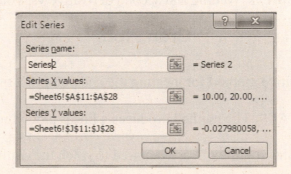

Figure 7-21 Edit Series window.

In the plot, right click on the second derivative series and chose **Format Data Series….**Under

Series Options, select **Plot Series On Secondary Axis**.

Display the combination plot, add axes labels and change the number formatting so that

the chart appears as shown in Figure 7-22.

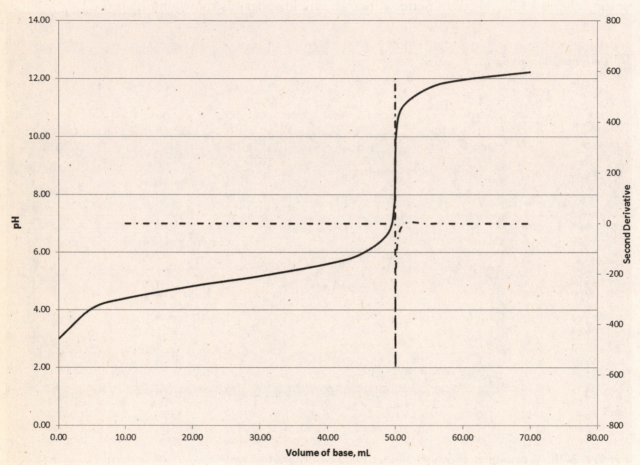

Figure 7-22 Combination pH vs. volume and second derivative vs. volume chart.

Gran Plots

The Gran plot is an alternative method for locating the end point in a titration. The principles are

given in Feature 14-6 of FAC9. The method produces a linear plot which can reveal both the acid

dissociation constant and the volume of base required to reach the equivalence point. Unlike the

normal titration curve and derivative curves, which find the end point only from data located in the end point region, the Gran plot uses data far away from the end point. This approach can

decrease the tedium of making many measurements after dispensing very small volumes of

titrant in the end point region. Gran plots can be used in acid-base, redox, precipitation and

complexation titrations. We'll consider only acid-base titrations here.

For the titration of a weak acid HA ($K_a = 1.00 \times 10^{-5}$) with NaOH, the applicable

equation for constructing a Gran plot is,

$$[H_3O^+]V_{NaOH} = K_aV_{eq} - K_aV_{NaOH}$$

where V_{eq} is the equivalence point volume. A plot of the left hand side of this equation vs. the

volume of titrant V_{NaOH} should yield a straight line with a slope of $-K_a$ and an intercept of K_aV_{eq}.

Usually, points in the mid stages of the titration are plotted and used to obtain these values. The

Gran plot can exhibit curvature in the early stages if K_a is too large, and it can curve near the

equivalence point.

We'll now prepare a Gran plot of the acid-base titration data used previously in the

worksheets shown in Figures 7-12 and 7-18. Begin a new worksheet as shown in Figure 7-23.

You can copy the volume and pH values from the worksheet for Figure 7-12.

	A	B	C	D
1	Weak Acid/Strong Base Titration, Gran Plot			
2	K_a	1.00E-05		
3	K_w	1.00E-14		
4	V_{HA}	50.00		
5	c^0_{HA}	0.1000		
6	c^0_{NaOH}	0.1000		
7				
8	V_{NaOH}	pH	$[H_3O^+]V_{NaOH}$	
9	20.00	4.82		
10	25.00	5.00		
11	30.00	5.18		
12	40.00	5.60		

Figure 7-23 Worksheet for Gran plot.

Note that volumes very early in the titration or very near the equivalence point are excluded. In cells C9:C12, the product of the H_3O^+ concentration and the volume of NaOH is calculated. Obtain the slope, the intercept, the value of K_a and the equivalence point volume. Plot the results as shown in Figure 7-24. You can also obtain the statistics of the slope and intercept and calculate the standard deviations in K_a and V_{eq}.

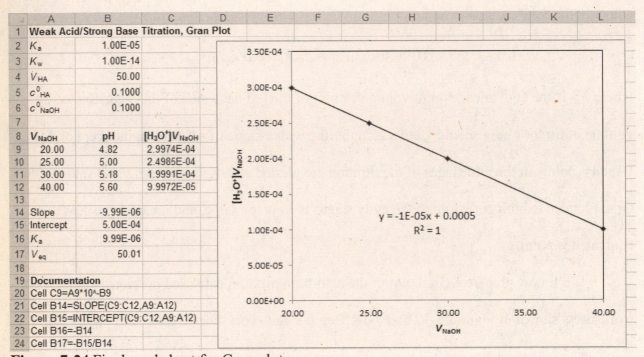

Figure 7-24 Final worksheet for Gran plot.

Summary

In this chapter, we have learned how to prepare spreadsheets for several types of acid/base titrations. These spreadsheets can be used to construct titration curves, derivative curves, or Gran plots. We have also seen how to obtain plots showing how different species vary in concentration as a function of a master variable such as pH. We have seen how to prepare a combination chart in which different axes are used. The next two chapters will extend these ideas to calculations of polyfunctional acids and bases and complexation equilibria.

Problems

1. A 25.00-mL aliquot of 0.0500 M HBr is titrated with 0.0250 M NaOH. Use the stoichiometric approach to calculate the pH of the solution after every 5.00 mL of titrant until you are within 1-mL of the equivalence point. In the equivalence point region, calculate the pH after the addition of every 0.100 mL. Continue until the titration is 1.00 mL beyond the equivalence point. Then calculate the pH every 5.00 mL until 70.00 mL have been added. Prepare a titration curve from these data.

2. Complete the calculations of problem 1 using the charge-balance equation method, and prepare a titration curve.

3. Find the pH of a solution that is prepared by adding 5.00 mL of 0.0500 M acetic acid to 10.00 mL of 0.0200 M sodium acetate.

4. Find the pH of a solution that is prepared by dissolving 3.56 g of lactic acid (90.08 g/mol) and 4.78 g of sodium lactate (112.06 g/mol) in water and diluting to 500.0 mL.

5. Find the pH of a solution prepared by dissolving 1.78 g of picric acid and 1.25 g of sodium picrate in water and diluting to 1.00 L.

6. A solution is 0.0500 M NH_4Cl and 0.036 M NH_3. What is the OH^- concentration and the pH (a) neglecting activity corrections, and (b) taking activity coefficients into account.

7. A 50.00-mL aliquot of 0.1000 M NaOH is titrated with 0.1000 M HCl. Calculate the pH of the solution after the addition of 0.00, 10.00, 25.00, 40.00, 45.00, 49.00, 50.00, 51.00, 55.00, and 60.00 mL of acid and prepare a titration curve from the data.

8. Use the stoichiometric approach to calculate the pH after the addition of 0.00, 5.00, 15.00, 25.00, 40.00, 45.00, 49.00, 50.00, 51.00, 55.00, and 60.00 mL of 0.1000 M NaOH in the titration of 50.00 mL of

   (a)   0.1000 M HBr

   (b)   0.1000 M HNO_2.

   (c)   0.1000 M lactic acid.

   (d)   0.1000 M pyridinium chloride.

9.   For each of the titrations in Problem 7 above, locate the end points by

   (a)   preparing a first- and second-derivative plot.

   (b)   preparing a Gran plot.

10.   Use the stoichiometric approach to calculate the pH after addition of 0.00, 5.00, 15.00,

        25.00, 40.00, 45.00, 49.00, 50.00, 51.00, 55.00, and 60.00 mL of 0.1000 M HCl in the

        titration of 50.00 mL of

   (a)   0.1000 M ammonia.

   (b)   0.1000 M hydrazine.

   (c)   0.1000 M sodium cyanide.

11.   For each of the titrations in Problem 10 above, locate the end points by:

   (a)   preparing a first- and second-derivative plot.

   (b)   preparing a Gran plot.

12.   Use the master equation approach for each of the titrations in Problem 10 and construct

        titration curves.

13.   Use the stoichiometric approach to calculate the pH after addition of 0.00, 5.00, 15.00,

        25.00, 40.00, 49.00, 50.00, 51.00, 55.00, and 60.00 mL of reagent in the titration of 50.0

        mL of:

   (a)   0.1000 M anilinium chloride with 0.1000 M NaOH.

   (b)   0.01000 M chloroacetic acid with 0.01000 M NaOH.

   (c)   0.1000 M hypochlorous acid with 0.1000 M NaOH.

   (d)   0.1000 M hydroxylamine with 0.1000 M HCl

   (e)   Construct titration curves for each of the above titrations.

14.   Use the master equation approach for each of the titrations in Problem 13 and construct

      titration curves for each.

15.   Prepare distribution diagrams of α_0 and α_1 over the pH range of 0 to 14. Calculate these

      values every 0.5 pH unit for:

   (a)   acetic acid   (d)   hydroxylamine

   (b)   picric acid   (e)   piperidine

   (c)   hypochlorous acid   (f)   ammonia

Chapter 8

Polyfunctional Acids and Bases

In Chapter 7, we explored the use of distribution diagrams to show the variation of species concentrations with pH. Distribution diagrams and α values are even more useful when polyfunctional acids and bases are considered as we show in this chapter. In addition, we'll investigate the construction of titration curves and logarithmic concentration diagrams for polyfunctional systems.

Distribution Diagrams

Distribution diagrams are very useful with polyprotic acids if we know the pH. We may know the pH, for example, if we have prepared or desire to prepare a buffer solution. In many situations, we want to know the concentrations of all species at a given pH. We can extend the diprotic acid system described here to more complicated systems, such as triprotic acids.

A General Expression for α Values

The alpha values for any weak acid with n ionizable protons, H_nA, can be written as follows;

$$\alpha_0 = \frac{[H_nA]}{c_T} = \frac{[H_3O^+]^n}{D}$$

$$\alpha_1 = \frac{[H_{n-1}A^-]}{c_T} = \frac{K_{a1}[H_3O^+]^{n-1}}{D}$$

$$\alpha_2 = \frac{[H_{n-2}A^{2-}]}{c_T} = \frac{K_{a1}K_{a2}[H_3O^+]^{n-2}}{D}$$

$$\vdots$$

214

$$\alpha_n = \frac{[A^{n-}]}{c_T} = \frac{K_{a1}K_{a2}\cdots K_{an}}{D}$$

where c_T is the analytical concentration of the acid and the denominator D is given by

$$D = [H_3O^+]^n + K_{a1}[H_3O^+]^{(n-1)} + K_{a1}K_{a2}[H_3O^+]^{(n-2)} + \cdots + K_{a1}K_{a2}\cdots K_{an}$$

Diprotic Acid System

The expressions for calculating α values look somewhat complex, but Excel and similar tools permit us to calculate the values with ease. We illustrate these calculations with a diprotic acid, maleic acid (H_2M), with $K_{a1} = 1.30 \times 10^{-2}$ and $K_{a2} = 5.90 \times 10^{-7}$. We'll calculate the alpha values every 0.5 pH unit from pH = 0 to pH = 10. The spreadsheet and its documentation are shown in Figure 8-1. Although it is not necessary to do so, we've entered a column to calculate the denominator D used in each α calculation. This step saves us time and lessens the chance for mistakes in entering the formulas for the α calculations.

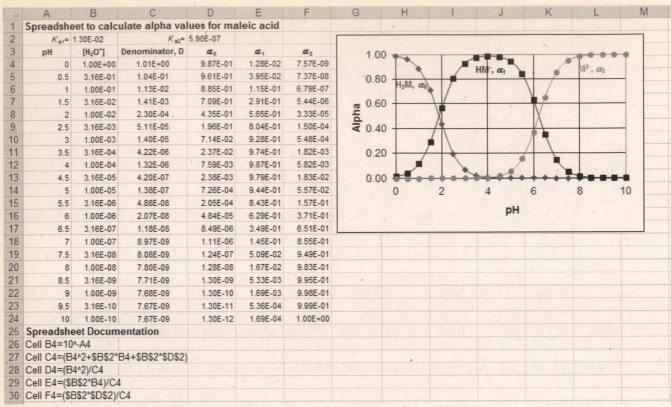

	A	B	C	D	E	F
1	Spreadsheet to calculate alpha values for maleic acid					
2	K_{a1}= 1.30E-02			K_{a2}= 5.90E-07		
3	pH	[H₃O⁺]	Denominator, D	α_0	α_1	α_2
4	0	1.00E+00	1.01E+00	9.87E-01	1.28E-02	7.57E-09
5	0.5	3.16E-01	1.04E-01	9.61E-01	3.95E-02	7.37E-08
6	1	1.00E-01	1.13E-02	8.85E-01	1.15E-01	6.79E-07
7	1.5	3.16E-02	1.41E-03	7.09E-01	2.91E-01	5.44E-06
8	2	1.00E-02	2.30E-04	4.35E-01	5.65E-01	3.33E-05
9	2.5	3.16E-03	5.11E-05	1.96E-01	8.04E-01	1.50E-04
10	3	1.00E-03	1.40E-05	7.14E-02	9.28E-01	5.48E-04
11	3.5	3.16E-04	4.22E-06	2.37E-02	9.74E-01	1.82E-03
12	4	1.00E-04	1.32E-06	7.59E-03	9.87E-01	5.82E-03
13	4.5	3.16E-05	4.20E-07	2.38E-03	9.79E-01	1.83E-02
14	5	1.00E-05	1.38E-07	7.26E-04	9.44E-01	5.57E-02
15	5.5	3.16E-06	4.88E-08	2.05E-04	8.43E-01	1.57E-01
16	6	1.00E-06	2.07E-08	4.84E-05	6.29E-01	3.71E-01
17	6.5	3.16E-07	1.18E-08	8.49E-06	3.49E-01	6.51E-01
18	7	1.00E-07	8.97E-09	1.11E-06	1.45E-01	8.55E-01
19	7.5	3.16E-08	8.08E-09	1.24E-07	5.09E-02	9.49E-01
20	8	1.00E-08	7.80E-09	1.28E-08	1.67E-02	9.83E-01
21	8.5	3.16E-09	7.71E-09	1.30E-09	5.33E-03	9.95E-01
22	9	1.00E-09	7.68E-09	1.30E-10	1.69E-03	9.98E-01
23	9.5	3.16E-10	7.67E-09	1.30E-11	5.36E-04	9.99E-01
24	10	1.00E-10	7.67E-09	1.30E-12	1.69E-04	1.00E+00
25	Spreadsheet Documentation					
26	Cell B4=10^-A4					
27	Cell C4=(B4^2+$B$2*B4+$B$2*$D$2)					
28	Cell D4=(B4^2)/C4					
29	Cell E4=($B$2*B4)/C4					
30	Cell F4=($B$2*$D$2)/C4					

Figure 8-1 Spreadsheet to calculate alpha values for maleic acid.

Note that we enter the dissociation constants into cells B2 and D2. The pH values are entered into cells A4 through A24. The H_3O^+ concentration is then calculated in cell B4 using the formula **=10^-A4**. This formula is copied into cells B5:B24. The value for D is calculated in cell C4 by the formula shown in documentation cell A27. The formulas to obtain α_0, α_1, and α_2 are shown in documentation cells A28, A29, and A30. On the right, the plot shows the composition of maleic acid solutions as a function of pH.

Distribution diagrams provide a wealth of information regarding the three species, H_2M, HM^-, and M^{2-}. For example, we might ask what the concentrations of these three species are at pH 5.0 in a solution whose molar analytical concentration of H_2M is 0.1 M. A glance at the plot tells us that there is essentially no H_2M at pH 5.0, that $\alpha_1 \approx 0.95$, and $\alpha_2 \approx 0.05$. If we require more accurate data, we consult the spreadsheet and conclude that $\alpha_0 = 7.26 \times 10^{-4}$, $\alpha_1 = 0.944$, and $\alpha_2 = 0.0557$. From these values we find that $[H_2M] = \alpha_0 c_T = 7.26 \times 10^{-5}$ M, $[HM^-] = \alpha_1 c_T =$ 0.0944 M, and $[M^{2-}] = \alpha_2 c_T = 0.00577$ M. Note that $\alpha_0 + \alpha_1 + \alpha_2 = 1$ as required by mass balance.

After entering the values for maleic acid, you can extend the calculations for several other diprotic acids, such as malonic acid and oxalic acid,. The K_a values can be found in the Appendixes of FAC9 and AC7. You can simply change the values for K_{a1} and K_{a2} in your spreadsheet for these additional diprotic acids. As a further exercise, prepare a spreadsheet to calculate α_0 through α_3 for a triprotic acid such as phosphoric acid.

What About Bases?

Thus far in our discussion, we have neglected bases. This apparent oversight occurred not because bases are unimportant, but because we treat bases in essentially the same way that we treat acids. Let's take ethylenediamine, $H_2NCH_2CH_2NH_2$, as an example. Ethylenediamine contains two amine groups, which can be protonated to give $^+H_3NCH_2CH_2NH_3^+$, the ethylene diammonium ion, which is analogous to H_2A. The basic ionization constants for ethylendiamine are: $K_{b1} = 8.5 \times 10^{-5}$ and $K_{b2} = 7.1 \times 10^{-8}$. Recall that $K_aK_b = K_w$, so we calculate

$$K_{a1} = \frac{K_w}{K_{b2}} = \frac{1.00 \times 10^{-14}}{7.1 \times 10^{-8}} = 1.4 \times 10^{-7} \text{ and}$$

$$K_{a2} = \frac{K_w}{K_{b1}} = \frac{1.00 \times 10^{-14}}{8.5 \times 10^{-5}} = 1.2 \times 10^{-10}$$

You can enter these values into your diprotic acid spreadsheet and observe the results. Which curves correspond to which ethylenediamine species? Compare the shapes and positions of the distribution diagrams for ethylenediamine with those for acidic species.

Logarithmic Concentration Diagrams

A logarithmic concentration diagram is a plot of log concentration vs. a master variable such as pH. The log concentration diagram only applies for a specific initial concentration of acid. Such diagrams can be obtained from the distribution diagrams previously discussed. The log concentration diagram is useful because it expresses the concentrations of all species in a polyprotic acid solution as a function of pH. This allows us to observe at a glance the species that are important at a particular pH. The logarithmic scale is used because the concentrations can vary over many orders of magnitude.

We'll modify our distribution diagram for maleic acid in the previous section to produce a logarithmic concentration diagram for an acid concentration $c_T = 0.10$ M. Prepare the worksheet shown in Figure 8-2. Begin by copying the worksheet of Figure 8-1 to a blank worksheet.

	A	B	C	D	E	F	G	H	I	J	K
1	Spreadsheet for logarithmic concentration diagram										
2	$K_{a1}=$	1.30E-02		$K_{a2}=$ 5.90E-07		$c_T=$ 0.1					
3	pH	[H_3O^+]	Denominator, D	[H_2M]	[M]	[M^2]	log [H_2M]	log [HM]	log [M^2]	log [H_3O^+]	log [OH]
4	0	1.00E+00	1.01E+00	9.87E-02	1.28E-03	7.57E-10	-1.00561	-2.89167	-9.12081	0	-14
5	0.5	3.16E-01	1.04E-01	9.61E-02	3.95E-03	7.37E-09	-1.0175	-2.40355	-8.1327	-0.5	-13.5
6	1	1.00E-01	1.13E-02	8.85E-02	1.15E-02	6.79E-08	-1.05308	-1.93914	-7.16828	-1	-13
7	1.5	3.16E-02	1.41E-03	7.09E-02	2.91E-02	5.44E-07	-1.14956	-1.53562	-6.26476	-1.5	-12.5
8	2	1.00E-02	2.30E-04	4.35E-02	5.65E-02	3.33E-06	-1.36174	-1.2478	-5.47695	-2	-12
9	2.5	3.16E-03	5.11E-05	1.96E-02	8.04E-02	1.50E-05	-1.70857	-1.09462	-4.82377	-2.5	-11.5
10	3	1.00E-03	1.40E-05	7.14E-03	9.28E-02	5.48E-05	-2.14637	-1.03242	-4.26157	-3	-11
11	3.5	3.16E-04	4.22E-06	2.37E-03	9.74E-02	1.82E-04	-2.62517	-1.01123	-3.74038	-3.5	-10.5
12	4	1.00E-04	1.32E-06	7.59E-04	9.87E-02	5.82E-04	-3.11981	-1.00586	-3.23501	-4	-10
13	4.5	3.16E-05	4.20E-07	2.38E-04	9.79E-02	1.83E-03	-3.62301	-1.00906	-2.73821	-4.5	-9.5
14	5	1.00E-05	1.38E-07	7.26E-05	9.44E-02	5.57E-03	-4.13915	-1.02521	-2.25436	-5	-9
15	5.5	3.16E-06	4.88E-08	2.05E-05	8.43E-02	1.57E-02	-4.68833	-1.07438	-1.80353	-5.5	-8.5
16	6	1.00E-06	2.07E-08	4.84E-06	6.29E-02	3.71E-02	-5.31536	-1.20142	-1.43057	-6	-8
17	6.5	3.16E-07	1.18E-08	8.49E-07	3.49E-02	6.51E-02	-6.07118	-1.45724	-1.18639	-6.5	-7.5
18	7	1.00E-07	8.97E-09	1.11E-07	1.45E-02	8.55E-02	-6.95279	-1.83885	-1.068	-7	-7
19	7.5	3.16E-08	8.08E-09	1.24E-08	5.09E-03	9.49E-02	-7.90747	-2.29353	-1.02267	-7.5	-6.5
20	8	1.00E-08	7.80E-09	1.28E-09	1.67E-03	9.83E-02	-8.89209	-2.77815	-1.0073	-8	-6
21	8.5	3.16E-09	7.71E-09	1.30E-10	5.33E-04	9.95E-02	-9.88712	-3.27317	-1.00232	-8.5	-5.5
22	9	1.00E-09	7.68E-09	1.30E-11	1.69E-04	9.98E-02	-10.8855	-3.77159	-1.00074	-9	-5
23	9.5	3.16E-10	7.67E-09	1.30E-12	5.36E-05	9.99E-02	-11.885	-4.27108	-1.00023	-9.5	-4.5
24	10	1.00E-10	7.67E-09	1.30E-13	1.69E-05	1.00E-01	-12.8849	-4.77093	-1.00007	-10	-4
25	Documentation										
26	Cell B4=10^-A4			Cell G4=LOG(D4)							
27	Cell C4=(B4^2+$B$2*B4+$B$2*$D$2)			Cell H4=LOG(E4)							
28	Cell D4=((B4^2)/C4)*$F$2			Cell I4=LOG(F4)							
29	Cell E4=(($B$2*B4)/C4)*$F$2			Cell J4=LOG(B4)							
30	Cell F4=(($B$2*$D$2)/C4)*$F$2			Cell K4=-14-J4							

Figure 8-2 Spreadsheet for constructing logarithmic concentration diagram.

In cell D4, multiply α_0 by the c_T value in cell F2. Make this an absolute reference to cell F2, and copy the resulting formula into cells D5:D24. Likewise in cells E4 and E5, multiply the α_1 and α_2 from the previous worksheet by the c_T value in cell F2. Copy these into cells E5:E24 and F5:F24. In columns G, H, and I, we take the logarithms of the values in columns D, E, and F. To complete the calculations we take the logarithm of the H_3O^+ concentration in column J and the logarithm of the OH^- concentration in column K. Since pOH = 14 – pH, –log [OH^-] = 14 + pH. Hence, log [OH^-] = – 14 – log [H_3O^+] as shown in the documentation for cell K4. Once the formulas have been entered correctly, the spreadsheet should appear as shown in Figure 8-2.

Next, the log concentrations will be plotted against the pH. Select cells A4:A24 and

columns G:K, rows 4:24. Remember to hold down the Ctrl key when selecting non-adjacent

ranges. Select the **Scatter** type (markers only). Your plot should look similar to that in Figure 8-

3 after formatting. The labels for each species were added as text boxes from the Layout tab.

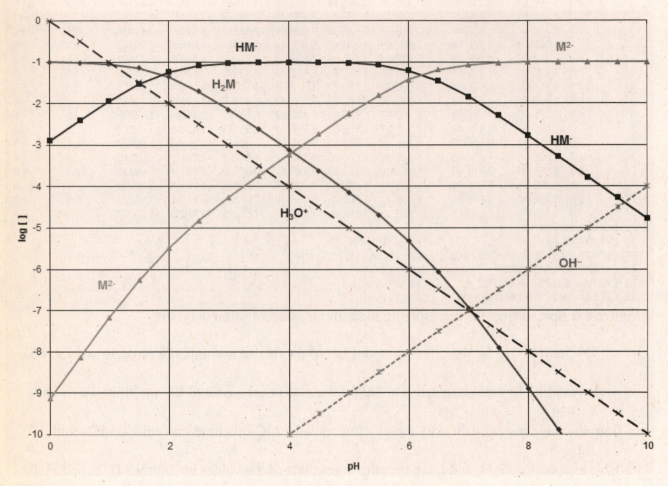

Figure 8-3 Logarithmic concentration diagram for 0.10 M maleic acid.

Note that the plot is specific for a total analytical concentration of 0.10 M and for maleic acid,

since the acid dissociation constants are included in the calculations. The log concentration

diagram is actually a graphical representation of the mass balance equation and the equilibrium

constants.

Estimating Concentrations at a Given pH Value

The logarithmic concentration diagram can be very useful in making more exact calculations than graphical approximations and in determining which species are important at a given pH. For example, if we are interested in calculating concentrations at pH 5.7, we can use the log concentration diagram to tell us which species to include in the calculation. At pH 5.7, the concentrations of the malate containing species are: $[H_2M] \approx 10^{-5}$ M, $[HM^-] \approx 0.07$ M, $[M^{2-}] \approx$ 0.02 M. Hence, the only maleate species of importance at this pH are HM^- and M^{2-}. Since $[OH^-]$ is four orders of magnitude lower than $[H_3O^+]$, we could carry out a more accurate calculation than the above estimates by considering only 3 species. Alternatively, we could insert pH 5.7 into the spreadsheet used to prepare Figure 8-2. If we do so we find the following concentrations: $[H_2M] \approx 1.18 \times 10^{-5}$ M, $[HM^-] \approx 0.077$ M, $[M^{2-}] = 0.023$ M. Note how close our original estimates were to these values.

Finding pH Values

If we don't know the pH, the logarithmic concentration diagram can also be used to give us an approximate pH value. For example, let's find the pH of a 0.1 M maleic acid solution. Since the log concentration diagram expresses mass balance and the equilibrium constants, we need only one additional equation such as charge balance to solve the problem exactly. The charge balance equation for this system is

$$[H_3O^+] = [HM^-] + 2[M^{2-}] + [OH^-]$$

The pH is found by graphically superimposing this equation on the log concentration diagram as described below. Beginning with a pH of 0, move from left to right along the H_3O^+ line until it intersects a line representing one of the species on the right hand side of the charge balance equation. We see that the H_3O^+ line first intersects the HM^- line at a pH of approximately 1.5. At this point, $[H_3O^+] = [HM^-]$. We also see that the concentrations of the other negatively charged species M^{2-} and OH^- are negligible compare to the HM^- concentration. Hence, the pH of a 0.1 M solution of maleic acid is approximately 1.5. A more accurate calculation using the quadratic formula gives pH = 1.52.

Let's now ask the question: "What is the pH of a 0.100 M solution of NaHM?" In this case, the charge balance equation is

$$[H_3O^+] + [Na^+] = [HM^-] + 2[M^{2-}] + [OH^-]$$

The Na^+ concentration is just the total concentration of maleate containing species

$$[Na^+] = c_T = [H_2M] + [HM^-] + [M^{2-}]$$

Substituting this latter equation into the charge balance equation gives

$$[H_3O^+] + [H_2M] = [M^{2-}] + [OH^-]$$

Now we superimpose this equation on the log concentration diagram. If we again begin on the left at pH 0 and move along either the H_3O^+ line or the H_2M line, we see that at pH values greater than about 2, the concentration of H_2M exceeds the H_3O^+ concentration by about an order of magnitude. Hence, we move along the H_2M line until it intersects either the M^{2-} line or the OH^- line. We see that it intersects the M^{2-} line first at pH $\approx$ 4.1. Because $[H_3O^+]$ and $[OH^-]$ are relatively small compared to $[H_2M]$ and $[M^{2-}]$, at this point, we can say that $[H_2M] \approx [M^{2-}]$. Thus, we conclude that the pH of a 0.100 M NaHM solution is approximately 4.1. A more exact calculation using the quadratic formula reveals that the pH of this solution is 4.08.

Finally, we let's find the pH of a 0.100 M solution of Na_2M. The charge balance equation is the same as before

$$[H_3O^+] + [Na^+] = [HM^-] + 2[M^{2-}] + [OH^-]$$

Now, however, the Na^+ concentration is given by

$$[Na^+] = 2c_T = 2[H_2M] + 2[HM^-] + 2[M^{2-}]$$

Substituting this into the charge balance equation gives

$$[H_3O^+] + 2[H_2M] + [HM^-] = [OH^-]$$

In this case, it is easier to find the OH^- concentration. Again, we move down the OH^- line now from right to left until it intersects the HM^- line at a pH of approximately 9.7. Since $[H_3O^+]$ and $[H_2M]$ are negligibly small at this intersection, $[HM^-] \approx [OH^-]$ we conclude that pH 9.7 is the approximate pH of a 0.100 M solution of Na_2M. A more exact calculation using the quadratic formula gives the pH as 9.61. Note that we have only plotted the values to pH 10. For more basic solutions we could readily insert higher pH values in the spreadsheet used to prepare Figure 8-2 and extend the plot. As an extension to this exercise, prepare a log concentration diagram for ethylenediamine, a base.

Titration Curves for Weak Acids

In this section, we extend the approaches used in Chapter 7 for constructing titration curves to polyfunctional acids and bases. We consider here both a stoichiometric approach and a master equation approach.

Stoichiometric Method

In the stoichiometric approach, we divide the titration curve up into regions, and use the stoichiometry of the reaction to find the concentrations. If the K_a values are not too large and the ratio of K_{a1}/K_{a2} is at least 10^3, we can treat the system by the techniques derived for monoprotic acids. However, if the K_a values are larger than about 10^{-4}, we may have to use a more complex scheme as described in FAC9 Section 15E and AC7 Section 13D.

As an example of a system with fairly large K_{a1}, let's take the titration of 25.00 mL of

0.1000 M maleic acid, H_2M, with 0.1000 M NaOH as was done in FAC9, Example 15-9 and

	A	B	C	D	E	F	G	H
1	Spreadsheet to calculate titration curve for maleic acid.							
2	K_{a1}	1.30E-02	K_{b1}	1.69E-08	Vol. H_2M	25.00		
3	K_{a2}	5.90E-07	Initial c_{H2M}	0.1000				
4	K_w	1.00E-14	Initial c_{NaOH}	0.1000				
5	Vol. NaOH, mL	c_{H2M}	c_{HM-}	c_{M2-}	$[H_3O^+]$	pH	[OH]	pOH

AC7, Example 13-4. We start by setting up a worksheet as illustrated in Figure 8-4. The volumes

for which we desire pH values are given on the final worksheet of Figure 8-5. Enter these

volumes in cells A6:A29. We have entered $K_{b1} = K_w/K_{a2}$ in cell D2 as shown.

Figure 8-4 Initial worksheet for titration of maleic acid with NaOH

Analytical Concentrations. We'll first write formulas to calculate the analytical concentrations

of H_2M, HM^-, and M^{2-} since we can do this directly from the stoichiometry. Since maleic acid is

a diprotic acid, there will be two equivalence points, the first at 25.00 mL of titrant, and the

second at 50.00 mL of titrant. In the initial part of the titration curve, we will have H_2M, which

we started with, and HM^- produced by the titration reaction. The analytical concentration of H_2M

is given by

$$c_{H_2M} = \frac{\text{no. mmoles } H_2M \text{ initially present} - \text{no. mmoles NaOH added}}{\text{total solution volume in mL}}$$

while the analytical concentration of HM^- is

$$c_{HM^-} = \frac{\text{no. of mmoles NaOH added}}{\text{total solution volume in mL}}$$

Hence, we can write the formulas equivalent to these equations in cells B6 and C6. These

formulas apply for all volumes before the first equivalence point, so we can select cells B6 and

C6 and use the fill handle to copy the formulas through row 17. The formulas are shown in the

documentation for Figure 8-5.

Between the first and second equivalence point, we have a solution composed of HM^-

produced prior to the first equivalence point and M^{2-} produced by the titration reaction after the

first equivalence point. Hence, we can write

$$c_{HM^-} = \frac{\text{no. of mmoles NaOH added to first eq. point } - \text{ no. of mmoles NaOH added beyond first eq. point}}{\text{total solution volume, mL}}$$

$$c_{M^{2-}} = \frac{\text{no. mmoles NaOH added beyond first eq. point}}{\text{total solution volume, mL}}$$

We can write formulas corresponding to these equations in cells C18 and D18 and copy them

into cells C19:C25 and D19:D25 respectively. The formulas are shown in the documentation for

cells C18 and D18 in Figure 8-5.

Beyond the second equivalence point we have a solution composed of M^{2-} produced by

the reaction and excess NaOH, which is diluted by the addition of the base. The M^{2-}

concentration is given by

$$c_{M^{2-}} = \frac{\text{no. of mmoles NaOH added between first and second eq. points}}{\text{total solution volume, mL}}$$

The formula corresponding to this equation is entered into cell D26 and copied into cells

D27:D29. The formula is given in the documentation for cell D26.

The OH^- concentration beyond the second equivalence point is given by

$$[OH^-] = \frac{\text{mmoles NaOH added beyond second eq. point}}{\text{total solution volume, mL}}$$

We write the formula corresponding to this equation in cell G26 and copy it into cells G27:G29

(see documentation for cell G26).

Now we are ready to calculate pH values in various regions of the titration curve.

Initial pH. Note here that K_{a1} is 1.3×10^{-2}, meaning that maleic acid is moderately dissociated even at the start of the titration. This can give us a clue that we may have to use the quadratic equation right from the beginning. As shown in Example 15-9 of FAC9 or Example 13-4 of AC7, the equation for the initial $[H_3O^+]$ is

$$[H_3O^+]^2 + K_{a1}[H_3O^+] - K_{a1}c^0_{H_2M} = 0$$

where $c^0_{H_2M}$ is the initial analytical concentration of maleic acid. Enter the formula corresponding to the quadratic equation for the $[H_3O^+]$ in cell E6. The formula is listed in the documentation for this cell in Figure 8-5. In cell F6, enter the formula to calculate pH from the $[H_3O^+]$. Copy this into cells F7:F24. The initial pH should be 1.52 as found earlier when discussing logarithmic concentration diagrams.

First Buffer Region. Addition of 5.00 mL of NaOH results in a buffer consisting of H_2M and HM^-. The equation that results for $[H_3O^+]$ is

$$[H_3O^+]^2 + (c_{HM^-} + K_{a1})[H_3O^+] + K_{a1}c_{H_2M} = 0$$

Enter the formula corresponding to the solution to this equation in cell E7. The pH found for this first buffer point should be 1.74. If correct, copy the formula in cell E7 into cells E8:E16. Very near the first equivalence point, the simple buffer formula begins to give large errors. You can see in cells B16 and C16, the ratio c_{HM^-}/c_{H_2M} becomes large indicating that the solution is no longer a good buffer. If we were to go closer to the equivalence point (24.90 mL, for example), a more complex expression would need to be developed.

First Equivalence Point. At the first equivalence point, we have converted all the H_2M to HM^- so that $c_{HM^-} \approx 0.05$ M. The concentration of H_3O^+ must be calculated from Equation 15-15 of FAC8 or 13-3 of AC7. That is,

$$[H_3O^+] = \sqrt{\frac{K_{a2}c_{HM^-} + K_w}{1 + c_{HM^-} / K_{a1}}}$$

Hence, in cell E17, we enter the formula shown in the documentation section for that cell.

Second Buffer Region. Addition of a small volume of base beyond that required to reach the first equivalence point creates a new buffer system consisting of HM^- and M^{2-}. When enough base has been added so that the reaction of HM^- with water to give OH^- can be neglected (a few tenths of a milliliter beyond the first equivalence point), the pH of the mixture is readily obtained from K_{a2} by the simplified buffer equation

	A	B	C	D	E	F	G	H	I
1	Spreadsheet to calculate titration curve for maleic acid.								
2	K_{a1}	1.30E-02	K_{b1}	1.69E-08	Vol. H₂M	25.00			
3	K_{a2}	5.90E-07	Initial c_{H2M}	0.1000					
4	K_w	1.00E-14	Initial c_{NaOH}	0.1000					
5	Vol. NaOH, mL	c_{H2M}	c_{HM-}	c_{M2-}	[H₃O⁺]	pH	[OH]	pOH	
6	0.00	0.10000	0.00000		3.01E-02	1.52			
7	5.00	0.06667	0.01667		1.81E-02	1.74			
8	7.50	0.05385	0.02308		1.40E-02	1.85			
9	10.00	0.04286	0.02857		1.07E-02	1.97			
10	12.50	0.03333	0.03333		7.98E-03	2.10			
11	15.00	0.02500	0.03750		5.78E-03	2.24			
12	18.00	0.01628	0.04186		3.62E-03	2.44			
13	20.00	0.01111	0.04444		2.41E-03	2.62			
14	22.50	0.00526	0.04737		1.11E-03	2.95			
15	24.00	2.041E-03	0.04898		4.25E-04	3.37			
16	24.50	1.010E-03	0.04949		2.09E-04	3.68			
17	25.00	0.0000	0.05		7.80E-05	4.11			
18	25.50		0.048515	9.9010E-04	2.89E-05	4.54			
19	26.00		0.047059	1.9608E-03	1.42E-05	4.85			
20	30.00		0.036364	9.0909E-03	2.36E-06	5.63			
21	40.00		0.015385	2.3077E-02	3.93E-07	6.41			
22	45.00		0.007143	2.8571E-02	1.48E-07	6.83			
23	49.00		0.001351	3.2432E-02	2.46E-08	7.61			
24	49.50		0.000671	3.2886E-02	1.20E-08	7.92			
25	50.00		0.000000	3.3333E-02	4.21E-10	9.38	2.38E-05	4.62	
26	50.50			3.3113E-02	1.51E-11	10.82	6.62E-04	3.18	
27	51.00			3.2895E-02	7.60E-12	11.12	1.32E-03	2.88	
28	55.00			3.1250E-02	1.60E-12	11.80	6.25E-03	2.20	
29	60.00			2.9412E-02	8.50E-13	12.07	1.18E-02	1.93	
30	Spreadsheet Documentation								
31	Cell B6=($D$3*$F$2-$D$4*A6)/($F$2+A6)				Cell E18=$B$3*C18/D18				
32	Cell C6=$D$4*A6/($F$2+A6)				Cell D26=(($A$25-$A$17)*$D$4)/(A26+$F$2)				
33	Cell E6=(-($B$2)+SQRT(($B$2)^2+(4*D3*$B$2)))/2				Cell G25=(-$D$2+SQRT($D$2^2+4*$D$2*D25))/2				
34	Cell E7=(-(C7+$B$2)+SQRT((C7+$B$2)^2+(4*B7*$B$2)))/2				Cell G26=(A26-2*$F$2)*$D$4/(A26+$F$2)				
35	Cell F6=-LOG10(E6)				Cell H25=-LOG(G25)				
36	Cell C18=(($D$4*$A$17)-((A18-$F$2)*$D$4))/($F$2+A18)				Cell F25=14-H25				
37	Cell D18=(($D$4*$A$17)-((A18-$F$2)*$D$4))/($F$2+A18)				Cell E25=10^-F25				
38	Cell E17=SQRT(($B$3*C17+$B$4)/(1+C17/$B$2))								

Figure 8-5 Final worksheet for maleic acid titration curve.

$$[H_3O^+] = \frac{K_{a2}c_{HM^-}}{c_{M^{2-}}}$$

For the 25.50 mL volume of NaOH, we enter into cell E18 the formula for $[H_3O^+]$ corresponding to the previous equation as shown in the documentation section for this cell. The other values in the second buffer region are calculated in a similar manner. Thus, the formula in cell E18 is then copied into cells E19 through E25. Just prior to the second equivalence point (49.90 mL), the

ratio c_{M2-}/c_{HM-} becomes large, and the simplified buffer equation is no longer valid.

Second Equivalence Point. After the addition of 50.00 mL of 0.1000 M sodium hydroxide, the

solution is 0.0333 M in Na_2M (2.5 mmol/75.00 mL) and there is no longer any HM^- left.

Reaction of the base M^{2-} with water is the predominant equilibrium in the system and the only

one that we need to take into account. Thus, we use the following equilibrium to find $[OH^-]$

$$M^{2-} + H_2O \rightleftharpoons HM^- + OH^-$$

$$K_{b1} = \frac{K_w}{K_{a2}} = \frac{[OH^-][HM^-]}{[M^{2-}]} \approx \frac{[OH^-]^2}{[M^{2-}]}$$

$$[OH^-]^2 + K_{b1}[OH^-] - K_{b1}c_{M^{2-}} = 0$$

We enter the quadratic formula solution to this equation in cell G25 as shown in the

documentation for that cell. In cell H25, we calculate $pOH = -\log[OH^-]$, and in cell F25, we

calculate pH from $pH = 14 - pOH$.

pH Beyond the Second Equivalence Point. In the region just beyond the second equivalence

point (50.01 mL, for example), we still need to take into account the reaction of M^{2-} with water

in addition to the excess OH^- that has been added. Further additions of OH^- suppress the basic

dissociation of M^{2-}. As long as we are a few tenths of a milliliter beyond the second equivalence

point, the $[OH^-]$ is calculated from the concentration of NaOH added in excess of that required

for the complete neutralization of H_2M. Thus, when 50.50 mL of NaOH have been added, we

have a 0.50-mL excess of 0.1000 M NaOH, and we enter into G26 the formula shown in the

documentation section for that cell. The formula in G26 is copied into the remaining column G

cells. Your final worksheet should appear similar to that shown in Figure 8-5 after adding

documentation.

We can now plot the titration curve of pH vs. volume of NaOH added. Your plot should

be similar to that shown in Figure 8-6. Note that we had Excel draw a smooth curve through the

data points of the titration curve.

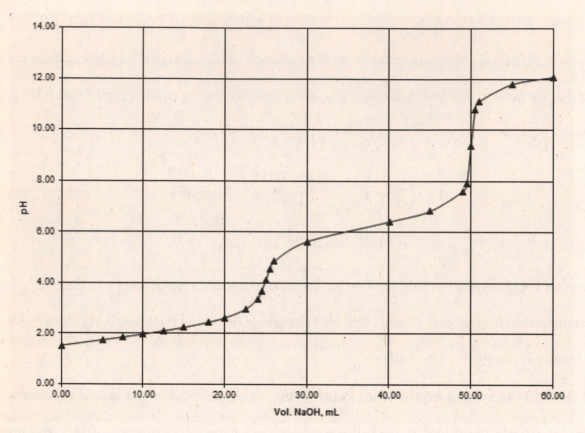

Figure 8-6 Titration curve for 25.00 mL of 0.100 M maleic acid with 0.1000 M NaOH.

Alpha Values During a Titration

Since we have now calculated pH as a function of volume of NaOH during the titration of maleic

acid, we can calculate the alpha values, which depend only on $[H_3O^+]$ and the equilibrium

constants. We can use the same formulas as we used in preparing the spreadsheet shown in

Figure 8-1.

Copy the constants, the volume data and the $[H_3O^+]$ data from the spreadsheet of Figure 8-5 to a new worksheet. Add labels for the denominator, D, and the alpha values. Add formulas for calculating these values. Plot the alpha values vs. the volume of NaOH, and include the chart in the spreadsheet. Your final spreadsheet should appear as shown in Figure 8-7.

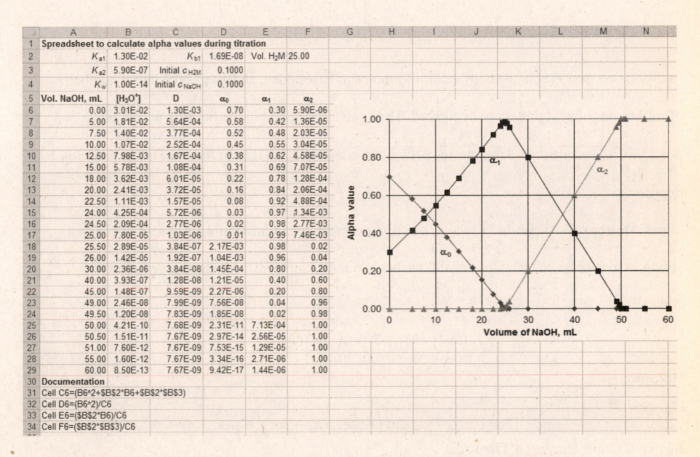

	A	B	C	D	E	F
1	Spreadsheet to calculate alpha values during titration					
2	K_{a1}	1.30E-02	K_{b1}	1.69E-08	Vol. H₂M	25.00
3	K_{a2}	5.90E-07	Initial c_{H2M}	0.1000		
4	K_w	1.00E-14	Initial c_{NaOH}	0.1000		
5	Vol. NaOH, mL	$[H_3O^+]$	D	α_0	α_1	α_2
6	0.00	3.01E-02	1.30E-03	0.70	0.30	5.90E-06
7	5.00	1.81E-02	5.64E-04	0.58	0.42	1.36E-05
8	7.50	1.40E-02	3.77E-04	0.52	0.48	2.03E-05
9	10.00	1.07E-02	2.52E-04	0.45	0.55	3.04E-05
10	12.50	7.98E-03	1.67E-04	0.38	0.62	4.58E-05
11	15.00	5.78E-03	1.08E-04	0.31	0.69	7.07E-05
12	18.00	3.62E-03	6.01E-05	0.22	0.78	1.28E-04
13	20.00	2.41E-03	3.72E-05	0.16	0.84	2.06E-04
14	22.50	1.11E-03	1.57E-05	0.08	0.92	4.88E-04
15	24.00	4.25E-04	5.72E-06	0.03	0.97	1.34E-03
16	24.50	2.09E-04	2.77E-06	0.02	0.98	2.77E-03
17	25.00	7.80E-05	1.03E-06	0.01	0.99	7.46E-03
18	25.50	2.89E-05	3.84E-07	2.17E-03	0.98	0.02
19	26.00	1.42E-05	1.92E-07	1.04E-03	0.96	0.04
20	30.00	2.36E-06	3.84E-08	1.45E-04	0.80	0.20
21	40.00	3.93E-07	1.28E-08	1.21E-05	0.40	0.60
22	45.00	1.48E-07	9.59E-09	2.27E-06	0.20	0.80
23	49.00	2.46E-08	7.99E-09	7.56E-08	0.04	0.96
24	49.50	1.20E-08	7.83E-09	1.85E-08	0.02	0.98
25	50.00	4.21E-10	7.68E-09	2.31E-11	7.13E-04	1.00
26	50.50	1.51E-11	7.67E-09	2.97E-14	2.56E-05	1.00
27	51.00	7.60E-12	7.67E-09	7.53E-15	1.29E-05	1.00
28	55.00	1.60E-12	7.67E-09	3.34E-16	2.71E-06	1.00
29	60.00	8.50E-13	7.67E-09	9.42E-17	1.44E-06	1.00
30	Documentation					
31	Cell C6=(B6^2+$B$2*B6+$B$2*$B$3)					
32	Cell D6=(B6^2)/C6					
33	Cell E6=($B$2*B6)/C6					
34	Cell F6=($B$2*$B$3)/C6					

Figure 8-7 Spreadsheet for calculating alpha values during maleic acid titration.

As a further exercise, superimpose the alpha plots on the titration curve to get a clear picture of the concentration changes that occur during the titration and the dominant species in each region.

An Inverse Master Equation Approach

Another way to obtain the titration curve is the master equation approach made popular by

Butler[1] and expanded by de Levie[2]. In this approach, the titration curve is developed as

a function of the extent of titration, or fraction titrated, ϕ, where

$$\phi = \frac{V_b c_b}{V_a c_a}$$

In this expression $V_b c_b$ is the number of millimoles of base added, and $V_a c_a$ is the initial number

of millimoles of acid present to be titrated. The equivalence point in the titration occurs when

$\phi = 1$. Since we calculate ϕ for various pH values, we should call this an *inverse master equation*

approach.

Weak Monoprotic Acid. Let's take as our first example, the titration of a weak acid HA with a

strong base such as NaOH. Once again, we begin our development with the charge balance

equation.

$$[Na^+] + [H_3O^+] = [A^-] + [OH^-]$$

But since,

$$[Na^+] = \frac{V_b c_b}{V_a + V_b} \text{ and } [A^-] = \alpha_1 c_T = \alpha_1 \left(\frac{V_a c_a}{V_a + V_b} \right)$$

the charge balance equation becomes

$$\frac{V_b c_b}{V_a + V_b} + [H_3O^+] = \alpha_1 \left(\frac{V_a c_a}{V_a + V_b} \right) + \frac{K_w}{[H_3O^+]}$$

By multiplying through by $V_a + V_b$, collecting terms and solving for ϕ, we can show that

---

[1] J. N. Butler, *Ionic Equilibrium: Solubility and pH Calculations*, (New York: Wiley, 1998) pp. 138-139 .
[2] R. de Levie, *J. Chem. Educ.*, **1993**, *70(3)*, 209-217.

$$\phi = \frac{V_b c_b}{V_a c_a} = \frac{\alpha_1 - \dfrac{[H_3O^+] - K_w/[H_3O^+]}{c_a}}{1 + \dfrac{[H_3O^+] - K_w/[H_3O^+]}{c_b}}$$

Before we proceed, let's examine some of the properties of this equation. First, ϕ is dimensionless, but if we know any three of the four variables V_a, V_b, c_a, or c_b, we can calculate the fourth from ϕ. Next, ϕ is expressed in terms of known constants, K_a and K_w, the molar analytical concentrations of acid and base, c_a and c_b, and the hydronium ion concentration (α_1 is calculated from K_a and $[H_3O^+]$). Thus, we can calculate a titration curve by substituting various values of $[H_3O^+]$ into this equation and calculating the corresponding value of ϕ. Note that this is precisely opposite the usual approach of calculating the pH for a given volume of added base. The calculation of ϕ is a more appropriate approach when we use the computer to perform the calculations. Furthermore, our expression for ϕ results in a single master equation for the entire titration curve instead of using different expressions for different regions as in the stoichiometric approach.

Next we'll calculate the titration curve of 0.100 M HA with 0.100 M NaOH. The K_a value for HA is 1×10^{-6}. The worksheet showing the appropriate calculations and the resulting plot is given in Figure 8-8. In order to get the appropriate x and y values, you'll have to switch axes. Excel always uses the leftmost of the columns selected for plotting as the x axis. First on the Insert tab select the **Scatter** chart with only markers. From the Design tab on the Chart Tools ribbon, choose **Select Data.** This opens the **Select Data Source** window. Click on Add to open the **Edit Series** window. You can define the x and y series by filling in the Series X values and Series Y values. If the x and y axes are reversed in a plot, you can also right click on the data series in the chart. Choose Select Data from the resulting drop-down menu. This opens the Select

Data Source window. Choose <u>E</u>dit to edit the series to interchange the axes. Note from the worksheet that prior to the equivalence point, α_1 and ϕ are essentially equal. This result makes good sense when we consider carefully the expression for ϕ. Under conditions where the quantity $([H_3O^+] - [OH^-])/c$ is small relative to α_1 or 1, $\phi \approx \alpha_1$. This occurs prior to the equivalence point because both $[H_3O^+]$ and $[OH^-]$ are relatively small during this part of the titration. Beyond the equivalence point, $[OH^-]$ becomes relatively large, and ϕ becomes significantly different from α_1 as we can see from the figure.

We now have a straightforward and elegant way to calculate the titration curve for *any* acid or base. In fact this ϕ-α approach works for all kinds of systems including strong acid-strong base titrations, weak acid-weak base titrations, complexometric titrations and redox titrations. Furthermore, no approximations are necessary in any of these situations. Except for activity effects, the *equations are exact*. Notice the relative simplicity of the spreadsheet of Figure 8-8 compared to the stoichiometric approach discussed in the previous section.

Now play with spreadsheet by changing the value of K_a and noting the effect. Also change the values of c_a and c_b. You will have to change the range to keep the ϕ values within the limits of 0 to 2. Negative ϕ values are meaningless. Note that as you decrease c_a and c_b, α begins to deviate more from ϕ. Explain why this is true in terms of the master equation.

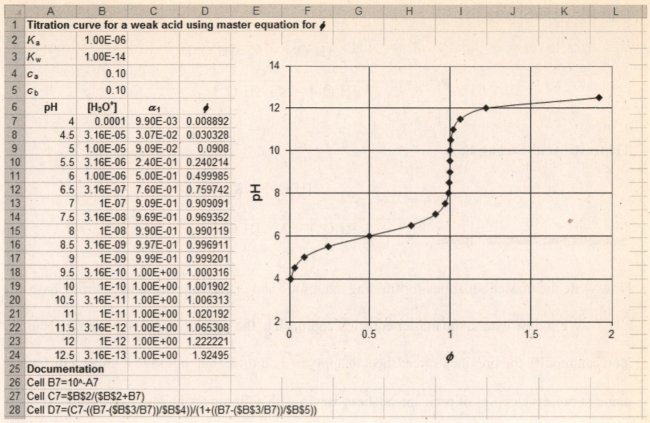

Figure 8-8 Spreadsheet for titration curve as a function of fraction titrated, ϕ.

The spreadsheet data:

	A	B	C	D
1	Titration curve for a weak acid using master equation for ϕ			
2	K_a	1.00E-06		
3	K_w	1.00E-14		
4	c_a	0.10		
5	c_b	0.10		
6	pH	[H_3O^+]	α_1	ϕ
7	4	0.0001	9.90E-03	0.008892
8	4.5	3.16E-05	3.07E-02	0.030328
9	5	1.00E-05	9.09E-02	0.0908
10	5.5	3.16E-06	2.40E-01	0.240214
11	6	1.00E-06	5.00E-01	0.499985
12	6.5	3.16E-07	7.60E-01	0.759742
13	7	1E-07	9.09E-01	0.909091
14	7.5	3.16E-08	9.69E-01	0.969352
15	8	1E-08	9.90E-01	0.990119
16	8.5	3.16E-09	9.97E-01	0.996911
17	9	1E-09	9.99E-01	0.999201
18	9.5	3.16E-10	1.00E+00	1.000316
19	10	1E-10	1.00E+00	1.001902
20	10.5	3.16E-11	1.00E+00	1.006313
21	11	1E-11	1.00E+00	1.020192
22	11.5	3.16E-12	1.00E+00	1.065308
23	12	1E-12	1.00E+00	1.222221
24	12.5	3.16E-13	1.00E+00	1.92495
25	Documentation			
26	Cell B7=10^-A7			
27	Cell C7=$B$2/($B$2+B7)			
28	Cell D7=(C7-((B7-($B$3/B7))/$B$4))/(1+((B7-($B$3/B7))/$B$5))			

Back to Polyprotic Acids. Let's begin our discussion of the master equation approach to polyfunctional acids and bases by writing down a series of master equations for titrations of various acids.

Strong Acid-Strong Base

$$\phi = \frac{1 - \dfrac{[H_3O^+] - K_w/[H_3O^+]}{c_a}}{1 + \dfrac{[H_3O^+] - K_w/[H_3O^+]}{c_b}}$$

Weak Monoprotic Acid-Strong Base

$$\phi = \frac{\alpha_1 - \dfrac{[H_3O^+] - K_w/[H_3O^+]}{c_a}}{1 + \dfrac{[H_3O^+] - K_w/[H_3O^+]}{c_b}}$$

Diprotic Acid-Strong Base

$$\phi = \frac{\alpha_1 + 2\alpha_2 - \dfrac{[H_3O^+] - K_w/[H_3O^+]}{c_a}}{1 + \dfrac{[H_3O^+] - K_w/[H_3O^+]}{c_b}}$$

Triprotic Acid-Strong Base

$$\phi = \frac{\alpha_1 + 2\alpha_2 + 3\alpha_3 - \dfrac{[H_3O^+] - K_w/[H_3O^+]}{c_a}}{1 + \dfrac{[H_3O^+] - K_w/[H_3O^+]}{c_b}}$$

Now write the master equations for titrating tetraprotic and pentaprotic acids with strong base.

We'll now extend the master equation approach to the titration curve of a diprotic acid. You can modify the previous spreadsheet to apply to a diprotic acid, but save it under a different name to preserve copies of each spreadsheet. Begin with a diprotic acid with $K_{a1} = 1.0 \times 10^{-4}$ and $K_{a2} = 1.0 \times 10^{-8}$. Use c_a and c_b values of 0.100 M. In the spreadsheet shown in Figure 8-9, separate columns have been added to calculate the numerator and denominator of ϕ. This makes it easier to enter these expressions without making errors or forgetting parenthetical expressions.

Now explore the diprotic acid system by systematically varying the K_a values. See what happens when the two constants are very close to one another. Begin with one K_a at say 1×10^{-2} and the other at 5×10^{-3}. Increase them both by two powers of ten and then four powers of ten, and so forth, and note the effect. Use the K_a values for maleic acid and compare the plot to that made previously. Modify your spreadsheet to include a column for $\alpha_1 + 2\alpha_2$. Where does the titration curve deviate from a plot of pH vs. $\alpha_1 + 2\alpha_2$. Why?

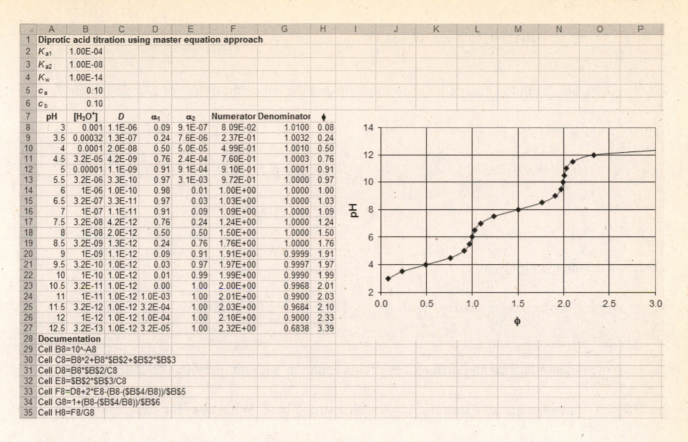

Figure 8-9. Spreadsheet for diprotic acid titration curve using master equation approach.

Titration Curves for Difunctional Bases

Now let's develop the equations for a difunctional base, ethylenediamine and plot its titration curve with hydrochloric acid. The pertinent equilibria are:

$$B + H_2O \rightleftharpoons BH^+ + OH^- \qquad K_{b1} = \frac{[BH^+][OH^-]}{[B]}$$

$$BH^+ + H_2O \rightleftharpoons BH_2^{2+} + OH^- \qquad K_{b2} = \frac{[BH_2^{2+}][OH^-]}{[BH^+]}$$

where ethylendiamine is abbreviated as B and its protonated forms as BH^+ and BH_2^{2+}.

The equilibria can also be written as acid dissociations as follows:

$$BH_2^{2+} + H_2O \rightleftharpoons BH^+ + H_3O^+ \qquad K_{a1} = \frac{K_w}{K_{b1}} = \frac{[BH^+][H_3O^+]}{[BH_2^{2+}]}$$

237

$$BH^+ + H_2O \rightleftharpoons B + H_2O^+ \qquad K_{a2} = \frac{K_w}{K_{b1}} = \frac{[B][H_3O^+]}{[BH^+]}$$

As before, from the charge balance equation and the concentrations, we can derive the following expression for the fraction titrated, ϕ:

$$\phi = \frac{V_a c_a}{V_b c_b} = \frac{\alpha_1 + 2\alpha_0 + \dfrac{[H_3O^+] - K_w/[H_3O^+]}{c_b}}{1 - \dfrac{[H_3O^+] - K_w/[H_3O^+]}{c_a}}$$

where we define the alphas as:

$$\alpha_0 = \frac{[BH_2^{2+}]}{c_T} \qquad \alpha_1 = \frac{[BH^+]}{c_T} \qquad \alpha_2 = \frac{[B]}{c_T}$$

Note the similarities and differences of this equation with our previous equations for ϕ for the titration of monoprotic and diprotic acids with a strong base. What does the expression for ϕ reduce to when $[H_3O^+]$ and $[OH^-]$ are both small or their difference is small relative to $\alpha_1 + 2\alpha_0$? Compare the signs of the last term in the numerator and in the denominator with our previous equations. Note where the molar analytical concentrations of the acid and base solutions appear in the last terms in the numerator and denominator. Do these differences make intuitive sense? Try to write down the master equation for the titration of a trifunctional base with strong acid by inspection. Do not be fooled here. Remember that α_0 now corresponds to the species having three protons, α_1 to the species with two protons, and so forth.

Now prepare a spreadsheet for the titration of ethylenediamine with HCl. You can find the acid dissociation constants in the appendices of FAC9 or AC7. Prepare a plot of pH vs. ϕ. Your final spreadsheet should be similar to that shown in Figure 8-10.

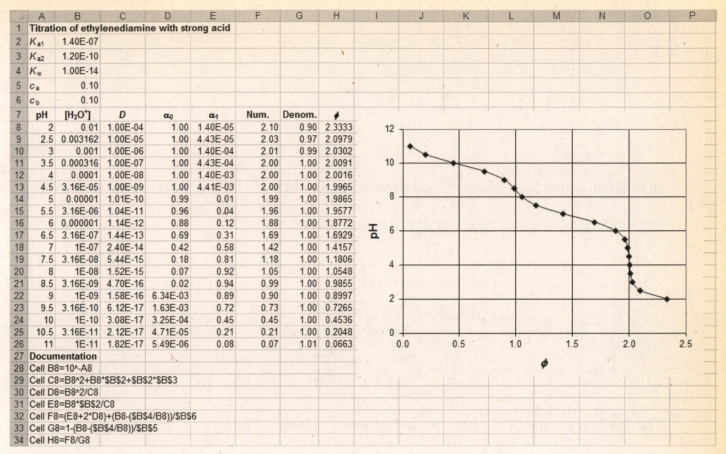

Figure 8-10 Spreadsheet for titration of ethylenediamine with strong acid.

You can modify the spreadsheet to include a column for $\alpha_1 + 2\alpha_0$ and compare this quantity to ϕ. Vary the K values as before and note the effects on the titration curves. As a further exercise, calculate the titration curve for sodium carbonate. You might also try calculating and plotting the titration curve for a trifunctional base.

Amphiprotic Species, Amino Acids, and the Isoelectric Point

Many compounds are *amphiprotic*; that is, they have both acidic and basic functional groups. The amino acids are important examples of amphiprotic compounds. Shown below is phenyalanine, a significant component of many proteins

Phe

In solution, amino acids dissociate and transfer a proton from the carboxylic acid group to the

amino group to form a *zwitterions*, similar to the structure below

PheH

The carboxyl group of the zwitter ion can protonate to form $PheH_2^+$, or it can lose a proton to

become Phe^-. In terms of acid dissociations, the pertinent equilibria can be written

$$PheH_2^+ + H_2O \rightleftharpoons H_3O^+ + PheH$$

$$PheH + H_2O \rightleftharpoons H_3O^+ + Phe^-$$

These equilibria are characterized by the two acid dissociation constants $K_{a1} = 6.3 \times 10^{-3}$ and

$K_{a2} = K_w/K_{b1} = 4.9 \times 10^{-10}$. We may then calculate the titration curve of phenylalanine exactly as

we did other diprotic acids. The completed spreadsheet and plot are shown in Figure 8-11.

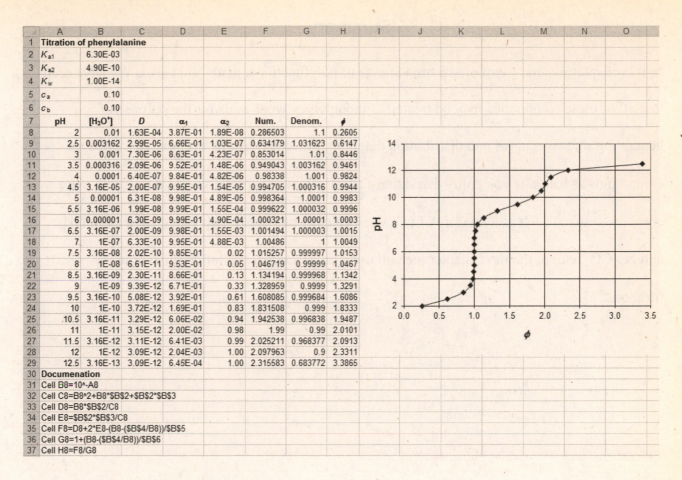

	A	B	C	D	E	F	G	H
1	Titration of phenylalanine							
2	K_{a1}	6.30E-03						
3	K_{a2}	4.90E-10						
4	K_w	1.00E-14						
5	c_a	0.10						
6	c_b	0.10						
7	pH	[H₃O⁺]	D	α_1	α_2	Num.	Denom.	ϕ
8	2	0.01	1.63E-04	3.87E-01	1.89E-08	0.286503	1.1	0.2605
9	2.5	0.003162	2.99E-05	6.66E-01	1.03E-07	0.634179	1.031623	0.6147
10	3	0.001	7.30E-06	8.63E-01	4.23E-07	0.853014	1.01	0.8446
11	3.5	0.000316	2.09E-06	9.52E-01	1.48E-06	0.949043	1.003162	0.9461
12	4	0.0001	6.40E-07	9.84E-01	4.82E-06	0.98338	1.001	0.9824
13	4.5	3.16E-05	2.00E-07	9.95E-01	1.54E-05	0.994705	1.000316	0.9944
14	5	0.00001	6.31E-08	9.98E-01	4.89E-05	0.998364	1.0001	0.9983
15	5.5	3.16E-06	1.99E-08	9.99E-01	1.55E-04	0.999622	1.000032	0.9996
16	6	0.000001	6.30E-09	9.99E-01	4.90E-04	1.000321	1.00001	1.0003
17	6.5	3.16E-07	2.00E-09	9.98E-01	1.55E-03	1.001494	1.000003	1.0015
18	7	1E-07	6.33E-10	9.95E-01	4.88E-03	1.00486	1	1.0049
19	7.5	3.16E-08	2.02E-10	9.85E-01	0.02	1.015257	0.999997	1.0153
20	8	1E-08	6.61E-11	9.53E-01	0.05	1.046719	0.99999	1.0467
21	8.5	3.16E-09	2.30E-11	8.66E-01	0.13	1.134194	0.999968	1.1342
22	9	1E-09	9.39E-12	6.71E-01	0.33	1.328959	0.9999	1.3291
23	9.5	3.16E-10	5.08E-12	3.92E-01	0.61	1.608085	0.999684	1.6086
24	10	1E-10	3.72E-12	1.69E-01	0.83	1.831508	0.999	1.8333
25	10.5	3.16E-11	3.29E-12	6.06E-02	0.94	1.942538	0.996838	1.9487
26	11	1E-11	3.15E-12	2.00E-02	0.98	1.99	0.99	2.0101
27	11.5	3.16E-12	3.11E-12	6.41E-03	0.99	2.025211	0.968377	2.0913
28	12	1E-12	3.09E-12	2.04E-03	1.00	2.097963	0.9	2.3311
29	12.5	3.16E-13	3.09E-12	6.45E-04	1.00	2.315583	0.683772	3.3865
30	Documenation							
31	Cell B8=10^-A8							
32	Cell C8=B8^2+B8*$B$2+$B$2*$B$3							
33	Cell D8=B8*$B$2/C8							
34	Cell E8=$B$2*$B$3/C8							
35	Cell F8=D8+2*E8-(B8-($B$4/B8))/$B$5							
36	Cell G8=1+(B8-($B$4/B8))/$B$6							
37	Cell H8=F8/G8							

Figure 8-11 Titration of phenylalanine.

As shown in Feature 15-2 of FAC9 and Feature 13-4 of AC7, there is a point in the titration of an amino acid called the *isoelectric point*. At this point, essentially all of the amino acid is in the zwitterionic form and thus, it will not migrate in an electric field. The pH at the isoelectric point for phenylalanine is calculated from

$$[H_3O^+] = \sqrt{\frac{K_{a1}K_w}{K_{b1}}} = \sqrt{K_{a1}K_{a2}} = \sqrt{6.3 \times 10^{-3} \times 4.9 \times 10^{-10}} = 1.7 \times 10^{-6}$$

and pH = $-\log(1.7 \times 10^{-6})$ = 5.75. Modify your titration curve for phenylalanine to include a horizontal dotted line at the isoelectric point pH.

Summary

In this chapter, we have explored a broad range of polyfunctional acids and bases. We have seen that all of these compounds may be treated as though they were acids for the purpose of generating titration curves and distribution diagrams. We could just as easily have treated all compounds as bases. If you enjoy derivations, try developing expressions for ϕ for some more complex systems such as a mixture of two weak acids, a mixture of two bases, such as NaOH and Na_2CO_3, etc. In the next chapter we will use similar approaches for complexometric titrations.

Problems

1.   Calculate alpha values for the following diprotic acids every 0.5 pH units from pH 0.0 to

14.0. Plot the distribution diagram for each of the acids labeling each species.

(a)   phthalic acid

(b)   malonic acid

(c)   tartaric acid

(d)   oxalic acid

2.   Calculate alpha values for the following triprotic acids every 0.5 pH units from pH 0.0 to

14.0. Plot the distribution diagram for each of the acids labeling each species.

(a)   citric acid

(b)   arsenic acid

(c)   phosphoric acid

3.   The pK values for the ethylene diammonium ion are pK_1 = 6.85 and pK_2 = 9.93. Plot the

distribution diagram and label each species. Find the alpha values every 0.5 pH units

from pH 0.0 to 14.0.

4.   (a)   Plot logarithmic concentration diagrams for 0.1000 M solutions of each of the

acids in Problem 1 above.

(b)   For phthalic acid, find the concentrations of all species at pH 4.8.

(c)   For tartaric acid, find the concentrations of all species at pH 4.3.

(d)   From the log concentration diagram, find the pH of a 0.1000 M solution of

phthalic acid, H_2P. Find the pH of a 0.100 M solution of HP^-.

5.   The following questions apply to a phosphoric acid solution.

(a)   Plot a logarithmic concentration diagram for 0.100 M H_3PO_4.

(b)   Find the concentrations of all species in a solution of 0.100 M H_3PO_4 at pH 6.2 .

(c)   Find the pH of a 0.100 M solution of $H_2PO_4^-$.

6.   Construct a logarithmic concentration diagram for 0.100 M ethylenediamine with the pK

        values given in Problem 3 above. Label the species. What are the concentrations of all

        species in a 0.100 M solution buffered to pH 5.5?

7.   Construct curves using the stoichiometric approach for the titration of 50.00 mL of a

        0.1000 M solution of compound A with a 0.2000 M solution of compound B in the

        following list. For each titration, calculate the pH after the addition of 0.00, 12.50, 20.00,

        24.00, 25.00, 26.00, 37.50, 45.00, 49.00, 50.00, 51.00, and 60.00 mL of compound B:

A	B
(a) Na_2CO_3	HCl
(b) ethylenediamine	HCl
(c) H_2SO_4	NaOH
(d) H_2SO_3	NaOH

8.   For the titration of each of the following acids with NaOH, plot the titration curves every

        0.5 pH unit using the inverse master equation approach.

(a)   malonic acid

(b)   oxalic acid

(c) o-phthalic acid

9.   If 25.00 mL of 0.1000 M solutions of the acids in Problem 8 are titrated with 0.2000 M

       NaOH, plot curves of pH vs. volume of titrant using the inverse master equation

       approach. Locate the end points using first- and second-derivative plots.

10.   For the titration of 25.00 mL of 0.1000 M phosphoric acid with 0.2500 M NaOH, plot the

       titration curve every 0.5 pH unit using the inverse master equation approach. Instead of

       fraction titrated ϕ, convert the x axis to volume of titrant. Locate the end points using

       first-and second-derivative plots.

11.   Prepare a titration curve plot of pH vs. ϕ for triethylenediamine being titrated with HCl.

       The pK values are given in problem 3 above. Find the end points by first- and second-

       derivative plots.

12.   Generate a curve using the stoichiometric approach for the titration of 50.00 mL of a

       solution in which the analytical concentration of NaOH is 0.1000 M and that for

       hydrazine is 0.0800 M. Calculate the pH after addition of 0.00, 10.00, 20.00, 24.00,

       25.00, 26.00, 35.00, 44.00, 45.00, 46.00, and 50.00 mL of 0.2000 M $HClO_4$.

13.   Generate a curve for the titration of 50.00 mL of a solution in which the analytical

       concentration of $HClO_4$ is 0.1000 M and that for formic acid is 0.0800 M. Calculate the

       pH after addition of 0.00, 10.00, 20.00, 24.00, 25.00, 26.00, 35.00, 44.00, 45.00, 46.00,

       and 50.00 mL of 0.2000 M KOH.

14.   Plot titration curves for the following amino acids and find the isoelectric point pH

       values.

       (a)   glycine (pK_1 = 2.34, pK_2 = 9.60).

(b)   β-alanine (pK_1 = 3.55, pK_2 = 10.24).

(c)   L-lysine (pK_1 = 2.18, pK_2 = 8.94, pK_3 = 10.63).

Chapter 9

Complexometric and Precipitation Titrations

In this chapter, we examine the equilibria associated with complexes formed between metals and various monodentate and polydentate ligands. These topics are covered in Chapter 17 of FAC9 and Chapter 15 of AC7. We begin our discussion by developing a general expression for the formation of complexes and follow with a description of the master equations for the titration of metals with ligands. We then consider several aspects of EDTA titrations. We conclude with a brief discussion of precipitation titrations.

Complexation Equilibria

When a metal M combines with a ligand L, we can write the following equilibria and the corresponding stepwise formation constants.

$$M + L \rightleftharpoons ML \qquad K_1 = \frac{[ML]}{[M][L]}$$

$$ML + L \rightleftharpoons ML_2 \qquad K_2 = \frac{[ML_2]}{[ML][L]}$$

$$ML_2 + L \rightleftharpoons ML_3 \qquad K_3 = \frac{[ML_3]}{[ML_2][L]}$$

$$\vdots$$

$$ML_{n-1} + L \rightleftharpoons ML_n \qquad K_n = \frac{[ML_n]}{[ML_{n-1}][L]}$$

We can also write reactions that are the sum of two or more steps and overall formation

constants designated by the symbol β_n. These overall formation constants are products of

stepwise constants with $\beta_1 = K_1$, $\beta_2 = K_1 K_2$, $\beta_3 = K_1 K_2 K_3$, *etc.*

$$M + 2L \rightleftharpoons ML_2 \qquad \beta_2 = \frac{[ML_2]}{[M][L]^2} = K_1 K_2$$

$$M + 3L \rightleftharpoons ML_3 \qquad \beta_3 = \frac{[ML_3]}{[M][L]^3} = K_1 K_2 K_3$$

$$\vdots$$

$$M + nL \rightleftharpoons ML_n \qquad \beta_n = \frac{[ML_n]}{[M][L]^n} = K_1 K_2 \cdots K_n$$

α-Values for Complexes

The fractions present in each form are given by

$$\alpha_M = \frac{1}{1 + \beta_1[L] + \beta_2[L]^2 + \beta_3[L]^3 + \cdots + \beta_n[L]^n}$$

$$\alpha_{ML} = \frac{\beta_1[L]}{1 + \beta_1[L] + \beta_2[L]^2 + \beta_3[L]^3 + \cdots + \beta_n[L]^n}$$

$$\alpha_{ML_2} = \frac{\beta_2[L]^2}{1 + \beta_1[L] + \beta_2[L]^2 + \beta_3[L]^3 + \cdots + \beta_n[L]^n}$$

$$\alpha_{ML_n} = \frac{\beta_n[L]^n}{1 + \beta_1[L] + \beta_2[L]^2 + \beta_3[L]^3 + \cdots + \beta_n[L]^n}$$

Note that these expressions are analogous to the α expressions we wrote for

polyfunctional acids and bases except that the equations here are written in terms of formation

equilibria while those for acids or bases are written in terms of dissociation equilibria. Also, the master variable is the ligand concentration [L] instead of the hydronium ion concentration $[H_3O^+]$. As an exercise, derive one or more of these α expressions.

As our first example, let's take the case of the ammonia complexes of Cu(II) with formulas $Cu(NH_3)^{2+}$, $Cu(NH_3)_2^{2+}$, $Cu(NH_3)_3^{2+}$ and $Cu(NH_3)_4^{2+}$. The logarithms of the stepwise formation constants are $\log K_1 = 4.04$, $\log K_2 = 3.43$, $\log K_3 = 2.80$ and $\log K_4 = 1.48$. The logarithms of the corresponding overall formation constants are $\log \beta_1 = 4.04$, $\log \beta_2 = 7.47$, $\log \beta_3 = 10.27$, and $\log \beta_4 = 11.75$.

The distribution diagram is a plot of the α values vs. pL = –log [L]. Prepare a spreadsheet to calculate the five α values. Use pL values from –1 to 6 in increments of 0.5 units. The final plot with appropriate labels is shown in Figure 9-1. Note the strong overlap of the α values in most regions.

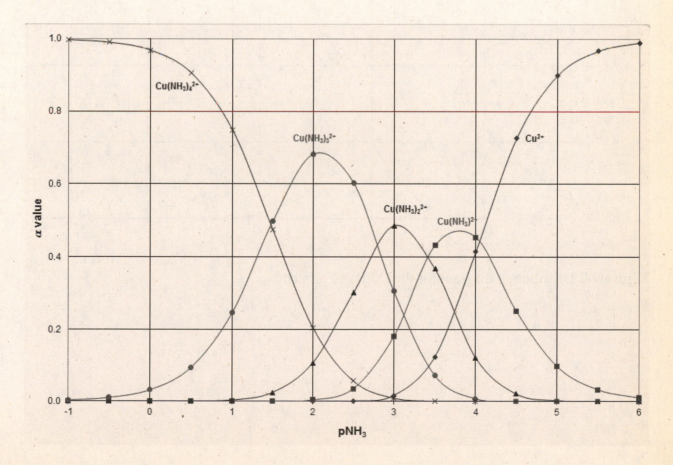

249

Figure 9-1 Alpha values for Cu(II)/NH₃ complexes.

Now prepare a spreadsheet for calculating the α values for the Cd(II)/Cl⁻ system. The logarithms of the stepwise formation constants for $CdCl^+$, $CdCl_2$, $CdCl_3^-$, and $CdCl_4^{2-}$ are 1.32, 0.90, 0.09, and –0.45 respectively. Use pL values from –1 to 4 in 0.5 increments. Figure 9-2 shows the normal distribution diagram. As you can see, even stronger overlap is seen than with the Cu(II)/NH₃ system.

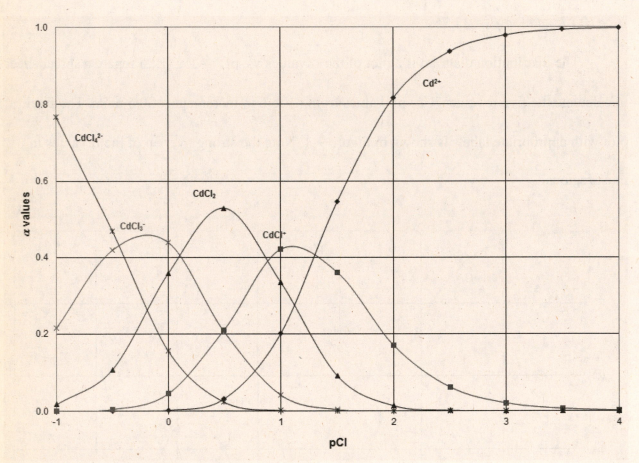

Figure 9-2 Distribution diagram for the Cd(II)/Cl⁻ system.

Next, we'll prepare a different type of distribution diagram. In a new column, calculate the sum of the α values for the 4 complexes, $\alpha_{CdCl^+} + \alpha_{CdCl_2} + \alpha_{CdCl_3^-} + \alpha_{CdCl_4^{2-}}$. In another column, calculate the sum of the α values for the complexes containing 2, 3, and 4 chlorides, $\alpha_{CdCl_2} + \alpha_{CdCl_3^-} + \alpha_{CdCl_4^{2-}}$. In another column, find $\alpha_{CdCl_3^-} + \alpha_{CdCl_4^{2-}}$, and finally in the last column, obtain $\alpha_{CdCl_4^{2-}}$. Plot these sums of α values vs. pCl$^-$ as shown in Figure 9-3.

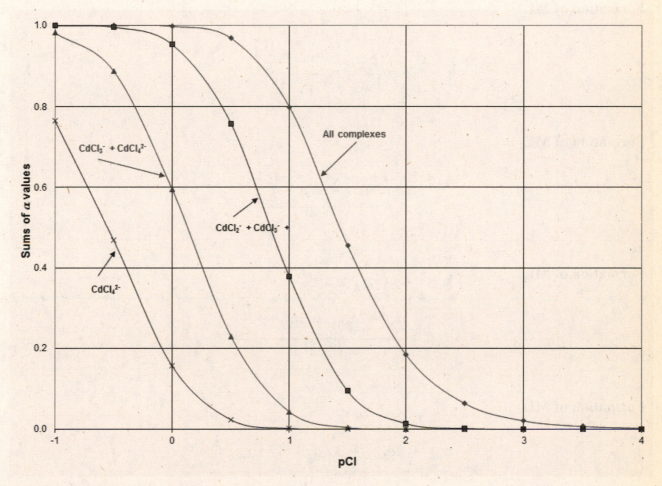

Figure 9-3 Sums of α values for Cd(II)/Cl$^-$ complexes.

The virtue of the sum type of distribution diagram is that you can easily see values of pL for which essentially all the metal is complexed by ligand. There are many cases in analytical chemistry where we would like to find the minimum concentration of ligand needed to complex a metal. For example, in Figure 9-3, we can see that for pCl values of 0 and lower ($[Cl^-] > 1.0$

M), essentially all the Cd^{2+} (99.8 % from the spreadsheet) is in the form of one of the complexes.

Inverse Master Equations for Complexometric Titrations

Inverse master equations for titrations of metals with ligands are analogous to those for acid-base

titrations.

Formation of ML

$$\phi = \frac{V_L c_L}{V_M c_M} = \frac{\alpha_{ML} + \dfrac{[L]}{c_M}}{1 - \dfrac{[L]}{c_L}}$$

Formation of ML$_2$

$$\phi = \frac{\alpha_{ML} + 2\alpha_{ML_2} + \dfrac{[L]}{c_M}}{1 - \dfrac{[L]}{c_L}}$$

Formation of ML$_3$

$$\phi = \frac{\alpha_{ML} + 2\alpha_{ML_2} + 3\alpha_{ML_3} + \dfrac{[L]}{c_M}}{1 - \dfrac{[L]}{c_L}}$$

Formation of ML$_n$

$$\phi = \frac{\alpha_{ML} + 2\alpha_{ML_2} + 3\alpha_{ML_3} + \cdots + n\alpha_{ML_n} + \dfrac{[L]}{c_M}}{1 - \dfrac{[L]}{c_L}}$$

Compare these with the analogous expressions for titrations of acids and bases, and note

the similarities and differences. Again, you may wish to derive the expression for the 1:1 metal-ligand complexometric titration. Begin with the mass balance expressions for both the metal and

the ligand.

Let's now construct the titration curve for the titration of 0.100 M Cd(II) with 0.300 M

Cl^-. Since Cd(II) forms complexes from $CdCl^+$ to $CdCl_4^{2-}$, we will need to calculate α_{ML},

α_{ML_2}, α_{ML_3}, and α_{ML_4}. The worksheet for the required calculations is shown in Figure 9-4.

Note we've also used the master equation for the formation of $CdCl_4^{2-}$ (ML_4) to find ϕ.

	A	B	C	D	E	F	G	H	I	J
1	Titration of Cd(II) with Cl⁻									
2	β_{f1}	2.089E+01								
3	β_{f2}	1.660E+02								
4	β_{f3}	2.042E+02								
5	β_{f4}	7.244E+01								
6	$c_{Cd(II)}$	0.100								
7	c_{Cl^-}	0.300								
8	pCl⁻	[Cl⁻]	D	α_{ML}	α_{ML2}	α_{ML3}	α_{ML4}	Numerator	Denominator	ϕ
9	1	0.1	4.960E+00	4.212E-01	3.346E-01	4.116E-02	1.460E-03	2.220E+00	0.666666667	3.33
10	1.1	0.07943282	3.812E+00	4.354E-01	2.747E-01	2.684E-02	7.566E-04	1.863E+00	0.735223922	2.53
11	1.2	0.06309573	3.031E+00	4.349E-01	2.180E-01	1.692E-02	3.788E-04	1.554E+00	0.789680885	1.97
12	1.3	0.05011872	2.490E+00	4.205E-01	1.674E-01	1.032E-02	1.836E-04	1.288E+00	0.832937589	1.55
13	1.4	0.03981072	2.108E+00	3.946E-01	1.248E-01	6.112E-03	8.633E-05	1.061E+00	0.86729761	1.22
14	1.5	0.03162278	1.833E+00	3.604E-01	9.053E-02	3.522E-03	3.952E-05	8.684E-01	0.894590745	0.97
15	2	0.01	1.226E+00	1.705E-01	1.354E-02	1.666E-04	5.910E-07	2.980E-01	0.966666667	0.31
16	2.5	0.00316228	1.068E+00	6.188E-02	1.554E-03	6.047E-06	6.785E-09	9.663E-02	0.989459074	0.10
17	3	0.001	1.021E+00	2.046E-02	1.625E-04	2.000E-07	7.095E-11	3.079E-02	0.996666667	0.03
18	3.5	0.00031623	1.007E+00	6.563E-03	1.649E-05	6.414E-09	7.197E-13	9.759E-03	0.998945907	9.77E-03
19	4	0.0001	1.002E+00	2.085E-03	1.656E-06	2.037E-10	7.229E-15	3.088E-03	0.999666667	3.09E-03
20	4.5	3.1623E-05	1.001E+00	6.603E-04	1.658E-07	6.452E-12	7.240E-17	9.768E-04	0.999894591	9.77E-04
21	5	0.00001	1.000E+00	2.089E-04	1.659E-08	2.041E-13	7.243E-19	3.089E-04	0.999966667	3.09E-04
22	Documentation									
23	Cell B9=10^-A9									
24	Cell C9=1+$B$2*B9+$B$3*B9^2+$B$4*B9^3+$B$5*B9^4									
25	Cell D9=$B$2*B9/C9									
26	Cell E9=$B$3*B9^2/C9									
27	Cell F9=$B$4*B9^3/C9									
28	Cell G9=$B$5*B9^4/C9									
29	Cell H9=D9+2*E9+3*F9+4*G9+B9/$B$6									
30	Cell I9=1-B9/$B$7									
31	Cell J9=H9/I9									

Figure 9-4 Worksheet for titration curve of Cd(II) being titrated with Cl⁻.

Now we can plot the titration curve of pCl vs. ϕ. The curve is shown in Figure 9-5. Note that

there is no useful inflection point. In other words, it is impossible to successfully titrate cadmium with chloride ion. This behavior is typical of most monodentate ligands such as the halogens and ammonia. The formation equilibria for the complexes overlap severely, and so there is no abrupt break in the titration curve. You can also see this from the distribution diagram of Figure 9-2.

Select a couple of different metals and ligands from the table of stepwise formation constants in the Appendices of FAC9 or AC7, enter the formation constants in a spreadsheet similar to Figure 9-4, and obtain the titration curves. Try $HgCl_4^{2-}$ and $HgBr_4^{2-}$. Can either of these complexes be used for successful titrations of the metals with the ligands? Why or why not?

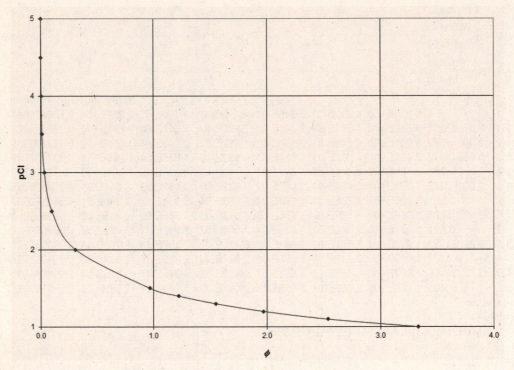

Figure 9-5 Titration curve for the Cd(II)-Cl⁻ system.

Ligands That Protonate

One complication that occurs in many complexation reactions is that the ligands can protonate. Of course, the ligands we have just considered, Cl⁻, Br⁻, and I⁻, are anions of strong acids, and we do not have to worry about protonation. However, many ligands, such as oxalate, cyanide, tartrate, salicylate, and phthalate, are anions of weak acids. Hence, we must consider protonation

side reactions.

Complexation Reactions with Protonating Ligands

Let's consider the formation of complexes between M and L, where L is the conjugate base of a

polyprotic acid, HL, H_2L, ... H_nL. Addition of acid to a solution containing M and L reduces the

concentration of free L available to complex with M. Hence, acidic solutions decrease the

effectiveness of L as a complexing agent. Controlling the pH of these solutions can control how

effective L is when complexing with different metals and thus the selectivity of L as a

complexing agent.

For a polyprotic acid, the fraction of the total ligand concentration present as L, HL, H_2L,

or H_nL is given by an α value. Since we're most often interested in the free ligand concentration,

[L], we're most concerned with the α_n value, where

$$\alpha_n = \frac{[\text{L}]}{c_\text{T}} = \frac{K_{a1}K_{a2}\dots K_{an}}{[\text{H}_3\text{O}^+]^n + K_{a1}[\text{H}_3\text{O}^+]^{n-1} + K_{a1}K_{a2}[\text{H}_3\text{O}^+]^{n-2} + \dots + K_{a1}K_{a2}\dots K_{an}}$$

From this equation the free ligand concentration can be calculated at any given pH as

$$[\text{L}] = \alpha_n c_\text{T}$$

As the solution gets more acidic, α_n becomes smaller and smaller, and the free ligand

concentration decreases. As the solution becomes basic, α_n increases until it approaches unity,

which indicates that all the ligand is present in the unprotonated L form.

Conditional Formation Constants

To take into account the effect of pH on the free ligand concentration in a complexation reaction,

it is useful to introduce a *conditional*, or *effective* formation constant. Such constants are pH-dependent equilibrium constants that apply at a single pH only. For the reaction of Fe^{3+} with oxalate, for example, we can write the formation constant K_1 for the first complex as

$$K_1 = \frac{[(FeOx)^+]}{[Fe^{3+}][Ox^{2-}]} = \frac{[(FeOx)^+]}{[Fe^{3+}]\alpha_2 c_T}$$

At a particular pH value, α_2 is constant and we can combine K_1 and α_2 to yield a new conditional constant K_1'

$$K_1' = \alpha_2 K_1 = \frac{[(FeOx)^+]}{[Fe^{3+}]c_T}$$

The use of these conditional constants greatly simplifies calculations because c_T is usually known or calculated. The overall formation constants, β values, for the higher complexes, $(FeOx_2)^-$ and $(FeOx_3)^{3-}$, can also be written as conditional constants.

EDTA: A Ubiquitous Ligand

Ethylenediaminetetraacetic acid, EDTA, has the structure shown below.

The EDTA molecule has six potential sites for binding a metal ion: the four carboxyl groups and the two amine groups, with each of the latter having an unshared pair of electrons. Thus, EDTA is a hexadentate ligand.

EDTA as a Tetraprotic Acid

The EDTA molecule is usually considered a tetraprotic acid corresponding to protonation of the four carboxyl groups. In acidic solutions, the amine groups can also protonate so that EDTA is sometimes considered as a hexaprotic acid. Here, we will neglect protonation of the amine groups and consider EDTA only as a tetraprotic acid. The dissociation constants for the acidic groups in EDTA are $K_1 = 1.02 \times 10^{-2}$, $K_2 = 2.14 \times 10^{-3}$, $K_3 = 6.92 \times 10^{-7}$, and $K_4 = 5.50 \times 10^{-11}$. The various EDTA species are usually abbreviated H_4Y, H_3Y^-, H_2Y^{2-}, HY^{3-}, and Y^{4-}. By the methods we developed in Chapter 8, we can readily find the fractions of the total EDTA concentration present in the 5 forms shown as a function of pH. The spreadsheet for the distribution diagram is shown in Figure 9-6 and the diagram itself in Figure 9-7. Note that the reactive form, Y^{4-} is not predominant until the pH gets more basic than 8. Metals that have very large formation constants with EDTA can form under less basic conditions, but those with moderate formation constants will not be very stable. The ability to control the concentration of the reactive form gives EDTA tremendous flexibility as a ligand.

	A	B	C	D	E	F	G	H
1	Spreadsheet to calculate α values for EDTA							
2	K_1	1.02E-02						
3	K_2	2.14E-03						
4	K_3	6.92E-07						
5	K_4	5.50E-11						
6	pH	[H_3O^+]	D	α_0	α_1	α_2	α_3	α_4
7	0.0	1	1.010	0.990	0.010	2.16E-05	1.50E-11	8.22E-22
8	0.5	0.316228	0.010	0.969	0.031	2.11E-04	4.63E-10	8.05E-20
9	1.0	0.1	1.10E-04	0.906	0.092	1.98E-03	1.37E-08	7.52E-18
10	1.5	0.031623	1.34E-06	0.744	0.240	0.016	3.55E-07	6.18E-16
11	2.0	0.01	2.24E-08	0.447	0.456	0.098	6.75E-06	3.71E-14
12	2.5	0.003162	6.41E-10	0.156	0.503	0.341	7.45E-05	1.30E-12
13	3.0	0.001	3.30E-11	0.030	0.309	0.661	4.57E-04	2.51E-11
14	3.5	0.000316	2.52E-12	3.97E-03	0.128	0.866	1.90E-03	3.30E-10
15	4.0	0.0001	2.30E-13	4.35E-04	0.044	0.949	6.56E-03	3.61E-09
16	4.5	3.16E-05	2.26E-14	4.42E-05	0.014	0.965	0.021	3.67E-08
17	5.0	0.00001	2.34E-15	4.27E-06	4.35E-03	0.931	0.064	3.54E-07
18	5.5	3.16E-06	2.66E-16	3.75E-07	1.21E-03	0.819	0.179	3.12E-06
19	6.0	0.000001	3.69E-17	2.71E-08	2.76E-04	0.591	0.409	2.25E-05
20	6.5	3.16E-07	6.96E-18	1.44E-09	4.63E-05	0.314	0.686	1.19E-04
21	7.0	1E-07	1.73E-18	5.78E-11	5.90E-06	0.126	0.873	4.80E-04
22	7.5	3.16E-08	5.00E-19	2.00E-12	6.45E-07	0.044	0.955	1.66E-03
23	8.0	1E-08	1.54E-19	6.49E-14	6.62E-08	0.014	0.980	5.39E-03
24	8.5	3.16E-09	4.88E-20	2.05E-15	6.61E-09	4.47E-03	0.979	0.017
25	9.0	1E-09	1.60E-20	6.27E-17	6.39E-10	1.37E-03	0.947	0.052
26	9.5	3.16E-10	5.61E-21	1.78E-18	5.75E-11	3.89E-04	0.852	0.148
27	10.0	1E-10	2.34E-21	4.27E-20	4.36E-12	9.32E-05	0.645	0.355
28	10.5	3.16E-11	1.31E-21	7.64E-22	2.47E-13	1.67E-05	0.365	0.635
29	11.0	1E-11	9.82E-22	1.02E-23	1.04E-14	2.22E-06	0.154	0.846
30	11.5	3.16E-12	8.79E-22	1.14E-25	3.67E-16	2.48E-07	0.054	0.946
31	12.0	1E-12	8.46E-22	1.18E-27	1.21E-17	2.58E-08	0.018	0.982
32	Documentation							
33	Cell B7=10^-A7							
34	Cell C7=B7^4+$B$2*B7^3+$B$2*$B$3*B7^2+$B$2*$B$3*$B$4*B7+$B$2*$B$3*$B$4*$B$5							
35	Cell D7=B7^4/C7							
36	Cell E7=$B$2*B7^3/C7							
37	Cell F7=$B$2*$B$3*B7^2/C7							
38	Cell G7=$B$2*$B$3*$B$4*B7/C7							
39	Cell H7=$B$2*$B$3*$B$4*$B$5/C7							

Figure 9-6 Spreadsheet for calculating distribution diagram for EDTA.

In the distribution diagram, note that the first two dissociation constants are very close to one another and the corresponding α values overlap considerably. You might find it useful to mark positions on the plot where pH = pK_a.

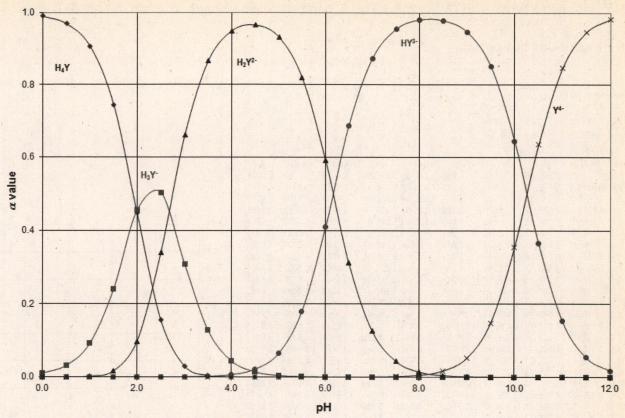

Figure 9-7 Distribution diagram for EDTA.

Titrating EDTA with Base

As a tetraprotic acid, EDTA can be titrated with base. We will now compute the titration curve

of 0.1000 M H_4Y with 0.1000 M NaOH using the inverse master equation approach. The

equation used is an extension of the equation for a triprotic acid given in Chapter 8.

$$\phi = \frac{\alpha_1 + 2\alpha_2 + 3\alpha_3 + 4\alpha_4 - \dfrac{[H_3O^+] - K_w/[H_3O^+]}{c_a}}{1 + \dfrac{[H_3O^+] - K_w/[H_3O^+]}{c_b}}$$

The spreadsheet and the titration curve are displayed in Figure 9-8. Note that there is no break

corresponding to loss of the first proton since K_{a1} and K_{a2} are very close to one another. A break

corresponding to the loss of two protons is observed at $\phi = 2.0$. Likewise, a break corresponding

to the loss of three protons is seen at $\phi = 3.0$. Only a buffering effect is seen at $\phi = 4.0$. Because

HY^{3-} is such a weak acid ($K_{a4} = 5.5 \times 10^{-11}$), comparable in strength to water, no break is seen

for loss of this last proton.

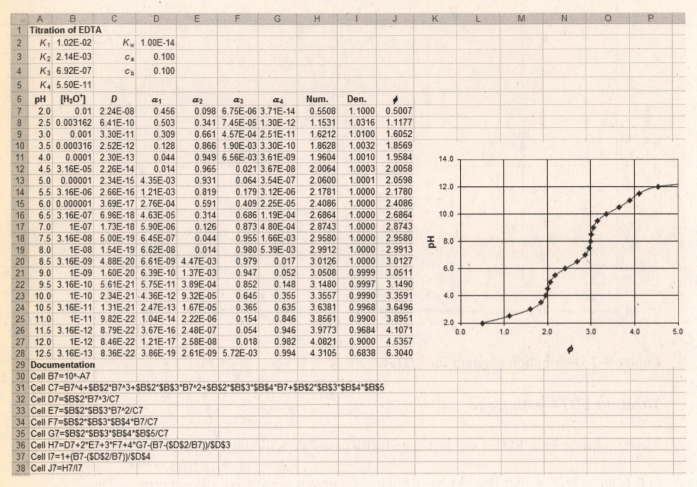

	A	B	C	D	E	F	G	H	I	J
1	Titration of EDTA									
2	K_1	1.02E-02	K_w	1.00E-14						
3	K_2	2.14E-03	c_a	0.100						
4	K_3	6.92E-07	c_b	0.100						
5	K_4	5.50E-11								
6	pH	[H_3O^+]	D	α_1	α_2	α_3	α_4	Num.	Den.	ϕ
7	2.0	0.01	2.24E-08	0.456	0.098	6.75E-06	3.71E-14	0.5508	1.1000	0.5007
8	2.5	0.003162	6.41E-10	0.503	0.341	7.45E-05	1.30E-12	1.1531	1.0316	1.1177
9	3.0	0.001	3.30E-11	0.309	0.661	4.57E-04	2.51E-11	1.6212	1.0100	1.6052
10	3.5	0.000316	2.52E-12	0.128	0.866	1.90E-03	3.30E-10	1.8628	1.0032	1.8569
11	4.0	0.0001	2.30E-13	0.044	0.949	6.56E-03	3.61E-09	1.9604	1.0010	1.9584
12	4.5	3.16E-05	2.26E-14	0.014	0.965	0.021	3.67E-08	2.0064	1.0003	2.0058
13	5.0	0.00001	2.34E-15	4.35E-03	0.931	0.064	3.54E-07	2.0600	1.0001	2.0598
14	5.5	3.16E-06	2.66E-16	1.21E-03	0.819	0.179	3.12E-06	2.1781	1.0000	2.1780
15	6.0	0.000001	3.69E-17	2.76E-04	0.591	0.409	2.25E-05	2.4086	1.0000	2.4086
16	6.5	3.16E-07	6.96E-18	4.63E-05	0.314	0.686	1.19E-04	2.6864	1.0000	2.6864
17	7.0	1E-07	1.73E-18	5.90E-06	0.126	0.873	4.80E-04	2.8743	1.0000	2.8743
18	7.5	3.16E-08	5.00E-19	6.45E-07	0.044	0.955	1.66E-03	2.9580	1.0000	2.9580
19	8.0	1E-08	1.54E-19	6.62E-08	0.014	0.980	5.39E-03	2.9912	1.0000	2.9913
20	8.5	3.16E-09	4.88E-20	6.61E-09	4.47E-03	0.979	0.017	3.0126	1.0000	3.0127
21	9.0	1E-09	1.60E-20	6.39E-10	1.37E-03	0.947	0.052	3.0508	0.9999	3.0511
22	9.5	3.16E-10	5.61E-21	5.75E-11	3.89E-04	0.852	0.148	3.1480	0.9997	3.1490
23	10.0	1E-10	2.34E-21	4.36E-12	9.32E-05	0.645	0.355	3.3557	0.9990	3.3591
24	10.5	3.16E-11	1.31E-21	2.47E-13	1.67E-05	0.365	0.635	3.6381	0.9968	3.6496
25	11.0	1E-11	9.82E-22	1.04E-14	2.22E-06	0.154	0.846	3.8561	0.9900	3.8951
26	11.5	3.16E-12	8.79E-22	3.67E-16	2.48E-07	0.054	0.946	3.9773	0.9684	4.1071
27	12.0	1E-12	8.46E-22	1.21E-17	2.58E-08	0.018	0.982	4.0821	0.9000	4.5357
28	12.5	3.16E-13	8.36E-22	3.86E-19	2.61E-09	5.72E-03	0.994	4.3105	0.6838	6.3040
29	Documentation									
30	Cell B7=10^-A7									
31	Cell C7=B7^4+$B$2*B7^3+$B$2*$B$3*B7^2+$B$2*$B$3*$B$4*B7+$B$2*$B$3*$B$4*$B$5									
32	Cell D7=$B$2*B7^3/C7									
33	Cell E7=$B$2*$B$3*B7^2/C7									
34	Cell F7=$B$2*$B$3*$B$4*B7/C7									
35	Cell G7=$B$2*$B$3*$B$4*$B$5/C7									
36	Cell H7=D7+2*E7+3*F7+4*G7-(B7-($D$2/B7))/$D$3									
37	Cell I7=1+(B7-($D$2/B7))/$D$4									
38	Cell J7=H7/I7									

Figure 9-8 Titration of H_4Y with strong base.

Titrating Metals with EDTA

EDTA invariably forms 1:1 complexes with metals. We will now develop the titration curve for

a metal such as Ca^{2+} with EDTA. In a complexometric titration curve, we can compute either pL

vs. titrant volume (or ϕ) or pM vs. volume. Because electrodes and indicators often respond to

the free metal, a plot of pM vs. volume is most useful. We present both the stoichiometric

approach and the inverse master equation approach using the titration of 50.00 mL of 0.00500 M

Ca^{2+} with 0.0100 M EDTA in a solution buffered to pH 10.0 as our example.

Stoichiometric Approach

In the stoichiometric approach, we divide the titration curve into 3 regions: the preequivalence-point region, the equivalence point, and the postequivalence point region. Our first task is to look up the conditional formation constant of the calcium-EDTA complex at pH 10.0. From Table 17-4 of FAC9 or 15-4 of AC7, the formation constant K_{CaY} is 5.0×10^{10}. The α_4 value from Figure 9-6 is 0.35. Hence, the conditional constant is

$$K'_{CaY} = \frac{[CaY^{2-}]}{[Ca^{2+}]c_T} = \alpha_4 K_{CaY}$$

$$= 0.35 \times 5.0 \times 10^{10} = 1.75 \times 10^{10}$$

The spreadsheet is shown in Figure 9-9. The various constants are entered at the top of the sheet. The volumes for which pCa is to be calculated are entered in cells A5:A19. We'll explain the other entries below.

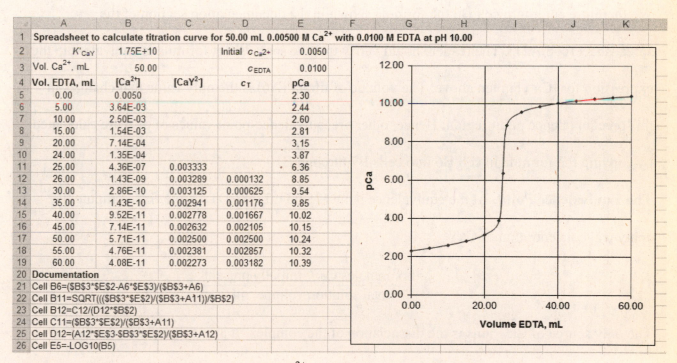

	A	B	C	D	E	F ... K
1	Spreadsheet to calculate titration curve for 50.00 mL 0.00500 M Ca^{2+} with 0.0100 M EDTA at pH 10.00					
2	K'_{CaY}	1.75E+10		Initial c_{Ca2+}	0.0050	
3	Vol. Ca^{2+}, mL	50.00		c_{EDTA}	0.0100	
4	Vol. EDTA, mL	$[Ca^{2+}]$	$[CaY^{2-}]$	c_T	pCa	
5	0.00	0.0050			2.30	
6	5.00	3.64E-03			2.44	
7	10.00	2.50E-03			2.60	
8	15.00	1.54E-03			2.81	
9	20.00	7.14E-04			3.15	
10	24.00	1.35E-04			3.87	
11	25.00	4.36E-07	0.003333		6.36	
12	26.00	1.43E-09	0.003289	0.000132	8.85	
13	30.00	2.86E-10	0.003125	0.000625	9.54	
14	35.00	1.43E-10	0.002941	0.001176	9.85	
15	40.00	9.52E-11	0.002778	0.001667	10.02	
16	45.00	7.14E-11	0.002632	0.002105	10.15	
17	50.00	5.71E-11	0.002500	0.002500	10.24	
18	55.00	4.76E-11	0.002381	0.002857	10.32	
19	60.00	4.08E-11	0.002273	0.003182	10.39	
20	Documentation					
21	Cell B6=($B$3*$E$2-A6*$E$3)/($B$3+A6)					
22	Cell B11=SQRT((($B$3*$E$2)/($B$3+A11))/$B$2)					
23	Cell B12=C12/(D12*$B$2)					
24	Cell C11=($B$3*$E$2)/($B$3+A11)					
25	Cell D12=(A12*$E$3-$B$3*$E$2)/($B$3+A12)					
26	Cell E5=-LOG10(B5)					

Figure 9-9 Spreadsheet for titration of Ca^{2+} with EDTA by stoichiometric approach.

Preequivalence Point Values for pCa. The initial $[Ca^{2+}]$ at 0.00 mL titrant is just the value in cell E2. The initial pCa is calculated from the initial $[Ca^{2+}]$ in the usual manner as shown in the documentation for cell E5 (cell A26). This formula is copied into cells E6 through E19. For the other entries prior to the equivalence point, the equilibrium concentration of Ca^{2+} is equal to the untitrated excess of the cation plus any calcium from the dissociation of the complex, the latter being equal numerically to c_T.

$$[Ca^{2+}] = \frac{\text{no. millimoles } Ca^{2+} \text{ initially present} - \text{no. millimoles EDTA added}}{\text{total solution volume, mL}} + c_T$$

Usually, c_T is small relative to the analytical concentration of the uncomplexed calcium ion. Thus, we can write,

$$[Ca^{2+}] \approx \frac{\text{no. millimoles } Ca^{2+} \text{ initially present} - \text{no. millimoles EDTA added}}{\text{total solution volume, mL}}$$

We therefore enter into cell B6 the formula shown in the documentation section of the spreadsheet (cell A21). The reader should verify that the spreadsheet formula is equivalent to the expression for $[Ca^{2+}]$ given above. The volume of titrant (A6) is the only value that changes in this preequivalence point region. Hence, other preequivalence-point values of pCa are calculated by copying the formula in cell B6 into cells B7 through B10.

The Equivalence Point. At the equivalence point (25.00 mL of EDTA), we first compute the analytical concentration of CaY^{2-}:

$$c_{CaY^{2-}} = \frac{\text{no. mmoles } Ca^{2+} \text{ initially present}}{\text{total solution volume, mL}}$$

The only source of Ca^{2+} ions is the dissociation of the complex. It also follows that the Ca^{2+} concentration must be equal to the sum of the concentrations of the uncomplexed EDTA, c_T. Thus,

$$[Ca^{2+}] = c_T \quad \text{and} \quad [CaY^{2-}] = c_{CaY^{2-}} - [Ca^{2+}] \cong c_{CaY^{2-}}$$

The formula for $[CaY^{2-}]$ is thus entered into cell C11 as shown in the documentation in cell A24.

Again, you should verify the formula and cell references in cell A 24. To obtain $[Ca^{2+}]$, we

substitute into the expression for K'_{CaY},

$$K'_{CaY} = \frac{[CaY^{2-}]}{[Ca^{2+}]c_T} \approx \frac{c_{CaY^{2-}}}{[Ca^{2+}]^2}$$

$$[Ca^{2+}] = \sqrt{\frac{c_{CaY^{2-}}}{K'_{CaY}}}$$

We thus enter into cell B11 the formula corresponding to this expression as shown in cell A22.

Postequivalence Point Region. Beyond the equivalence point, analytical concentrations of

CaY^{2-} and EDTA are obtained directly from the stoichiometric data. Since there is now excess

EDTA, we can write

$$c_{CaY^{2-}} = \frac{\text{no. mmoles } Ca^{2+} \text{ initially present}}{\text{total solution volume, mL}}$$

and

$$c_{EDTA} = \frac{\text{no. mmoles EDTA added} - \text{no. mmoles } Ca^{2+} \text{ initially present}}{\text{total solution volume, mL}}$$

As an approximation,

$$[CaY^{2-}] = c_{CaY^{2-}} - [Ca^{2+}] \approx c_{CaY^{2-}} \approx \frac{\text{no. mmole } Ca^{2+} \text{ initally present}}{\text{total solution volume, mL}}$$

Since this expression is the same as that previously entered into cell C11, we copy that equation

into cell C12. We also note that $[CaY^{2-}]$ will be given by this same expression (with the volume

varied) throughout the remainder of the titration. Hence the formula in cell C12 is copied into

cells C13 through C19.

Also we approximate

$$c_T = c_{EDTA} + [Ca^{2+}] \approx c_{EDTA} = \frac{\text{no. mmoles EDTA added } - \text{ no. mmoles Ca}^{2+} \text{ initially present}}{\text{total solution volume, mL}}$$

We enter this formula into cell D12 as shown in the documentation (cell A25) and copy it into

cells D13 through D16. To calculate $[Ca^{2+}]$, we then substitute into the conditional formation-

constant expression, and obtain

$$K'_{CaY} = \frac{[CaY^{2-}]}{[Ca^{2+}]c_T} \approx \frac{c_{CaY^{2-}}}{[Ca^{2+}]c_{EDTA}}$$

$$[Ca^{2+}] = \frac{c_{CaY^{2-}}}{c_{EDTA}K'_{CaY}}$$

Hence, the $[Ca^{2+}]$ in cell B12 is computed from the values in cells C12 and D12 as shown in cell

A23. We copy this formula into cells B13 through B19, and plot the titration curve shown in

Figure 9-9.

Inverse Master Equation Approach

Since we have a 1:1 complex, the master equation is given by

$$\phi = \frac{V_L c_L}{V_M c_M} = \frac{\alpha_{ML} + \dfrac{[L]}{c_M}}{1 - \dfrac{[L]}{c_L}}$$

You can show that the master equation for a 1:1 complex can also be written in terms of the

equilibrium concentration of metal ion as follows:

$$\phi = \frac{1 - \dfrac{[M]}{c_M}}{1 - \alpha_L + \dfrac{[M]}{c_L}}$$

where

$$\alpha_L = \frac{[L]}{c_T} = \frac{1}{1 + K'_{CaY}[Ca^{2+}]}$$

Here, c_T is the molar analytical concentration of ligand, and α_L is the fraction of *uncomplexed*

ligand.

The spreadsheet for the titration of Ca^{2+} with EDTA is shown in Figure 9-10. Note that

the equivalence point pCa^{2+} is approximately the same as that found by the stoichiometric

approach of Figure 9-9 (6.36).

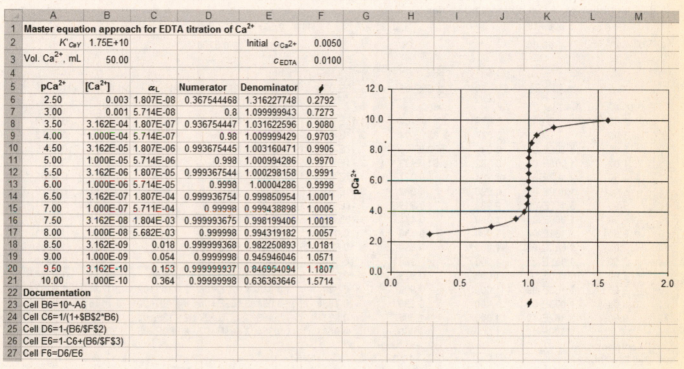

	A	B	C	D	E	F
1	Master equation approach for EDTA titration of Ca^{2+}					
2	K'_{CaY}	1.75E+10			Initial c_{Ca2+}	0.0050
3	Vol. Ca^{2+}, mL	50.00			c_{EDTA}	0.0100
4						
5	pCa^{2+}	$[Ca^{2+}]$	α_L	Numerator	Denominator	ϕ
6	2.50	0.003	1.807E-08	0.367544468	1.316227748	0.2792
7	3.00	0.001	5.714E-08	0.8	1.099999943	0.7273
8	3.50	3.162E-04	1.807E-07	0.936754447	1.031622596	0.9080
9	4.00	1.000E-04	5.714E-07	0.98	1.009999429	0.9703
10	4.50	3.162E-05	1.807E-06	0.993675445	1.003160471	0.9905
11	5.00	1.000E-05	5.714E-06	0.998	1.000994286	0.9970
12	5.50	3.162E-06	1.807E-05	0.999367544	1.000298158	0.9991
13	6.00	1.000E-06	5.714E-05	0.9998	1.00004286	0.9998
14	6.50	3.162E-07	1.807E-04	0.999936754	0.999850954	1.0001
15	7.00	1.000E-07	5.711E-04	0.99998	0.999438898	1.0005
16	7.50	3.162E-08	1.804E-03	0.999993675	0.998199406	1.0018
17	8.00	1.000E-08	5.682E-03	0.999998	0.994319182	1.0057
18	8.50	3.162E-09	0.018	0.999999368	0.982250893	1.0181
19	9.00	1.000E-09	0.054	0.9999998	0.945946046	1.0571
20	9.50	3.162E-10	0.153	0.999999937	0.846954094	1.1807
21	10.00	1.000E-10	0.364	0.99999998	0.636363646	1.5714
22	Documentation					
23	Cell B6=10^-A6					
24	Cell C6=1/(1+$B$2*B6)					
25	Cell D6=1-(B6/$F$2)					
26	Cell E6=1-C6+(B6/$F$3)					
27	Cell F6=D6/E6					

Figure 9-10 Inverse master equation approach for EDTA titration of Ca^{2+}.

Note also how much easier it is to do the calculations by the inverse master equation approach.

Next we'll see what effect pH has on the shape and end point of the titration curve by

computing the titration curves for pH values of 4.0, 6.0, 8.0, and 10.0 and plotting them on the

same chart. The setup will be a little different from that in Figure 9-10 since α_L changes with pH.

We enter in the spreadsheet the formation constant K_{CaY} and compute the conditional constant in

the calculation of α_L. The spreadsheet and chart are shown in Figure 9-11. Note that at pH 4.0, no discernible break occurs at the stoichiometric value ($\phi = 1.0$). At pH 6.0, a small break occurs.

Only pH 8.0 and 10.0 give satisfactory breaks at the equivalence point.

As an extension, look up the formation constants of the EDTA complexes for some other

metals, such as Zn^{2+}, Mg^{2+}, and Fe^{2+}. Compute the titration curves. A considerably more

challenging exercise is to plot these titration curves for the situation in which there is an auxiliary

complexing agent such as ammonia present.

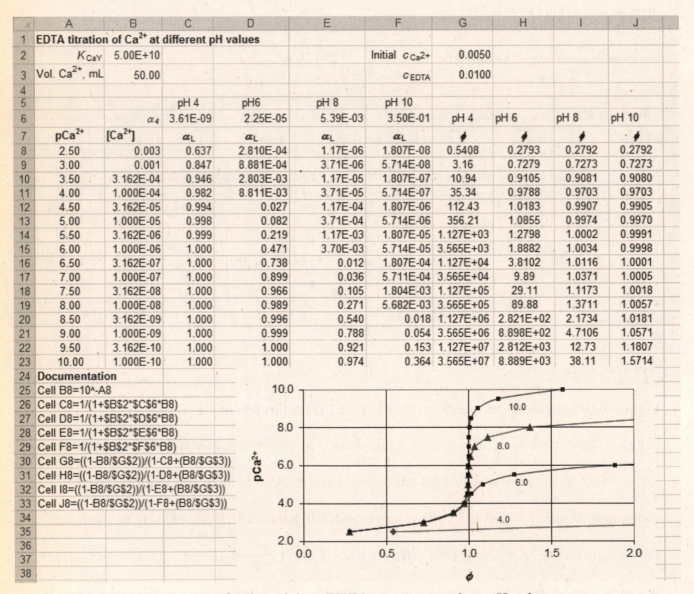

	A	B	C	D	E	F	G	H	I	J
1	EDTA titration of Ca^{2+} at different pH values									
2	K_{CaY}	5.00E+10				Initial c_{Ca2+}	0.0050			
3	Vol. Ca^{2+}, mL	50.00				c_{EDTA}	0.0100			
4										
5			pH 4	pH6	pH 8	pH 10				
6		α_4	3.61E-09	2.25E-05	5.39E-03	3.50E-01	pH 4	pH 6	pH 8	pH 10
7	pCa^{2+}	$[Ca^{2+}]$	α_L	α_L	α_L	α_L	ϕ	ϕ	ϕ	ϕ
8	2.50	0.003	0.637	2.810E-04	1.17E-06	1.807E-08	0.5408	0.2793	0.2792	0.2792
9	3.00	0.001	0.847	8.881E-04	3.71E-06	5.714E-08	3.16	0.7279	0.7273	0.7273
10	3.50	3.162E-04	0.946	2.803E-03	1.17E-05	1.807E-07	10.94	0.9105	0.9081	0.9080
11	4.00	1.000E-04	0.982	8.811E-03	3.71E-05	5.714E-07	35.34	0.9788	0.9703	0.9703
12	4.50	3.162E-05	0.994	0.027	1.17E-04	1.807E-06	112.43	1.0183	0.9907	0.9905
13	5.00	1.000E-05	0.998	0.082	3.71E-04	5.714E-06	356.21	1.0855	0.9974	0.9970
14	5.50	3.162E-06	0.999	0.219	1.17E-03	1.807E-05	1.127E+03	1.2798	1.0002	0.9991
15	6.00	1.000E-06	1.000	0.471	3.70E-03	5.714E-05	3.565E+03	1.8882	1.0034	0.9998
16	6.50	3.162E-07	1.000	0.738	0.012	1.807E-04	1.127E+04	3.8102	1.0116	1.0001
17	7.00	1.000E-07	1.000	0.899	0.036	5.711E-04	3.565E+04	9.89	1.0371	1.0005
18	7.50	3.162E-08	1.000	0.966	0.105	1.804E-03	1.127E+05	29.11	1.1173	1.0018
19	8.00	1.000E-08	1.000	0.989	0.271	5.682E-03	3.565E+05	89.88	1.3711	1.0057
20	8.50	3.162E-09	1.000	0.996	0.540	0.018	1.127E+06	2.821E+02	2.1734	1.0181
21	9.00	1.000E-09	1.000	0.999	0.788	0.054	3.565E+06	8.898E+02	4.7106	1.0571
22	9.50	3.162E-10	1.000	1.000	0.921	0.153	1.127E+07	2.812E+03	12.73	1.1807
23	10.00	1.000E-10	1.000	1.000	0.974	0.364	3.565E+07	8.889E+03	38.11	1.5714
24	Documentation									
25	Cell B8=10^-A8									
26	Cell C8=1/(1+$B$2*$C$6*B8)									
27	Cell D8=1/(1+$B$2*$D$6*B8)									
28	Cell E8=1/(1+$B$2*$E$6*B8)									
29	Cell F8=1/(1+$B$2*$F$6*B8)									
30	Cell G8=((1-B8/$G$2))/(1-C8+(B8/$G$3))									
31	Cell H8=((1-B8/$G$2))/(1-D8+(B8/$G$3))									
32	Cell I8=((1-B8/$G$2))/(1-E8+(B8/$G$3))									
33	Cell J8=((1-B8/$G$2))/(1-F8+(B8/$G$3))									
34										
35										
36										
37										
38										

Figure 9-11 Titration curves for the calcium-EDTA system at various pH values.

Precipitation Titrations

In Example 17-1 of FAC9 and Example 15-2 of AC7 we constructed a curve for the titration of

$NaCl$ with $AgNO_3$. In some ways, precipitation titrations are similar to the acid-base titrations

discussed in Chapter 7. Like we did there, we split the titration curve into three parts: points prior

to the equivalence point, the equivalence point, and points beyond the equivalence point. In each

of these regions of the curve, we made certain assumptions regarding relative concentrations of

solution species to simplify the calculations. Here, we construct titration curves for this same

example by three different methods. We first use a stoichiometric method, identical to that used

in Example 17-1. In the second method, we use a single master equation and the quadratic

formula to obtain the pAg for different known volumes. The last method uses a single equation

and inverts the problem to have Excel solve for the volume of reagent required to produce given

values of pAg and pCl.

Stoichiometric Method

The titration reaction is:

$$Ag^+ + Cl^- \rightarrow AgCl(s)$$

Let's begin by labeling a blank worksheet and entering the known quantities as shown in Figure

9-12. We'll obtain the pAg values for each volume of $AgNO_3$ shown in cells A8 through A28.

	A	B	C	D
1	Precipitation Titration, Stoichiometric Method			
2	K_{sp}	1.82E-10		
3	V_{NaCl}	50.00		
4	c_{NaCl}	0.0500		
5	c_{AgNO3}	0.1000		
6				
7	V_{AgNO3}	[Cl$^-$]	[Ag$^+$]	pAg
8	1.00			
9	5.00			
10	10.00			
11	15.00			
12	20.00			
13	22.00			
14	23.00			
15	24.00			
16	24.50			
17	24.75			
18	24.90			
19	25.00			
20	25.10			
21	25.25			
22	25.50			
23	26.00			
24	27.00			
25	28.00			
26	30.00			
27	35.00			
28	40.00			

Figure 9-12 Spreadsheet for precipitation titration.

In the pre-equivalence point region, we haven't added enough $AgNO_3$ to react with all of the NaCl and hence have unreacted Cl$^-$. In this region, it is relatively easy to find the [Cl$^-$] from the stoichiometry and then the [Ag$^+$] from the solubility product. From mass balance considerations, the total solution chloride ion concentration comes from unreacted NaCl and any AgCl that dissociates. Except for volumes extremely close to the equivalence point (e.g., within 0.01 mL), it is reasonable to assume that [Cl$^-$] from this latter source is negligible. Hence, we can write

$$[\text{Cl}^-] \approx c_{\text{NaCl}} = \frac{\text{original no. mmol NaCl present} - \text{no. mmol } AgNO_3 \text{ added}}{\text{total volume of solution, mL}}$$

For the first volume of 1.00 mL of titrant, we can enter into cell B8 the formula corresponding to this equation:

$$\texttt{=(\$B\$3*\$B\$4-\$B\$5*A8)/(\$B\$3+A8)} [\hookleftarrow]$$

Note here that the product of cells B3 and B4 is the original number of millimoles of chloride present. Since this is fixed, we make the cell references absolute references by adding a dollar sign (**$**) before the letter and number. The product of cells B5 and A8 is the number of millimoles of titrant added, while the denominator is the total volume of solution after the addition. Note that since the volume of $AgNO_3$ is variable, we make this cell reference a relative reference by not adding the dollar sign. As we fill in the entries for other volumes, this reference will change.

In cell C8, we can calculate the $[Ag^+]$ from K_{sp} by

$$[Ag^+] = \frac{K_{sp}}{[Cl^-]}$$

We then enter the formula **=$B$2/B8[↵]** in cell C8. In cell D8, we calculate pAg from $pAg = -\log [Cl^-]$. Enter into cell D8, the formula **=-LOG(C8)[↵]**. Now, fill in the remaining calculations for volumes prior to the equivalence point by selecting cells B8:D8 and using the fill handle to complete the entries. Your worksheet should then appear as shown in Figure 9-13.

	A	B	C	D
1	**Precipitation Titration, Stoichiometric Method**			
2	K_{sp}	1.82E-10		
3	V_{NaCl}	50.00		
4	c_{NaCl}	0.0500		
5	c_{AgNO3}	0.1000		
6				
7	V_{AgNO3}	[Cl⁻]	[Ag⁺]	pAg
8	1.00	4.71E-02	3.87E-09	8.41
9	5.00	3.64E-02	5.01E-09	8.30
10	10.00	2.50E-02	7.28E-09	8.14
11	15.00	1.54E-02	1.18E-08	7.93
12	20.00	7.14E-03	2.55E-08	7.59
13	22.00	4.17E-03	4.37E-08	7.36
14	23.00	2.74E-03	6.64E-08	7.18
15	24.00	1.35E-03	1.35E-07	6.87
16	24.50	6.71E-04	2.71E-07	6.57
17	24.75	3.34E-04	5.44E-07	6.26
18	24.90	1.34E-04	1.36E-06	5.87
19	25.00			

Figure 9-13 Worksheet completing the pre-equivalance point calculations.

At the equivalence point, we've added enough $AgNO_3$ to react completely with the NaCl initially present. The only Cl^- comes from the dissociation of AgCl. Hence, at the equivalence point $[Ag^+] = [Cl^-]$ and

$$[Ag^+] = \sqrt{K_{sp}}$$

In cell C19, we thus enter the formula **=SQRT (B2) [↵]**.

In the post-equivalence point region, we calculate the $[Ag^+]$ from the excess $AgNO_3$ added. Except for the region very near the equivalence point (0.01 mL), we can neglect any Ag^+ from the dissociation of AgCl. Thus,

$$[Ag^+] \approx c_{AgNO_3} = \frac{\text{no. of mmol } AgNO_3 \text{ added} - \text{original no. of mmol NaCl present}}{\text{total volume of solution, mL}}$$

In cell C20, we enter the formula corresponding to this equation shown in the documentation

section. Use the fill handle the complete the entries for column D. Your worksheet should appear

as shown in Figure 9-14 after completing the documentation.

	A	B	C	D
1	**Precipitation Titration, Stoichiometric Method**			
2	K_{sp}	1.82E-10		
3	V_{NaCl}	50.00		
4	c_{NaCl}	0.0500		
5	c_{AgNO3}	0.1000		
6				
7	V_{AgNO3}	[Cl⁻]	[Ag⁺]	pAg
8	1.00	4.71E-02	3.87E-09	8.41
9	5.00	3.64E-02	5.01E-09	8.30
10	10.00	2.50E-02	7.28E-09	8.14
11	15.00	1.54E-02	1.18E-08	7.93
12	20.00	7.14E-03	2.55E-08	7.59
13	22.00	4.17E-03	4.37E-08	7.36
14	23.00	2.74E-03	6.64E-08	7.18
15	24.00	1.35E-03	1.35E-07	6.87
16	24.50	6.71E-04	2.71E-07	6.57
17	24.75	3.34E-04	5.44E-07	6.26
18	24.90	1.34E-04	1.36E-06	5.87
19	25.00		1.35E-05	4.87
20	25.10		1.33E-04	3.88
21	25.25		3.32E-04	3.48
22	25.50		6.62E-04	3.18
23	26.00		1.32E-03	2.88
24	27.00		2.60E-03	2.59
25	28.00		3.85E-03	2.41
26	30.00		6.25E-03	2.20
27	35.00		1.18E-02	1.93
28	40.00		1.67E-02	1.78
29	**Documentation**			
30	Cell B8=($B$3*$B$4-$B$5*A8)/($B$3+A8)			
31	Cell C8=$B$2/B8			
32	Cell D8=-LOG(C8)			
33	Cell C19=SQRT(B2)			
34	Cell C20=($B$5*A20-$B$3*$B$4)/($B$3+A20)			

Figure 9-14 Final worksheet for calculating pAg for varies volumes of $AgNO_3$ added.

Now we're ready to plot the data. Select cells A8:A28 and D8:D28. Select an

XY(Scatter) plot. Format the chart and locate it as a new sheet. Add a smooth line so that the

titration curve appears as shown in Figure 9-15.

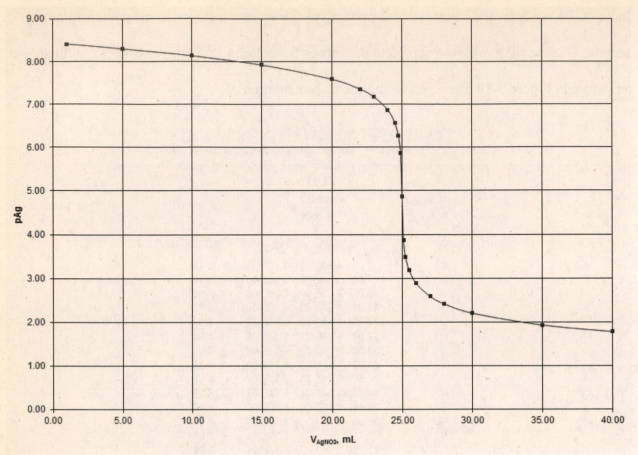

Figure 9-15 A plot of the precipitation titration curve.

Direct Master Equation Approach

In our second approach, a master equation is written to find $[Ag^+]$ throughout the titration curve.

The equation is a quadratic, but as we know from Chapter 5, quadratic equations are easily

solved with Excel. We will consider again Example 17-1 of FAC9 and Example 15-2 of AC7.

For this example, we can write the charge balance expression as

$$[Na^+] + [Ag^+] = [Cl^-] + [NO_3^-]$$

We can express the concentrations of Na^+ and NO_3^- by

$$[Na^+] = \frac{c_{NaCl}V_{NaCl}}{V_{NaCl} + V_{AgNO_3}} \text{ and } [NO_3^-] = \frac{c_{AgNO_3}V_{AgNO_3}}{V_{NaCl} + V_{AgNO_3}}$$

Substituting these into the mass balance expression and expressing the chloride ion concentration

by $K_{sp}/[Ag^+]$, we can obtain, after some algebraic manipulation, the following quadratic equation

in terms of the single unknown quantity, $[Ag^+]$

$$(V_{NaCl}+V_{AgNO_3})[Ag^+]^2 + (c_{NaCl}V_{NaCl}-c_{AgNO_3}V_{AgNO_3})[Ag^+] - (V_{NaCl}+V_{AgNO_3})K_{sp} = 0$$

This equation can be solved with the aid of our quadratic equation solver (Figure 5-2).

Your final worksheet after documentation should appear as shown in Figure 9-16. Note that the

values calculated are the same, within the number of figures shown, as those from the

stoichiometric method. One advantage of the quadratic equation approach is that it can be used

very close to the equivalence point since it is an exact solution. You can try this by inserting the

volumes 24.99 mL and 24.999 mL prior to the equivalence point. These should return the values

5.08 and 4.89 for pAg. You should also note that we could obtain these results using Excel's

Solver as was done in Chapter 5 or Excel's Goal Seek as was done in Chapter 7.

Finally, use the results of Figure 9-16 and plot the titration curve as before. The curve

should be identical to Figure 9-15.

	A	B	C	D	E	F
1	Precipitation Titration, Master Equation Approach					
2	K_{sp}	1.82E-10				
3	V_{NaCl}	50.00				
4	c_{NaCl}	0.0500				
5	c_{AgNO3}	0.1000				
6						
7	a	b	c	V_{AgNO3}	$[Ag^+]$	pAg
8	51.00	2.4	-9.28E-09	1.00	3.87E-09	8.41
9	55.00	2	-1.00E-08	5.00	5.00E-09	8.30
10	60.00	1.5	-1.09E-08	10.00	7.28E-09	8.14
11	65.00	1	-1.18E-08	15.00	1.18E-08	7.93
12	70.00	0.5	-1.27E-08	20.00	2.55E-08	7.59
13	72.00	0.3	-1.31E-08	22.00	4.37E-08	7.36
14	73.00	0.2	-1.33E-08	23.00	6.64E-08	7.18
15	74.00	0.1	-1.35E-08	24.00	1.35E-07	6.87
16	74.50	0.05	-1.36E-08	24.50	2.71E-07	6.57
17	74.75	0.025	-1.36E-08	24.75	5.43E-07	6.26
18	74.90	0.01	-1.36E-08	24.90	1.35E-06	5.87
19	75.00	0	-1.37E-08	25.00	1.35E-05	4.87
20	75.10	-0.01	-1.37E-08	25.10	1.35E-04	3.87
21	75.25	-0.025	-1.37E-08	25.25	3.33E-04	3.48
22	75.50	-0.05	-1.37E-08	25.50	6.63E-04	3.18
23	76.00	-0.1	-1.38E-08	26.00	1.32E-03	2.88
24	77.00	-0.2	4.19E-09	27.00	2.60E-03	2.59
25	78.00	-0.3	-1.42E-08	28.00	3.85E-03	2.41
26	80.00	-0.5	-1.46E-08	30.00	6.25E-03	2.20
27	85.00	-1	-1.55E-08	35.00	1.18E-02	1.93
28	90.00	-1.5	-1.64E-08	40.00	1.67E-02	1.78
29	Documentation					
30	Cell A8=$B$3+D8					
31	Cell B8=$B$3*$B$4-$B$5*D8					
32	Cell C8=-$B$2*($B$3+D8)					
33	Cell E8=(-B8+SQRT(B8^2-4*A8*C8))/(2*A8)					
34	Cell F8=-LOG(E8)					

Figure 9-16 Final worksheet for master equations approach.

Calculation of Volumes Needed to Produce a Given pAg

In this last method, we invert the problem and calculate the volume of reagent needed to produce

a given pAg value. This is sometimes called the inverse, master equation approach.

We begin with the charge balance equation as before. After substituting values for $[Na^+]$,

$[NO_3^-]$ and $[Cl^-]$, we obtain

$$\frac{c_{NaCl}V_{NaCl}}{V_{NaCl}+V_{AgNO_3}} + [Ag^+] = \frac{K_{sp}}{[Ag^+]} + \frac{c_{AgNO_3}V_{AgNO_3}}{V_{NaCl}+V_{AgNO_3}}$$

Solving this equation for V_{AgNO_3} yields

$$V_{AgNO_3} = \frac{\frac{K_{sp}}{[Ag^+]}V_{NaCl} - c_{NaCl}V_{NaCl} - [Ag^+]V_{NaCl}}{[Ag^+] - \frac{K_{sp}}{[Ag^+]} - c_{AgNO_3}}$$

This is now a master equation to calculate the volume needed to achieve a given $[Ag^+]$.

We can now enter values for pAg, calculate $[Ag^+]$, and find the volumes needed to obtain these

values since all other values are constant in the problem. Begin with a worksheet as shown in

Figure 9-17. To calculate $[Ag^+]$ enter into cell B8 the formula =10^-A8.

	A	B	C	D
1	**Precipitation Titration, Inverse Method**			
2	K_{sp}	1.82E-10		
3	V_{NaCl}	50.00		
4	c_{NaCl}	0.0500		
5	c_{AgNO3}	0.1000		
6				
7	pAg	$[Ag^+]$	V	
8	1.8			
9	2.0			
10	2.5			
11	3.0			
12	3.5			
13	4.0			
14	4.5			
15	5.0			
16	5.5			
17	6.0			
18	6.5			
19	7.0			
20	7.5			
21	8.0			
22	8.2			
23	8.4			

Figure 9-17 Worksheet for titration curve by the inverse master equation method.

In cell C8 enter the formula corresponding to the equation for V_{AgNO_3}. Select cells B8:C8 and

drag the fill handle to complete the calculations. Your final worksheet should appear as shown in

Figure 9-18 after adding documentation and inserting a chart of the data.

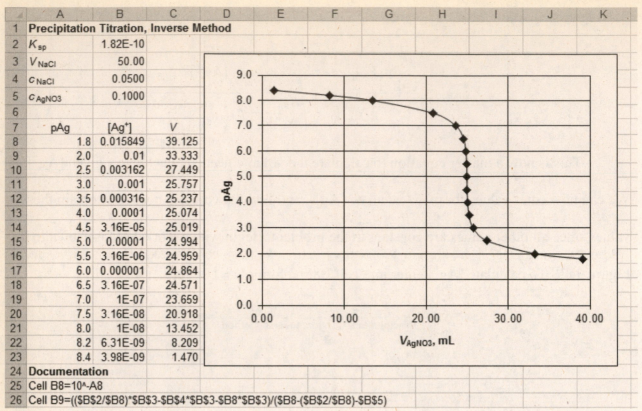

	A	B	C	D	E	F	G	H	I	J	K
1	**Precipitation Titration, Inverse Method**										
2	K_{sp}	1.82E-10									
3	V_{NaCl}	50.00									
4	c_{NaCl}	0.0500									
5	c_{AgNO3}	0.1000									
6											
7	pAg	[Ag⁺]	V								
8	1.8	0.015849	39.125								
9	2.0	0.01	33.333								
10	2.5	0.003162	27.449								
11	3.0	0.001	25.757								
12	3.5	0.000316	25.237								
13	4.0	0.0001	25.074								
14	4.5	3.16E-05	25.019								
15	5.0	0.00001	24.994								
16	5.5	3.16E-06	24.959								
17	6.0	0.000001	24.864								
18	6.5	3.16E-07	24.571								
19	7.0	1E-07	23.659								
20	7.5	3.16E-08	20.918								
21	8.0	1E-08	13.452								
22	8.2	6.31E-09	8.209								
23	8.4	3.98E-09	1.470								
24	**Documentation**										
25	Cell B8=10^-A8										
26	Cell B9=(($B$2/$B8)*$B$3-$B$4*$B$3-$B8*$B$3)/($B8-($B$2/$B8)-$B$5)										

Figure 9-18 Final worksheet for inverse method showing documentation and plot of the data.

As a further exercise, you may wish to calculate titration curves for the silver/bromide system ($K_{sp} = 5.2 \times 10^{-13}$) or another precipitation titration of interest. As an added challenge, have Excel calculate titration curves for two different concentrations of $AgNO_3$ as in FAC9, Figure 17-2 or AC7 Figure 15-2.

Summary

In this chapter we have seen how to calculate the distribution of species in solutions containing metals and ligands. We have also explored the computation of titration curves and the influence of pH on the titration of Ca^{2+} with EDTA. Once again we have seen the power and efficiency of Excel in making these calculations. The computation of α values has again given us a very useful tool in the visualization of solution equilibria. Finally, we have constructed titration curves for precipitation titrations by three different methods.

Problems

1.   The four complexes of Cd^{2+} with CN^- have stepwise formation constants of $K_1 = 3.02 \times 10^5$, $K_2 = 1.32 \times 10^5$, $K_3 = 4.27 \times 10^4$, $K_4 = 3.54 \times 10^3$. Prepare a distribution diagram that ranges from $pCN^- = 1.0$ to 8.0 in steps of 0.5 units. Label each of the species. Determine from the chart the minimum pCN^- that gives 99% $Cd(CN)_4^{2-}$?

2.   There are six complexes of Ni^{2+} with NH_3. The logarithms of the overall formation constants are $\log \beta_1 = 2.80$, $\log \beta_2 = 5.04$, $\log \beta_3 = 6.77$, $\log \beta_4 = 7.96$, $\log \beta_6 = 8.71$, and $\log \beta_7 = 8.74$. Prepare a distribution diagram from $pNH_3 = -1$ to 5 in steps of 0.5 units. Label each species. What pNH_3 value maximizes the $Ni(NH_3)_3^{2+}$ concentration?

3.   For the Cd^{2+}/CN^- system in Problem 1 above, prepare a sum plot similar to Figure 9-3 showing the α_4, the sum of the α_4 and α_3, up to the fraction representing the sum of all complexes.

4.   For the Ni/NH_3 system in Problem 2 above, prepare a sum plot similar to Figure 9-3 showing the $Ni(NH_3)_6^{2+}$ fraction, the sum of the $Ni(NH_3)_6^{2+}$ and $Ni(NH_3)_5^{2+}$ fractions, up to the fraction representing the sum of all complexes. Label the plot.

5.   Use the inverse master equation approach to prepare a titration curve for the titration of 0.0900 M Cd^{2+} with 0.2500 M I^-. Do you see any breaks in the titration curve? Would a titration be successful?

6.   Use the inverse master equation approach to prepare a titration curve for the titration of 0.1250 M Cd^{2+} with 0.3500 M CN^-. Use the formation constants from Problem 1 above. Are there discernible breaks in the curve? Would a titration be successful?

7.   Triethylenetetramine (trien, T) is a quadridentate ligand and can protonate to form HT^+, H_2T^{2+}, H_3T^{3+} and H_4T^{4+}. The pK values for the dissociation of these protonated forms are p$K_1 = 3.32$, p$K_2 = 6.67$, p$K_3 = 9.20$ and p$K_4 = 9.92$. The pK_1 value refers to the loss of a proton from H_4T^{4+}. Prepare a spreadsheet to calculate the α values for the various protonated and unprotonated forms of trien. Plot the distribution of these species versus pH from pH 0.0 to 14.0 in 0.5 pH increments. Label the plot with the appropriate species.

8.   Copper(II) forms a strong complex with triethylenetetramine (trien, T). The formation constant of the 1:1 complex with the unprotonated T molecule is 2.51×10^{20}. Calculate the conditional formation constant for the Cu(II)/T complex every 0.5 pH units from pH 0.0 to 14.0. For what range of pH values is the conditional formation constant at least 10^{10}?

9.   Calculate conditional constants for the formation of the EDTA complex of Fe^{2+} every 0.5 pH units from pH 4.0 to 10.0. What pH range will give a conditional formation constant of at least 10^{10}?

10.   Calculate conditional constants for the formation of the EDTA complex of Ba^{2+} every 0.5 pH units from pH 4.0 to 10.0. What pH range will give a conditional formation constant of 10^4 or lower?

11.   Use the stoichiometric approach and construct a titration curve for 50.00 mL of 0.01000 M Sr^{2+} titrated with 0.02000 M EDTA in a solution buffered to pH 11.0. Calculate pSr values after the addition of 0.00, 10.00, 24.00, 24.90, 25.00, 25.10, 26.00, and 30.00 mL of titrant.

12.   Use the stoichiometric approach and construct a titration curve for 50.00 mL of 0.0150 M

Fe^{2+} titrated with 0.0300 M EDTA in a solution buffered to pH 7.0. Calculate pFe values

after the addition of 0.00, 10.00, 24.00, 24.90, 25.00, 25.10, 26.00, and 30.00 mL of

titrant.

13.   Use the inverse master equation approach and construct titration curves for 25.00 mL of

0.0100 M Sr^{2+} titrated with 0.0150 M EDTA in solutions buffered at pH. Plot the curves

together as in Figure 9-11.

(a)   6.0   (c)   10.0

(b)   8.0   (d)   12.0

14.   Use the inverse master equation approach and construct titration curves for 35.00 mL of

0.0200 M Fe^{2+} titrated with 0.015 M EDTA in a solution buffered at pH 7.5. Plot the

curves on the same graph as was done in Figure 9-11.

(a)   5.0   (c)   9.0

(b)   7.0   (d)   11.0

15.   Plot the titration curve for the titration of 50.00 mL of 0.0500 NaBr with 0.1000 M

AgNO$_3$. Use (a) the stoichiometric method, (b) the direct master equation approach, and

(c) the inverse master equation approach.

16.   Plot the titration curve for the titration of 50.00 mL of 0.050M NaI with 0.1000 M

AgNO$_3$. Use (a) the stoichiometric method, (b) the direct master equation approach, and

(c) the inverse master equation approach.

17.   Calculate the silver ion concentration after the addition of 5.00, 10.00, 15.00, 20.00,

25.00, 30.00, 35.00, 39.00, 39.50, 39.90, 39.99, 40.00, 40.01, 40.10, 40.50, 41.00, 45.00,

and 50.00 mL of 0.05000 M AgNO$_3$ to 50.0 mL of 0.0400 M KSCN. Construct a titration curve from these data plotting pAg as a function of titrant volume. Use the stoichiometric

approach.

18.   Repeat the calculations in Problem 17 using the direct master equation approach.

Chapter 10

Potentiometry and Redox Titrations

In this chapter we explore the use of Excel for calculating electrode potentials and applying the

mathematical expressions used in redox equilibria. We'll also discussoxidation-reduction

titrations. For calculating and displaying titration curves, we'll use both the stoichiometric

approach and the master equation approach. The topics discussed here are covered in Chapters

18-21 of FAC9 and Chapters 16-19 of AC7.

Calculating Electrode Potentials

We begin by writing a general redox half reaction and the corresponding Nernst equation.

$$O + ne^- \rightleftharpoons R$$

$$E = E^0 - \frac{RT}{nF} \ln \frac{[R]}{[O]} = E^0 - \frac{2.303RT}{nF} \log \frac{[R]}{[O]}$$

At 25° C, the factor $2.303RT/F = 0.0592$ so that the Nernst equation becomes

$$E = E^0 - \frac{0.0592}{n} \log \frac{[R]}{[O]}$$

We first take the case of a platinum electrode immersed in a solution containing Sn^{2+} and

Sn^{4+} and find the potential for different ratios of [R]/[O]. Set up a spreadsheet like that of Figure

10-1. Write formulas to calculate [R]/[O], log([R]/[O]) and electrode potential E in columns C-E.

Plot E vs. [R]/[O] and E vs. log([R]/[O]). Determine the slope and intercept of the plot of E vs.

log([R]/[O]). Your plots should appear as shown in Figure 10-2. What is the slope of the plot?

How does it relate to the Nernst factor $2.303RT/nF$? What is the y-intercept and what is its

significance? How might you determine the standard potential E^0 from such a plot? How much

does the potential change for a one decade change in [R]/[O]? Explain chemically from the half reaction and LeChatlier's principle, why the electrode potential is less positive as [R]/[O]

increases.

	A	B	C	D	E
1	**Half cell potential**				
2	R=	Sn^{2+}			
3	O=	Sn^{4+}			
4	n=	2			
5	E^0=	0.154			
6	[R]	[O]	[R]/[O]	log([R]/[O])	E
7	1.00E-05				
8	2.50E-05				
9	5.00E-05				
10	7.50E-05				
11	1.00E-04				
12	2.50E-04				
13	5.00E-04				
14	7.50E-04				
15	0.0010				
16	0.0025				
17	0.0050				
18	0.0075				
19	0.010				
20	0.025				
21	0.050				
22	0.075				
23	0.10				
24	0.25				
25	0.50				
26	0.75				
27	1.00				

Figure 10-1 Spreadsheet setup for calculating electrode potential.

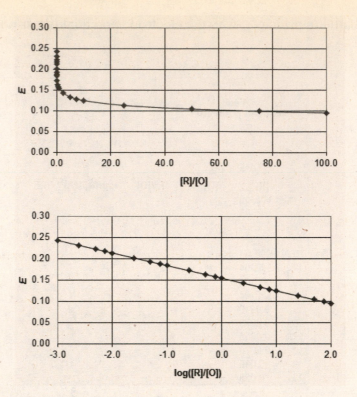

Figure 10-2 Plots of electrode potential vs. [R]/[O] and log([R]/[O]).

As an extension to this exercise, change the spreadsheet so that it describes the Fe^{3+}/Fe^{2+}

couple and repeat the calculations. Modify the spreadsheet to calculate electrode potentials for

metal ion/metal systems, such as Cu^{2+}/Cu or Zn^{2+}/Zn. For these systems, remember that the

activity of the metal is unity since it is in its standard state.

Cell Potentials and Equilibrium Constants

We can also use Excel to calculate cell potentials and equilibrium constants. For simple

processes involving only a single oxidant and reductant, a general spreadsheet such as that

shown in Figure 10-3 can be constructed for the calculation. The figure shows the results of

calculations for the cell given in Example 19-1 of FAC9 and Example 17-1 of AC7:

$$Cu \,|\, Cu^{2+}(0.0200 \text{ M})\,||\, Ag^{+}(0.0200 \text{ M})\,|\, Ag$$

This approach works only for simple reactions and not for those involving protons, hydroxide or

extra species other than O and R. The user enters oxidant and reductant concentrations, numbers of electrons, and standard electrode potentials, E^0 values. The program calculates E_{left}, E_{right}, E^0_{cell}, $\log K_{eq}$ and K_{eq}. You can fill in values for the other entries. Be sure to use 1 for the activity of solids or pure liquids. If you wish, include a column to calculate ΔG^0 for the cell reaction. Write down the specific reactions for each of the equilibrium constants.

It is more difficult to generalize about reactions involving extra species since many of these are individual situations. However, several reactions are of the type $O + nH^+ + ne^- \rightleftharpoons R$. Hence, you could modify the spreadsheet of Figure 10-3 to include reactions of this type.

	A	B	C	D	E	F	G	H	I	J	K	L	M	N	O	P
1	Cell potentials and equilibrium constants for simple reactions															
2	Reactions are O + ne- → R															
3	no protons or extra species															
4																
5																
6	Cell	$[R]_{left}$	$[O]_{left}$	E^0_{left}	n_{left}	E_{left}	$[O]_{right}$	$[R]_{right}$	E^0_{right}	n_{right}	E_{right}	E_{cell}	E^0_{cell}	n_{cell}	$\log K_{eq}$	K_{eq}
7	Cu/Cu2+//Ag+/Ag	1	0.0200	0.337	2	0.2867	0.0200	1	0.799	1	0.6984	0.412	0.462	2	15.61	4.06E+15
8	Pb/Pb2+//Cd2+/Cd			-0.126					-0.403				-0.277	2	-9.36	4.38E-10
9	Zn/Zn2+//Co2+/Co			-0.763					-0.277				0.486	2	16.42	2.62E+16
10	Pt/Fe3+,Fe2+//Hg2+/Hg			0.771					0.854				0.083	2	2.80	6.37E+02
11	Al/Al3+//Cu2+/Cu			-1.662					0.337				1.999	6	202.60	3.99E+202
12																
13	Documentation															
14	Cell F7=D7-(0.0592/E7)*LOG(B7/C7)															
15	Cell K7=I7-(0.0592/J7)*LOG(H7/G7)															
16	Cell L7=K7-F7															
17	Cell M7=I7-D7															
18	Cell O7=N7*M7/0.0592															
19	Cell P7=10^O7															
20	User enters values in cells B7:E7, G7:J7, and N7															

Figure 10-3 Spreadsheet to calculate cell potentials and equilibrium constants.

α-Values for Redox Species

The α values that we used for acid-base and complexation equilibria are also useful in redox equilibria. To calculate redox α's, we must solve the Nernst equation for the ratio of the concentration of the reduced species to the oxidized species. Here, we will use an approach similar to that of de Levie.[1] Since,

$$E = E^0 - \frac{2.303RT}{nF} \log \frac{[R]}{[O]}$$

---

[1] R. de Levie, *J. Electroanal. Chem.*, **1992**, *323*, 347-355.

we can write,

$$\frac{[R]}{[O]} = 10^{-\frac{nF(E-E^0)}{2.303RT}} = 10^{-nf(E-E^0)}$$

where at 25° C,

$$f = \frac{F}{2.303RT} = \frac{1}{0.0592}$$

We can find the fractions α of the total $[R] + [O]$ as follows:

$$\alpha_R = \frac{[R]}{[R] + [O]} = \frac{[R]/[O]}{[R]/[O] + 1} = \frac{10^{-nf(E-E^0)}}{10^{-nf(E-E^0)} + 1}$$

As an exercise, show that α_R is given by

$$\alpha_R = \frac{1}{10^{-nf(E^0-E)} + 1}$$

and that α_O is

$$\alpha_O = 1 - \alpha_R = \frac{1}{10^{-nf(E-E^0)} + 1}$$

Furthermore, you can rearrange the equations as follows:

$$\alpha_R = \frac{10^{-nfE}}{10^{-nfE} + 10^{-nfE^0}} \qquad \alpha_O = \frac{10^{-nfE^0}}{10^{-nfE} + 10^{-nfE^0}}$$

The reason for expressing the α's in this way is so that they are in a similar form to those for a

weak monoprotic acid presented in Chapter 7.

$$\alpha_0 = \frac{[H_3O^+]}{[H_3O^+] + K_a} \qquad \alpha_1 = \frac{K_a}{[H_3O^+] + K_a}$$

or, alternatively,

$$\alpha_0 = \frac{10^{-pH}}{10^{-pH} + 10^{-pK_a}} \qquad \alpha_1 = \frac{10^{-pK_a}}{10^{-pH} + 10^{-pK_a}}$$

Notice the very similar forms of the α's for redox species and those for the weak monoprotic acid. The term 10^{-nfE} in the redox expression is analogous to 10^{-pH} in the acid-base case, and the term 10^{-nfE^0} is analogous to 10^{-pK_a}. These analogies will become more apparent when we plot α_O and α_R versus E in the same way that we plotted α_0 and α_1 versus pH. It is important to recognize here that we obtain these relatively straightforward expressions for the redox alphas only for redox half-reactions that have 1:1 stoichiometry. For other stoichiometries, which we will not treat here, the expressions become considerably more complex. For simple cases, these equations provide us with a nice way to visualize redox chemistry and to calculate redox titration curves. If we have formal potential data in a constant ionic strength medium, we can use the $E^{0'}$ values in place of the E^0 values in the α expressions

Now we'll examine graphically the dependence of the redox α's on the potential E. We'll determine this dependence for both the Fe^{3+}/Fe^{2+} and the Ce^{4+}/Ce^{3+} couples in 1 M H_2SO_4, where the formal potentials are known. For these two couples, the α's are given by

$$\alpha_{Fe^{2+}} = \frac{10^{-fE}}{10^{-fE} + 10^{-fE_{Fe}^{0'}}} \qquad \alpha_{Fe^{2+}} = \frac{10^{-fE_{Fe}^{0'}}}{10^{-fE} + 10^{-fE_{Fe}^{0'}}}$$

$$\alpha_{Ce^{3+}} = \frac{10^{-fE}}{10^{-fE} + 10^{-fE_{Ce}^{0'}}} \qquad \alpha_{Ce^{4+}} = \frac{10^{-fE_{Ce}^{0'}}}{10^{-fE} + 10^{-fE_{Ce}^{0'}}}$$

Note that the *only* difference in the expressions for the two sets of α's is the two different formal potentials $E_{Fe}^{0'} = 0.68$ V and $E_{Ce}^{0'} = 1.44$ V in 1 M H_2SO_4. The effect of this difference will be apparent in the α-plots that we'll construct. Since $n = 1$ for both couples, it does not appear in these equations for α.

The spreadsheet and plot for α values are shown in Figure 10-4. In cell B6, we've computed f at 25° C from 2/0.0592. We've calculated the α's every 0.05 V from 0.50 V to 1.75

V. The shapes of the α-plots are identical to those for acid-base systems as you might expect

from the form of the analogous expressions. At what potentials are $\alpha_O = \alpha_R$? How do the α-plots

change for different 1:1 redox systems. You can easily calculate α for a given electrode potential

E, but how would you find E for a given value of α? Try to develop an Excel spreadsheet to

perform this task.

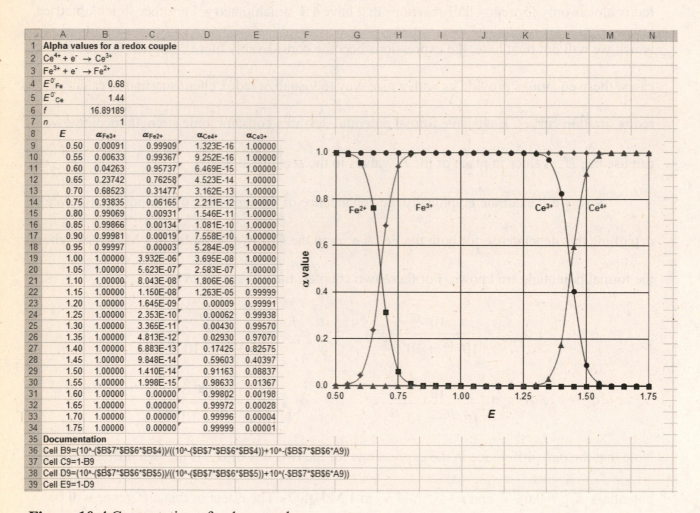

	A	B	C	D	E
1	Alpha values for a redox couple				
2	$Ce^{4+} + e^- \rightarrow Ce^{3+}$				
3	$Fe^{3+} + e^- \rightarrow Fe^{2+}$				
4	$E^0{}_{Fe}$	0.68			
5	$E^0{}_{Ce}$	1.44			
6	f	16.89189			
7	n	1			
8	E	α_{Fe3+}	α_{Fe2+}	α_{Ce4+}	α_{Ce3+}
9	0.50	0.00091	0.99909	1.323E-16	1.00000
10	0.55	0.00633	0.99367	9.252E-16	1.00000
11	0.60	0.04263	0.95737	6.469E-15	1.00000
12	0.65	0.23742	0.76258	4.523E-14	1.00000
13	0.70	0.68523	0.31477	3.162E-13	1.00000
14	0.75	0.93835	0.06165	2.211E-12	1.00000
15	0.80	0.99069	0.00931	1.546E-11	1.00000
16	0.85	0.99866	0.00134	1.081E-10	1.00000
17	0.90	0.99981	0.00019	7.558E-10	1.00000
18	0.95	0.99997	0.00003	5.284E-09	1.00000
19	1.00	1.00000	3.932E-06	3.695E-08	1.00000
20	1.05	1.00000	5.623E-07	2.583E-07	1.00000
21	1.10	1.00000	8.043E-08	1.806E-06	1.00000
22	1.15	1.00000	1.150E-08	1.263E-05	0.99999
23	1.20	1.00000	1.645E-09	0.00009	0.99991
24	1.25	1.00000	2.353E-10	0.00062	0.99938
25	1.30	1.00000	3.365E-11	0.00430	0.99570
26	1.35	1.00000	4.813E-12	0.02930	0.97070
27	1.40	1.00000	6.883E-13	0.17425	0.82575
28	1.45	1.00000	9.848E-14	0.59603	0.40397
29	1.50	1.00000	1.410E-14	0.91163	0.08837
30	1.55	1.00000	1.998E-15	0.98633	0.01367
31	1.60	1.00000	0.00000	0.99802	0.00198
32	1.65	1.00000	0.00000	0.99972	0.00028
33	1.70	1.00000	0.00000	0.99996	0.00004
34	1.75	1.00000	0.00000	0.99999	0.00001
35	Documentation				
36	Cell B9=(10^(-($B$7*$B$6*$B$4))/((10^(-($B$7*$B$6*$B$4))+10^(-($B$7*$B$6*A9))				
37	Cell C9=1-B9				
38	Cell D9=(10^(-($B$7*$B$6*$B$5))/((10^(-($B$7*$B$6*$B$5))+10^(-($B$7*$B$6*A9))				
39	Cell E9=1-D9				

Figure 10-4 Computation of redox α values.

It is worth mentioning that we normally think of calculating the potential of an electrode

for a redox system in terms of concentration rather than the other way around. Just as pH is the

independent variable in our α calculations with acid-base systems, potential is the independent variable in redox calculations. It's much easier to calculate α for a series of potential values than to solve the expressions for potential given various values of α.

Redox Titrations

Now that we've explored the relationship between α and potential for redox systems, we can calculate titration curves using the master equation approach. Before doing that, however, we'll use the stoichiometric approach to construct the titration curve for the titration of iron(II) with cerium(IV) which is covered in FAC9, Section 19D and AC7 Section 17C. The titration reaction is

$$Fe^{2+} + Ce^{4+} \rightleftharpoons Fe^{3+} + Ce^{3+}$$

Recall that for such a rapid and reversible reaction, the titration reaction *is at equilibrium throughout the titration*. Thus the electrode potentials for the two half-reactions are equal throughout. Or,

$$E_{Ce^{4+}/Ce^{3+}} = E_{Fe^{3+}/Fe^{2+}} = E_{system}$$

where E_{system} is the *potential of the system*.

Stoichiometric Method

As our example we'll consider the titration of 50.00 mL of 0.0500 M Fe^{2+} with 0.1000 M Ce^{4+} in 1 M H_2SO_4. To use the stoichiometric method, we divide the titration curve into the preequivalence point region, the equivalence point, and the postequivalence point region. In each region we calculate the system potential E_{system}. Since there are $50.00 \times 0.0500 = 2.500$ mmoles of Fe^{2+}, the equivalence point will occur when 2.500 mmole/0.1000 mmol/mL = 25.00 mL of Ce^{4+} has been added. The spreadsheet and plot are shown in Figure 10-**5**. We describe the calculations for each region below.

Preequivalence Point Region. From the stoichiometry, we can compute the concentrations of

Fe^{2+} remaining and that of Fe^{3+} produced. Hence, it is easiest to use the Fe^{3+}/Fe^{2+} couple to

calculate the system potential in this region. The concentrations are calculated as follows

$$[Fe^{3+}] = \frac{\text{mmoles } Ce^{4+} \text{ added}}{\text{total solution volume, mL}}$$

$$[Fe^{2+}] = \frac{\text{no. of mmoles } Fe^{2+} \text{ initially present } - \text{ no. of mmoles } Ce^{4+} \text{ added}}{\text{total solution volume, mL}}$$

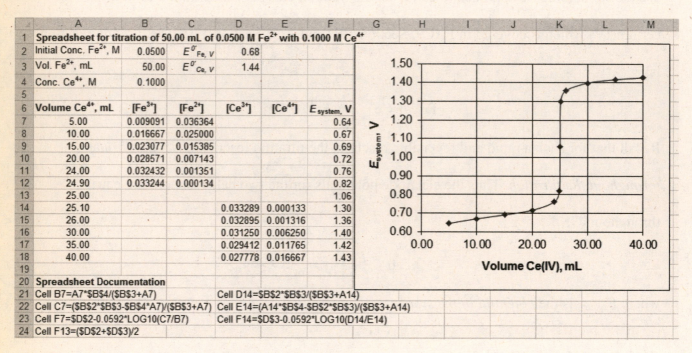

	A	B	C	D	E	F
1	Spreadsheet for titration of 50.00 mL of 0.0500 M Fe²⁺ with 0.1000 M Ce⁴⁺					
2	Initial Conc. Fe²⁺, M	0.0500	E⁰Fe, V	0.68		
3	Vol. Fe²⁺, mL	50.00	E⁰Ce, V	1.44		
4	Conc. Ce⁴⁺, M	0.1000				
5						
6	Volume Ce⁴⁺, mL	[Fe³⁺]	[Fe²⁺]	[Ce³⁺]	[Ce⁴⁺]	Esystem, V
7	5.00	0.009091	0.036364			0.64
8	10.00	0.016667	0.025000			0.67
9	15.00	0.023077	0.015385			0.69
10	20.00	0.028571	0.007143			0.72
11	24.00	0.032432	0.001351			0.76
12	24.90	0.033244	0.000134			0.82
13	25.00					1.06
14	25.10			0.033289	0.000133	1.30
15	26.00			0.032895	0.001316	1.36
16	30.00			0.031250	0.006250	1.40
17	35.00			0.029412	0.011765	1.42
18	40.00			0.027778	0.016667	1.43
19						
20	Spreadsheet Documentation					
21	Cell B7=A7*$B$4/($B$3+A7)			Cell D14=$B$2*$B$3/($B$3+A14)		
22	Cell C7=($B$2*$B$3-$B$4*A7)/($B$3+A7)			Cell E14=(A14*$B$4-$B$2*$B$3)/($B$3+A14)		
23	Cell F7=$D$2-0.0592*LOG10(C7/B7)			Cell F14=$D$3-0.0592*LOG10(D14/E14)		
24	Cell F13=($D$2+$D$3)/2					

Figure 10-6 Spreadsheet and plot for titration of Fe^{2+} with Ce^{4+}.

These concentrations are computed in cells B7 and C7 and the formulas copied into the

remaining preequivalence point cells. The cell potential is computed from the Nernst equation in

cell F7 and copied into the remaining preequivalence point cells.

Equivalence Point. The equivalence point potential is just one-half of the sum of the formal

potentials. This is computed in cell F13.

Postequivalence Point Region. In this region, the concentrations of Ce^{3+} and Ce^{4+} are more easily calculated. Hence, we use the Ce^{4+}/Ce^{3+} couple to find the system potential. The concentrations are found from

$$[Ce^{3+}] = \frac{\text{no. of mmoles } Fe^{2+} \text{ initially present}}{\text{total solution volume, mL}}$$

$$[Ce^{4+}] = \frac{\text{no. of mmoles } Ce^{4+} \text{ added} - \text{no. of mmoles } Fe^{2+} \text{ initially present}}{\text{total solution volume, mL}}$$

The formulas corresponding to these expressions are entered into cells D14 and E14 and copied into the remaining cells. The system potential is calculated in cell F14 from the Nernst equation for the Ce^{4+}/Ce^{3+} couple. The formula is copied into the remaining cells. From the plot, we can notice the sharp break in the equivalence point region.

Inverse Master Equation Approach

At all points during the titration, the concentrations of Fe^{3+} and Ce^{3+} are equal from the stoichiometry. Or

$$[Fe^{3+}] = [Ce^{3+}]$$

From the α's and the concentrations and volumes of the reagents, we can write

$$\alpha_{Fe^{3+}} \frac{V_{Fe} c_{Fe}}{V_{Fe} + V_{Ce}} = \alpha_{Ce^{3+}} \frac{V_{Ce} c_{Ce}}{V_{Fe} + V_{Ce}}$$

where V_{Fe} and c_{Fe} are the initial volume and concentration of Fe^{2+} present, and V_{Ce} and c_{Ce} are the volume and concentration of the titrant. By multiplying both sides of the equation by $V_{Fe} + V_{Ce}$ and dividing both sides by $V_{Fe} c_{Fe} \alpha_{Ce^{3+}}$, we find that

$$\phi = \frac{V_{Ce} c_{Ce}}{V_{Fe} c_{Fe}} = \frac{\alpha_{Fe^{3+}}}{\alpha_{Ce^{3+}}}$$

where ϕ is the extent of the titration (fraction titrated). We then substitute the expressions

previously derived for the α's and obtain

$$\phi = \frac{\alpha_{Fe^{3+}}}{\alpha_{Ce^{3+}}} = \frac{1 + 10^{-f(E_{Ce}^{0'}-E)}}{1 + 10^{-f(E-E_{Fe}^{0'})}}$$

where E is now the system potential. We then enter values of E in 0.5 V increments from 0.5 to

1.40 V in the spreadsheet and calculate ϕ as shown in Figure 10-7. An additional point at 1.42 V

was added since 1.45 V gave a ϕ value of more than 2.

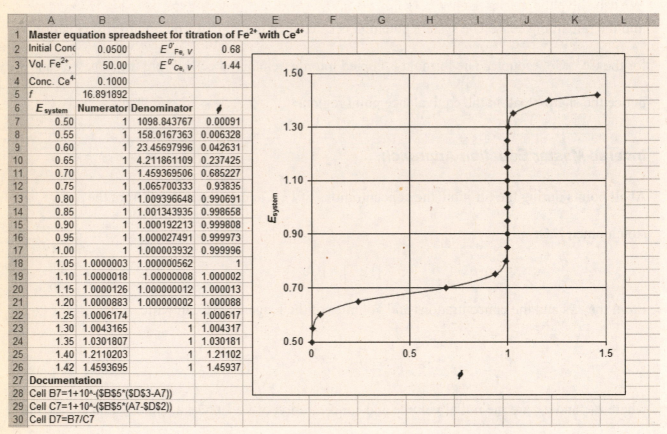

	A	B	C	D	E	F	G	H	I	J	K	L
1	Master equation spreadsheet for titration of Fe^{2+} with Ce^{4+}											
2	Initial Conc	0.0500	$E^{0'}_{Fe, V}$	0.68								
3	Vol. Fe^{2+},	50.00	$E^{0'}_{Ce, V}$	1.44								
4	Conc. Ce^{4+}	0.1000										
5	f	16.891892										
6	E_{system}	Numerator	Denominator	ϕ								
7	0.50	1	1098.843767	0.00091								
8	0.55	1	158.0167363	0.006328								
9	0.60	1	23.45697996	0.042631								
10	0.65	1	4.211861109	0.237425								
11	0.70	1	1.459369506	0.685227								
12	0.75	1	1.065700333	0.93835								
13	0.80	1	1.009396648	0.990691								
14	0.85	1	1.001343935	0.998658								
15	0.90	1	1.000192213	0.999808								
16	0.95	1	1.000027491	0.999973								
17	1.00	1	1.000003932	0.999996								
18	1.05	1.0000003	1.000000562	1								
19	1.10	1.0000018	1.00000008	1.000002								
20	1.15	1.0000126	1.000000012	1.000013								
21	1.20	1.0000883	1.000000002	1.000088								
22	1.25	1.0006174	1	1.000617								
23	1.30	1.0043165	1	1.004317								
24	1.35	1.0301807	1	1.030181								
25	1.40	1.2110203	1	1.21102								
26	1.42	1.4593695	1	1.45937								
27	Documentation											
28	Cell B7=1+10^-($B$5*($D$3-A7))											
29	Cell C7=1+10^-($B$5*(A7-$D$2))											
30	Cell D7=B7/C7											

Figure 10-7 Spreadsheet for titration of Fe^{2+} with Ce^{4+} by the master equation approach.

At this point we should mention that some redox titration expressions are more complex

than those presented here for a basic 1:1 situation. The next section considers a more complex

titration by the stoichiometric approach. If you are interested in exploring the master equation

approach for pH dependent redox titrations or other situations, consult the paper by de Levie.[1]

Calculating System Potentials From the Master Equation

Because we know the concentrations of the analyte and titrant and the volume of analyte, we might well ask what system potential corresponds to a given value of the titrant volume. In our previous example the concentration of Fe^{2+} is 0.05000 M and its volume is 50.00 mL, while the concentration of the titrant Ce^{4+} is 0.1000 M. We know that

$$\phi = \frac{V_{Ce} c_{Ce}}{V_{Fe} c_{Fe}}$$

We can solve the master equation for E if we know the extent of titration ϕ by using Excel's Solver or Goal Seek. For example, find the system potential if the titrant volume is 10.00 mL. The spreadsheet shown in Figure 10-8 will solve the equation for E given a known value of ϕ. Have Solver change cell A8 to reach a value in cell B8 of 0.25, which corresponds to a volume of 10.00 mL. Start with an initial estimate of 0.50 as shown. The value should be 0.652 when Solver has completed its search.

	A	B	C	D	E	F
1	Solver setup for finding E given a value of ϕ					
2	Initial Conc	0.0500	$E^{0'}{}_{Fe, v}$	0.68		
3	Vol. Fe^{2+},	50.00	$E^{0'}{}_{Ce, v}$	1.44		
4	Conc. Ce^{4+}	0.1000				
5	f	16.89189				
6						
7	E	ϕ				
8	0.5	0.00091				
9	Documentation					
10	Cell A8= Initial estimate or Solver result					
11	Cell B8=(1+10^-($B$5*($D$3-A8)))/(1+10^-($B$5*(A8-$D$2)))					

Figure 10-8 Spreadsheet to find E given a value of ϕ.

Try several other values of ϕ and calculate potentials. Could you calculate a titration curve this way?

A More Complex Titration Example

We'll next construct a spreadsheet and plot a titration curve for the titration of 50.00 mL of

0.02500 M U^{4+} with 0.1000 M Ce^{4+} using the stoichiometric approach. The solution is 1.0 M in

H_2SO_4. For simplicity, we will assume that the H_3O^+ concentration is also 1.0 M throughout the

titration. The titration reaction is

$$U^{4+} + 2\,Ce^{4+} + 2H_2O \rightleftharpoons UO_2^{2+} + 2Ce^{3+} + 4H^+$$

From Appendix 5 of FAC9 or Appendix 4 of AC7, we find that

$$UO_2^{2+} + 4H^+ + 2e^- \rightleftharpoons U^{4+} + 2H_2O \qquad E^0 = 0.334\ V$$

$$Ce^{4+} + e^- \rightleftharpoons Ce^{3+} \qquad\qquad E^{0'} = 1.44\ V$$

The spreadsheet and plot are shown in Figure 10-9.

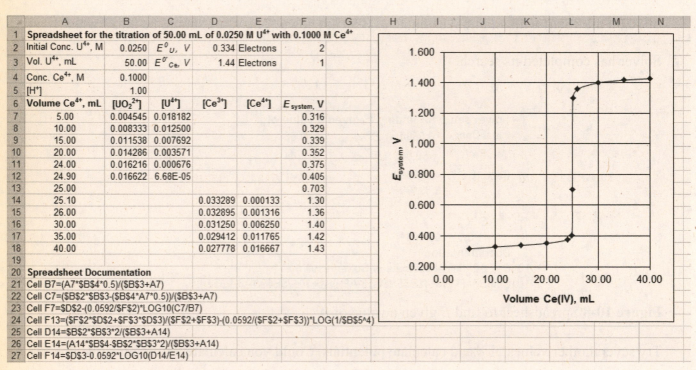

Figure 10-9 Spreadsheet and plot for titration of U^{4+} with Ce^{4+}.

Preequivalence Point Potential. Before the equivalence point we use the UO_2^{2+}/U^{4+} couple. The

concentration of UO_2^{2+} can be found from

$$[UO_2^{2+}] = \frac{\text{mmol Ce}^{4+} \text{ added} \times \dfrac{1 \text{ mmol UO}_2^{2+}}{2 \text{ mmol Ce}^{4+}}}{\text{total solution volume, mL}}$$

The formula corresponding to this equation is entered into cell B7 and copied into the remaining

preequivalence point cells (B8:B12). The concentration of U^{4+} is given by

$$[U^{4+}] = \frac{\text{mmol U}^{4+} \text{ initially present} - \text{mmol Ce}^{4+} \text{ added} \times \dfrac{1 \text{ mmol U}^{4+} \text{ reacted}}{2 \text{ mmol Ce}^{4+}}}{\text{total solution volume, mL}}$$

The formula for this expression is entered into cell C7 and copied into the rest of the

preequivalence point cells. The potential is calculated from the Nernst equation:

$$E = E_U^0 - \frac{0.0592}{2} \log \frac{[U^{4+}]}{[UO_2^{2+}][H^+]^4}$$

This formula is entered into cell F7 and copied into cells F8:F12.

Equivalence Point. The equivalence point potential E_{eq} is obtained from

$$E_{eq} = \frac{2E_U^0 + E_{Ce}^{0'}}{3} - \frac{0.0592}{3} \log \frac{1}{[H^+]^4}$$

The formula corresponding to this equation is entered into cell F13.

Postequivalence Point Region. In this region we use the Ce^{4+}/Ce^{3+} couple to calculate the

system potential. The concentration of Ce^{4+} is given by

$$[Ce^{4+}] = \frac{\text{mmol Ce}^{4+} \text{ added} - \text{mmol U}^{4+} \text{ initially present} \times \dfrac{2 \text{ mmol Ce}^{4+} \text{ reacted}}{\text{mmol U}^{4+}}}{\text{total solution volume, mL}}$$

The concentration of Ce^{3+} formed is given by

$$[Ce^{3+}] = \frac{\text{mmol U}^{4+} \text{ initially present} \times \dfrac{2 \text{ mmol Ce}^{3+} \text{ formed}}{\text{mmol U}^{4+}}}{\text{total solution volume, mL}}$$

The formulas for these concentrations are entered into cells D14 and E14 and copied into the remaining cells. The potential is calculated from the Nernst equation for the Ce^{4+}/Ce^{3+} couple as shown in the documentation for cell F14.

Summary

In this chapter we've explored using Excel for calculations of electrode potentials. We've seen how to calculate α values for redox equilibria and how to use two approaches for computing titration curves. The stoichiometric approach was used with a pH-dependent system and the master equation approach was used for a simple 1:1 redox titration.

Problems

1. Use a spreadsheet, such as that shown in Figure 10-1, to find the electrode potential as a function of the ratio of [R] to [O] for the following couples.

    (a) Ce^{4+}/Ce^{3+} (d) Ti^{3+}/Ti^{2+}

    (b) Co^{3+}/Co^{2+} (e) V^{3+}/V^{2+}

    (c) Tl^{3+}/Tl^{+} (f) Hg^{2+}/Hg_2^{2+}

2. Modify the electrode potential spreadsheet of Figure 10-1 to include metal ion/metal systems and find the electrode potential as a function of metal ion concentration over the range of 10^{-7} M to 1 M for the following systems. Plot E vs the logarithm of the metal ion concentration.

    (a) Cd^{2+}/Cd (d) Ag^{+}/Ag

    (b) Pb^{2+}/Pb (e) Sn^{2+}/Sn

    (c) $Hg_2^{2+}/Hg(l)$ (f) Al^{3+}/Al

3. Modify the electrode potential spreadsheet of Figure 10-1 to include systems with pH dependent half-reactions. For equal concentrations of [R] and [O], find the electrode potentials of the following systems as a function of pH over the range pH 1.0 to pH 7.0. Plot E vs. pH. What is the numerical value of the slope? What should the slope be according to the Nernst equation?

    (a) $Cr_2O_7^{2-}/Cr^{3+}$ (d) BrO_3^{-}/Br^{-}

    (b) MnO_4^{-}/Mn^{2+} (e) VO^{2+}/V^{3+}

    (c) H_3AsO_4/H_3AsO_3 (f) TiO^{2+}/Ti^{3+}

4. Use a spreadsheet, such as that shown in Figure 10-3, to calculate equilibrium constants for the following reactions.

(a) $Fe^{3+} + V^{2+} \rightleftharpoons Fe^{2+} + V^{3+}$

(b) $Fe(CN)_6^{3-} + Cr^{2+} \rightleftharpoons Fe(CN)_6^{4-} + Cr^{3+}$

(c) $2V(OH)_4^+ + U^{4+} \rightleftharpoons 2VO^{2+} + UO_2^{2+} + 4H_2O$

(d) $Tl^{3+} + 2Fe^{2+} \rightleftharpoons Tl^+ + 2Fe^{3+}$

(e) $Pb^{2+} + 2Ag(s) \rightleftharpoons 2Ag^+ + Pb(s)$

5.   Modify the spreadsheet of Figure 10-3 to include extra species other than O and R. Find

the equilibrium constants for the following reactions.

(a) $2Ce^{4+} + H_3AsO_3 + H_2O \rightleftharpoons 2Ce^{3+} + H_3AsO_4 + 2H^+$ (1 M $HClO_4$)

(b) $2V(OH)_4^+ + H_2SO_3 \rightleftharpoons SO_4^{2-} + 2VO^{2+} + 5H_2O$

(c) $VO^{2+} + V^{2+} + 2H^+ \rightleftharpoons 2V^{3+} + H_2O$

(d) $TiO^{2+} + Ti^{2+} + 2H^+ \rightleftharpoons 2Ti^{3+} + H_2O$

6.   Calculate α values for the following redox systems over the potential range given

(a) V^{2+} being titrated with Ce^{4+} to form V^{3+} and Ce^{3+} (range from -0.4 V to $+1.8$ V)

(b) V^{2+} being titrated with Fe^{3+} to form V^{3+} and Fe^{2+} (range from -0.4 V to $+0.9$ V)

(c) Ti^{2+} being titrated with Ce^{4+} to form Ti^{3+} and Ce^{3+} (range -0.5V to $+1.8$ V)

7.   Use the stoichiometric approach to construct curves for the following titrations. Calculate

potentials after the addition of 10.00, 25.00, 49.00, 49.90, 50.00, 50.10, 51.00, and 60.00

mL of the reagent. Where necessary, assume that $[H^+] = 1.00$ throughout.

(a)   50.00 mL of 0.1000 V^{2+} with 0.05000 M Sn^{4+}.

(b)   50.00 mL of 0.1000 M $Fe(CN)_6^{3-}$ with 0.1000 M Cr^{2+}.

(c)   50.00 mL of 0.1000 M $Fe(CN)_6^{4-}$ with 0.05000 M Tl^{3+}.

(d)   50.00 mL of 0.1000 M Fe^{3+} with 0.05000 M Sn^{2+}.

(e)   50.00 mL of 0.05000 M U^{4+} with 0.02000 M MnO_4^-.

8.   Use the inverse master equation approach to construct titration curves for 25.00 mL of

       0.1000 M analyte being titrated with 0.1000 M titrant for each of the systems in Problem

       6.

9.   For each of the titrations in Problem 8, use Solver to find the system potentials after

       10.00 mL of titrant as was done in Figure 10-8.

10.   For the titration of 50.00 mL of 0.0800 M Fe^{2+} with 0.1000 M Ce^{4+} use Solver to find the

       system potentials after 10.00 mL, 20.00 mL, 30.00 mL, 35.00 mL and 39.90 mL of

       titrant. Can you complete the titration curve in this manner? Explain.

Chapter 11

Dynamic Electrochemistry

In this chapter, we consider electrochemical methods in which net chemical reactions occur. In Chapter 10, we discussed potentiometric measurements where the net current was zero, and no net chemical changes took place. We'll focus here on coulometric methods, including coulometric titrations, and voltammetric methods. Dynamic electrochemical methods are discussed in Chapters 22 and 23 of FAC9 and Chapter 20 of AC7.

Controlled-Potential Coulometry

Coulometric methods are performed by measuring the quantity of electrical charge (electrons) required to convert a sample of an analyte quantitatively to a different oxidation state. Coulometric methods are as accurate as conventional gravimetric and volumetric procedures. In addition, they are easily automated. In controlled-potential coulometry, the potential of the working electrode is maintained at a constant value so that only the analyte is responsible for charge conduction across the electrode/solution interface.

Basic Principles

In controlled-potential coulometry, the current i can vary with time. Hence, the total quantity of charge measured in coulombs is given by the integral

$$Q = \int_0^t i\, dt$$

If we know the total charge required to completely reduce or oxidize the analyte, we can find the amount of analyte from Faraday's law

$$Q = nFn_A$$

where n is the number of moles of electrons in the analyte half-reaction, F is Faraday's constant, and n_A is the number of moles of analyte initially present. The number of coulombs can be obtained in several different ways, but with digital data collection, the current-time curve can be integrated numerically as shown next.

Numerical Integration

As an example, let's consider the controlled-potential coulometric determination of lead at a mercury cathode. The cathode potential was −0.50 V vs. the SCE. The current-time curve was recorded until the current fell to less than 0.1% of its initial value, as shown in Figure 11-1.

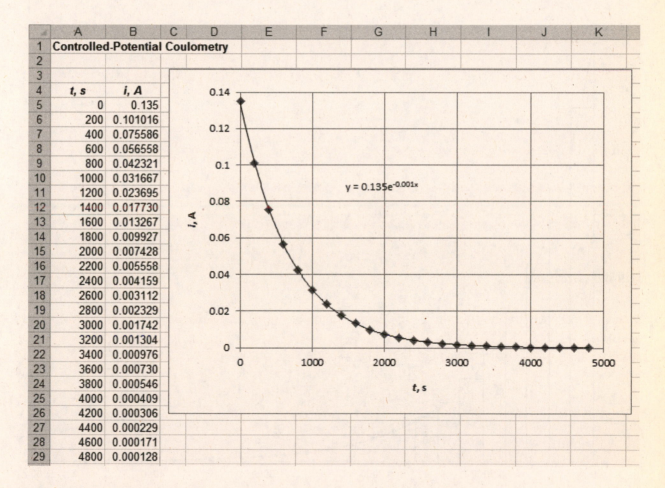

Figure 11-1 Current-time data and plot for coulometric determination of lead.

We'll show here two methods for doing numerical integration: the trapezoidal method and a Simpson's rule method.

Trapazoidal Method. In this method, the curve to be integrated is divided into rectangular

segments. The area of each segment is calculated, and the segment areas are summed to give the

total area. The trapezoidal approximation uses the average of the two y values as the height of the

rectangle, which is equivalent to using a trapezoid to approximate the rectangle. In our case, the

height of each rectangle is $\dfrac{y_n + y_{n+1}}{2}$ and the width is $x_{n+1} - x_n$. In our example, the width of

each rectangle is 200 s. The height of the first rectangle is $(0.135 + 0.101016)/2 = 0.118008$. The

area of the rectangle is the height times the width, so the first segment has an area of 23.6016.

The process is illustrated for the first 8 segments in Figure 11-2.

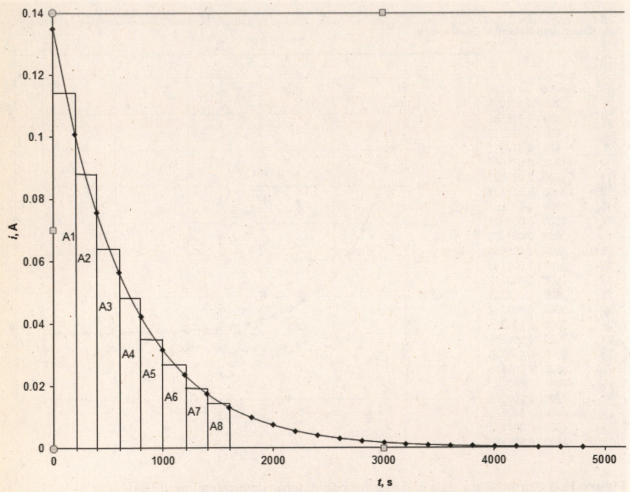

Figure 11-2 Illustration of numerical integration by trapezoidal method for 8 area segments.

The area found by summing all the segments is 93.6661 as shown in Figure 11-3.

	A	B	C
1	Controlled-Potential Coulometry		
2			
3			
4	*t, s*	*i, A*	Trapezoidal
5	0	0.135	
6	200	0.101016	23.60156
7	400	0.075586	17.66019
8	600	0.056558	13.21447
9	800	0.042321	9.88791
10	1000	0.031667	7.39876
11	1200	0.023695	5.53622
12	1400	0.017730	4.14255
13	1600	0.013267	3.09972
14	1800	0.009927	2.31941
15	2000	0.007428	1.73553
16	2200	0.005558	1.29863
17	2400	0.004159	0.97172
18	2600	0.003112	0.72710
19	2800	0.002329	0.54406
20	3000	0.001742	0.40710
21	3200	0.001304	0.30462
22	3400	0.000976	0.22794
23	3600	0.000730	0.17056
24	3800	0.000546	0.12762
25	4000	0.000409	0.09549
26	4200	0.000306	0.07146
27	4400	0.000229	0.05347
28	4600	0.000171	0.04001
29	4800	0.000128	0.02994
30		Area	93.6661
31		True Area	93.0151
32	Documentation		
33	Cell C6=200*(B5+B6)/2		
34	Cell C30=SUM(C6:C29)		

Figure 11-3 Area measurement by trapezoidal method of numerical integration.

The true area found by integration of the mathematical function used to generate the data was

93.0151. You can see that the trapezoidal approximation method overestimates the true area, in

this case by about 0.7%.

Simpson's Rule. A more accurate method estimates the curvature of the function by a parabola

(quadratic equation) over each subinterval.[1] For example, to find the first sub-area for our

coulometry experiment, shown in Figure 11-4, a quadratic equation of the form

$$A_{0-2} = \Delta x \left[\frac{y_0 + 4y_1 + y_2}{3} \right]$$

---

[1] G. B. Thomas, R. L. Finney, M. D. Weir, F. Giordano, *Thomas' Calculus*, 10th ed., (Reading, MA: Addison-Wesley Publishing Co., 2000), Ch. 7.

is used. This computes the exact area under the parabola from x_0 to x_2.

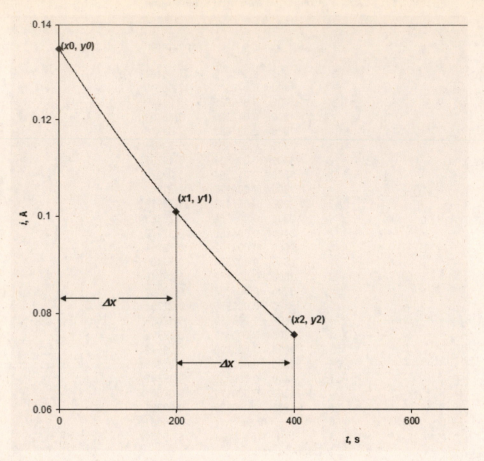

Figure 11-4 Fitting a parabola through three points.

   To find the total area from the lower limit to the upper limit, we must sum the areas of

each subinterval remembering to count each subinterval only once. For three intervals, the

equation for the total area from x_0 to x_6, A_{0-6}, is

$$A_{0-6} = \Delta x \left[\frac{y_0 + 4y_1 + y_2 \;+\; y_2 + 4y_3 + y_4 \;+\; y_4 + 4y_5 + y_6}{3} \right]$$

where each subinterval is shown in parentheses in the summation. The final equation for the total

area A_t is then

$$A_t = \Delta x \left[\frac{y_0 + 4y_1 + 2y_2 + 4y_3 + \cdots + 4y_{n-1} + y_n}{3} \right]$$

Here n must be an even number and each subinterval Δx must be identical.

With Excel, the easiest way to set up a Simpson's rule integration is to make a column of

coefficients, 1, 4, 2, 4, 2, 4 …4, 1. Each data value is then multiplied by the appropriate

coefficient. The resulting values are then summed and multiplied by $\Delta x/3$ to give the total area as

shown in Figure 11-5. Note that the area by Simpson's rule, 93.0187, is much closer to the true

area of 93.0151 (0.004% difference) than that calculated by the trapezoidal method.

	A	B	C	D	E	F	G	H
1	Controlled-Potential Coulometry							
2	i_0	0.135	F	96485				
3	AM Pb	207.2	n	2				
4	t	i_t	Trapezoidal	Simpson's	Coefficient			
5	0	0.135		0.135	1			
6	200	0.101016	23.60156	0.40406	4			
7	400	0.075586	17.66019	0.15117	2			
8	600	0.056558	13.21447	0.22623	4			
9	800	0.042321	9.88791	0.08464	2			
10	1000	0.031667	7.39876	0.12667	4			
11	1200	0.023695	5.53622	0.04739	2			
12	1400	0.017730	4.14255	0.07092	4			
13	1600	0.013267	3.09972	0.02653	2			
14	1800	0.009927	2.31941	0.03971	4			
15	2000	0.007428	1.73553	0.01486	2			
16	2200	0.005558	1.29863	0.02223	4			
17	2400	0.004159	0.97172	0.00832	2			
18	2600	0.003112	0.72710	0.01245	4			
19	2800	0.002329	0.54406	0.00466	2			
20	3000	0.001742	0.40710	0.00697	4			
21	3200	0.001304	0.30462	0.00261	2			
22	3400	0.000976	0.22794	0.00390	4			
23	3600	0.000730	0.17056	0.00146	2			
24	3800	0.000546	0.12762	0.00218	4			
25	4000	0.000409	0.09549	0.00082	2			
26	4200	0.000306	0.07146	0.00122	4			
27	4400	0.000229	0.05347	0.00046	2			
28	4600	0.000171	0.04001	0.00068	4		Amount of Pb	
29	4800	0.000128	0.02994	0.00013	1		no. mmole	0.4820
30		Area	93.6661	93.0187			no. mg	99.88
31		True Area	93.0151					
32	Documentation							
33	Cell C6=200*(B5+B6)/2							
34	Cell C30=SUM(C6:C29)							
35	Cell D5=E5*B5							
36	Cell D30=SUM(D5:D29)*200/3							
37	Cell H29=1000*D30/($D$3*$D$2)							
38	Cell H30=H29*$B$3							

Figure 11-5 Controlled-potential coulometry integration results for lead determination.

In cells H29 and H30, we calculate the amount (in mmoles) and the mass (mg) of Pb

present from the charge (coulombs) determined by Simpson's rule integration. The amount of Pb

is determined from the charge Q by Faraday's law

$$\text{amount of Pb} = \frac{1000 \times Q}{nF} \quad \text{in} \quad \frac{\text{mmol Pb/mol Pb} \times \cancel{C}}{\cancel{C}/\text{mol Pb}} = \text{mmol Pb}$$

and the mass of Pb is determined in the usual manner.

Coulometric Titrations

In contrast to controlled-potential coulometry, coulometric titrations are carried out under constant current conditions. In these titrations, a reagent, generated by electrolysis, reacts stoichiometrically with the analyte. The fundamental requirement is that the reaction that generates the reagent must proceed with a current efficiency of close to 100%. Under such conditions, we can monitor the amount of titrant added by monitoring the time, since the charge Q is given by

$$Q = I \times t$$

where I is the current. The number of moles of titrant added, n_t is then given by Faraday's law

$$n_t = \frac{Q}{nF}$$

As an example, we'll consider the determination of Fe(II) by coulometric titration with electrogenerated Ce(IV). An excess of Ce(III) is added at the start of the titration. Since the titration occurs under constant current conditions, we cannot control the electrode reaction as we could under constant potential conditions. At the very start of the titration, there is some direct oxidation of Fe(II) to Fe(III) at the Pt anode. Soon, however, the potential drifts to a value sufficiently positive that oxidation of Ce(III) to Ce(IV) occurs. The Ce(IV) is transported into the bulk of solution where it reacts with Fe(II). The presence of relatively large amounts of Ce(III), stabilizes the anode potential and prevents oxygen evolution. Thus, even though the electrolysis occurs under constant current conditions, interfering electrode reactions are prevented by the stabilizing influence of the reagent generating reaction.

The end point in a coulometric titration can be detected by any of the usual methods including visual indicators, potentiometry, spectrophotometry, and amperometry. Here, we'll use potentiometry to detect the end point. Coulometric titrations can be quite accurate and precise. Several applications are presented in FAC9, Chapter 22 or AC7, Chapter 20.

In our example, we'll consider the titration of 100.0 mL of a 1 M H_2SO_4 solution containing Fe(II) with Ce(IV) generated from 0.1 M Ce(III). A constant current of 25.0 mA is used, and the system potential E_{system} is monitored with the results shown in Figure 11-6.

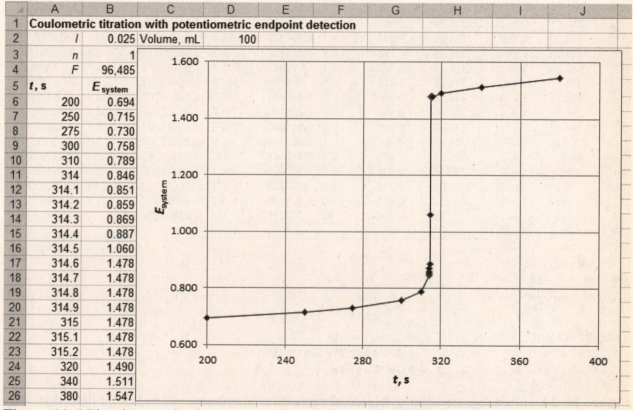

Figure 11-6 Titration results for coulometric titration of Fe(II) with Ce(IV).

Note here that the titration curve is somewhat asymmetrical compared to the normal titration curve (see Figure 10-5), because we have a large excess of Ce(III) throughout the titration. After the end point, this excess causes the system potential to be almost constant.

Now we need to find the end point and to calculate the amount of Fe(II) present. To find the end point, we'll use the second derivative method as discussed in Chapter 7. The final spreadsheet is shown in Figure 11-7, and an expanded-scale second derivative plot is shown in

Figure 11-8. From the second derivative, the end point occurs at ≈ 314.53 s.

	A	B	C	D	E	F	G	H	I	J
1	**Coulometric titration with potentiometric endpoint detection**									
2	I	0.025	Volume, mL	100						
3	n	1	Initial $[Ce^{3+}]$	0.1000						
4	F	96,485								
5	t, s	E_{system}	**Midpoint t**	ΔE	Δt	$\Delta E/\Delta t$	Midpoint t	$\Delta(\Delta E/\Delta t)$	Δt	$\Delta^2 E/\Delta t^2$
6	200	0.694								
7	250	0.715	225	0.020	50	0.00041				
8	275	0.730	262.5	0.015	25	0.000602	243.75	0.00019246	37.5	5.1324E-06
9	300	0.758	287.5	0.028	25	0.00112	275	0.00051779	25	2.0712E-05
10	310	0.789	305	0.031	10	0.003093	296.25	0.00197248	17.5	0.00011271
11	314	0.846	312	0.057	4	0.014205	308.5	0.01111261	7	0.00158752
12	314.1	0.851	314.05	0.006	0.1	0.057453	313.025	0.0432474	2.05	0.02109629
13	314.2	0.859	314.15	0.007	0.1	0.074046	314.1	0.01659298	0.1	0.16592978
14	314.3	0.869	314.25	0.010	0.1	0.104328	314.2	0.03028227	0.1	0.30282267
15	314.4	0.887	314.35	0.018	0.1	0.178292	314.3	0.07396371	0.1	0.73963706
16	314.5	1.060	314.45	0.173	0.1	1.72819	314.4	1.54989879	0.1	15.4989879
17	314.6	1.478	314.55	0.418	0.1	4.1828	314.5	2.45460963	0.1	24.5460963
18	314.7	1.478	314.65	0.000	0.1	0.000442	314.6	-4.18235761	0.1	-41.8235761
19	314.8	1.478	314.75	0.000	0.1	0.000443	314.7	4.8063E-07	0.1	4.8063E-06
20	314.9	1.478	314.85	0.000	0.1	0.000443	314.8	4.8207E-07	0.1	4.8207E-06
21	315	1.478	314.95	0.000	0.1	0.000444	314.9	4.8352E-07	0.1	4.8352E-06
22	315.1	1.478	315.05	0.000	0.1	0.000444	315	4.8498E-07	0.1	4.8498E-06
23	315.2	1.478	315.15	0.000	0.1	0.000445	315.1	4.8643E-07	0.1	4.8643E-06
24	320	1.490	317.6	0.012	4.8	0.00248	316.375	0.0020352	2.45	0.00083069
25	340	1.511	330	0.021	20	0.001036	323.8	-0.00144397	12.4	-0.00011645
26	380	1.547	360	0.036	40	0.000897	345	-0.00013912	30	-4.6375E-06
27	End point	314.53								
28	Coulombs	7.86325								
29	mmol Fe^{2+}	0.081497								
30	conc. Fe^{2+}	0.000815								
31	**Documentation**									
32	Cell C7=(A7+A6)/2			Cell I8=C8-C7						
33	Cell D7=B7-B6			Cell J8=H8/I8						
34	Cell E7=A7-A6			Cell B28=B27*$B$2						
35	Cell F7=D7/E7			Cell B29=1000*B28/($B$3*$B$4)						
36	Cell G8=(C8+C7)/2			Cell B30=B29/$D$2						
37	Cell H8=F8-F7									

Figure 11-7 Coulometric titration with second derivative potentiometric end point detection.

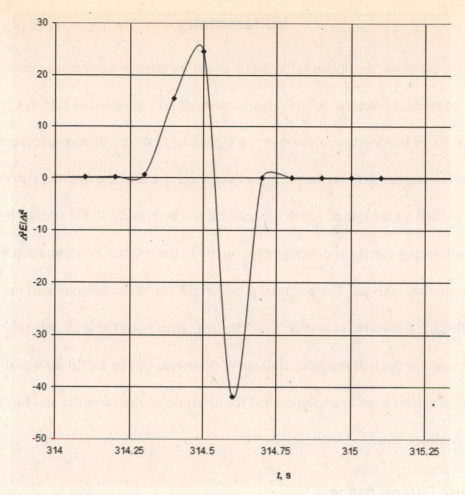

Figure 11-8 Second derivative end point detection.

From the time, we can calculate the amount of Fe(II) present from Faraday's law as shown in cells B28, 29, and 30 of Figure 11-7.

A useful exercise is to generate the titration curve of Figure 11-6. Prior to the equivalence point, for each time listed, calculate the amount of Fe^{3+} produced and the amount of Fe^{2+} remaining, knowing that the initial amount of Fe^{2+} is 0.0815 mmol. Use the Nernst equation to find the system potential. Find the equivalence point potential in the usual manner for a redox titration. After the equivalence point, calculate the amount of Ce^{4+} produced from the electrolysis and the amount of Ce^{3+} remaining. You could also use the inverse master equation approach. How do you define ϕ for this titration in terms of the generation time, t?

Voltammetry

In voltammetric methods, we obtain information about the analyte from measurements of current as a function of applied potential. Voltammetric methods are discussed in FAC9, Chapter 23, and AC7, Chapter 20. In linear-scan voltammetry, a sigmoidal shaped voltammetric wave is obtained. The constant current on the plateau of the wave is called the *limiting current*, i_l, because it is limited by the rate at which the reactant can be brought to the surface by mass transport. The limiting current is directly proportional to the reactant concentration and is the basis for quantitative analysis. The potential at which the current equals one-half the limiting current is called the *half-wave potential*, $E_{1/2}$. The half-wave potential is closely related to the standard potential for the half reaction. Half-wave potentials can be useful for qualitative identification of constituents in a solution and for determining the formulas and formation constants of complex ions.

Voltammetric Determination

In most quantitative applications of voltammetry, a calibration curve is used to evaluate the proportionality between the limiting current and concentration. We will take as an example, the polarographic determination of copper in an ammonia/ammonium chloride buffer. The limiting current was measured on the plateau of the polarographic wave and the results shown in Figure 11-9 obtained. The residual current was obtained on a blank containing no copper. The raw limiting currents are shown in cells C5:C10, and the values are corrected by subtracting the residual current in cells D5:D10. The concentration of the solution containing an unknown amount of copper is found in cell B14. The statistical treatment follows the procedures described previously in Chapter 4.

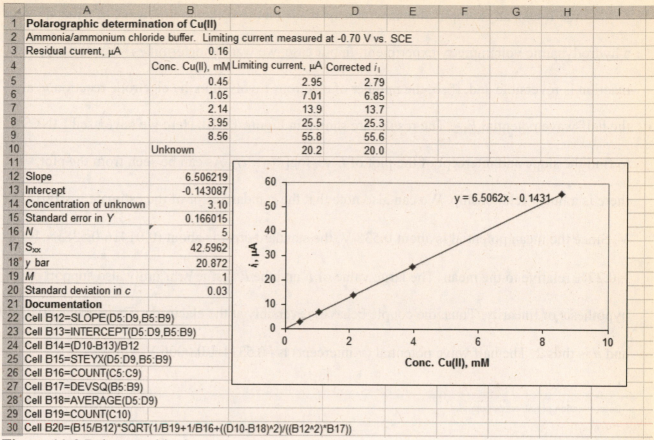

Figure 11-9 Polarographic determination of copper.

The spreadsheet (cells A1 through B30) contains:

	A	B
1	Polarographic determination of Cu(II)	
2	Ammonia/ammonium chloride buffer. Limiting current measured at -0.70 V vs. SCE	
3	Residual current, µA	0.16
4		Conc. Cu(II), mM Limiting current, µA Corrected i_l
5		0.45 2.95 2.79
6		1.05 7.01 6.85
7		2.14 13.9 13.7
8		3.95 25.5 25.3
9		8.56 55.8 55.6
10	Unknown	20.2 20.0
11		
12	Slope	6.506219
13	Intercept	-0.143087
14	Concentration of unknown	3.10
15	Standard error in Y	0.166015
16	N	5
17	S_{xx}	42.5962
18	y bar	20.872
19	M	1
20	Standard deviation in c	0.03
21	**Documentation**	
22	Cell B12=SLOPE(D5:D9,B5:B9)	
23	Cell B13=INTERCEPT(D5:D9,B5:B9)	
24	Cell B14=(D10-B13)/B12	
25	Cell B15=STEYX(D5:D9,B5:B9)	
26	Cell B16=COUNT(C5:C9)	
27	Cell B17=DEVSQ(B5:B9)	
28	Cell B18=AVERAGE(D5:D9)	
29	Cell B19=COUNT(C10)	
30	Cell B20=(B15/B12)*SQRT(1/B19+1/B16+((D10-B18)^2)/((B12^2)*B17))	

The chart shows i_l, µA (y-axis) vs. Conc. Cu(II), mM (x-axis) with the line $y = 6.5062x - 0.1431$.

Determination of Half-Wave Potentials

In voltammetry, half-wave potentials are often determined from the following equation for the voltammetric wave

$$E_{appl} = E_{1/2} - \frac{0.0592}{n}\log\frac{i}{i_l - i}$$

where i_l is the limiting current. Here, currents on the voltammetric wave are measured as a function of the applied potential, E_{appl}, and a plot is made of E_{appl} vs. $\log[i/(i_l - i)]$. The slope of the plot is $-0.0592/n$, and the y-intercept is $E_{1/2}$. Hence, the value of n can be determined from the slope and $E_{1/2}$ from the intercept. This plot is also a test for the reversibility of the couple. If the plot is linear and the slope is $0.0592/n$, the couple is said to be reversible. Irreversible couples will either give nonlinear plots, or a slope that is larger than predicted.

We'll take an example of measurements made on the voltammetric wave of a metal ion in a hydrodynamic voltammetry experiment. In our case, we want to determine if the electrode reaction is reversible and, if so, the number of electrons involved in the electrode reaction n, and the half-wave potential $E_{1/2}$. The results are shown in Figure 11-10. Here we have used LINEST to find the slope and intercept of the plot of E vs. $\log[i/(i_1 - i)]$. As can be seen from the plot, there is a linear relationship. We can also note that the standard error of the estimate is 0.001416 V. Since the mean potential is about 0.538 V, the standard error is about $(0.01416/0.538) \times 100\%$ = 0.27% relative to the mean. The large value of F and the R^2 value near unity also support our hypothesis of linearity. Thus, the couple behaves reversibly at the electrode. The slope is –0.0289 and n is thus 2. The half wave potential (y-intercept) is –0.5311 ± 0.0006 V.

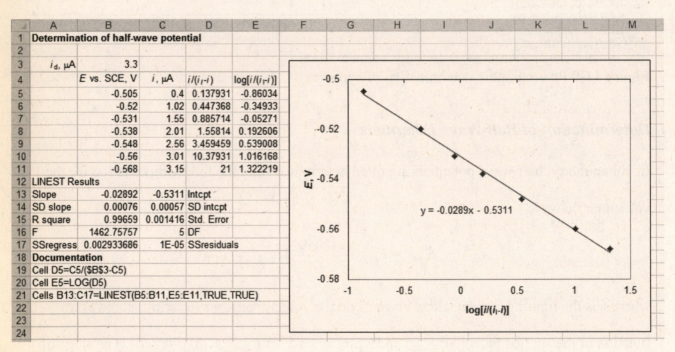

Figure 11-10 Determination of half-wave potential from polarographic data.

Determination of Formation Constant and Formula of a Complex

Voltammetry provides a method for determining formation constants and formulas for complexes of metals. When a ligand complexes a metal, the voltammetric half-wave potential shifts to more negative values than that of the simple metal ion itself. The shift in half-wave potentials, $\Delta E_{1/2}$, can be related to the formation constant K_f and the number of ligands x bound to the metal by

$$\Delta E_{1/2} = (E_{1/2})_c - E_{1/2} = -\frac{0.0592}{n}\log K_f - \frac{0.0592x}{n}\log c_L$$

where $(E_{1/2})_c$ and $E_{1/2}$ are the half-wave potentials for the complexed and uncomplexed metal ions, x is the combining ratio of complexing ligand to metal, and c_L is the ligand concentration. If measurements are made of the shift in half-wave potential vs. $\log c_L$, a straight line should be obtained with a slope of $-0.0592x/n$. The intercept should be $-(0.0592/n)\log K_f$. If n is known, x and K_f can be determined.

Let's take an example of the two-electron reduction of Cd(II) at a hanging mercury drop electrode. To determine both the formula and the formation constant of the Cd(II)/NH$_3$ complex, the voltammetric half-wave potentials were measured, as described in the previous section, for a series of ammonia concentrations. The results are shown in Figure 11-11. The raw data are in cells B4:B12 for the ligand concentration and cells D4:D12 for the $E_{1/2}$ values. We calculate log c_L in cells C4:C12 and $\Delta E_{1/2}$ in cells E4:E12. Again LINEST is used to obtain the slope, the intercept, and the statistics. The combining ratio for the ligand is found from the slope ($-0.117611 = -0.0592x/2$ or $x = 3.97 \approx 4$), and the formation constant is found from the intercept ($-0.206788 = -(0.0592/2)\log K_f$). From these values, the formula of the complex is Cd(NH$_3$)$_4^{2+}$ and log $K_f = 6.99 \pm 0.03$ or $K_f = 9.7 \pm 0.7 \times 10^6$.

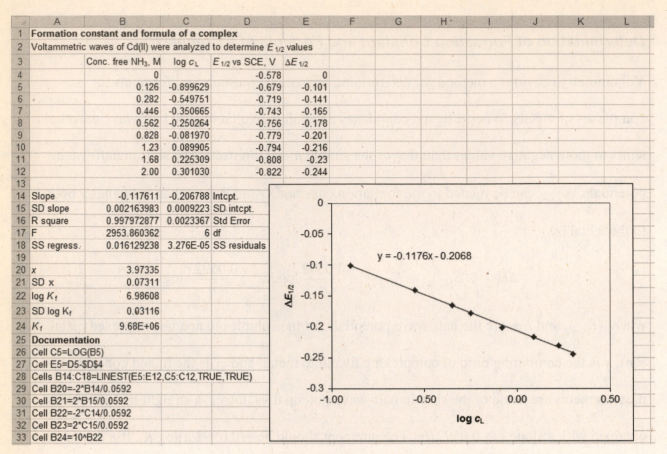

The spreadsheet shown in the figure contains the following:

	A	B	C	D	E
1	Formation constant and formula of a complex				
2	Voltammetric waves of Cd(II) were analyzed to determine $E_{1/2}$ values				
3		Conc. free NH_3, M	log c_L	$E_{1/2}$ vs SCE, V	$\Delta E_{1/2}$
4		0		-0.578	0
5		0.126	-0.899629	-0.679	-0.101
6		0.282	-0.549751	-0.719	-0.141
7		0.446	-0.350665	-0.743	-0.165
8		0.562	-0.250264	-0.756	-0.178
9		0.828	-0.081970	-0.779	-0.201
10		1.23	0.089905	-0.794	-0.216
11		1.68	0.225309	-0.808	-0.23
12		2.00	0.301030	-0.822	-0.244
13					
14	Slope	-0.117611	-0.206788	Intcpt.	
15	SD slope	0.002163983	0.0009223	SD intcpt.	
16	R square	0.997972877	0.0023367	Std Error	
17	F	2953.860362	6	df	
18	SS regress.	0.016129238	3.276E-05	SS residuals	
19					
20	x	3.97335			
21	SD x	0.07311			
22	log K_f	6.98608			
23	SD log K_f	0.03116			
24	K_f	9.68E+06			
25	Documentation				
26	Cell C5=LOG(B5)				
27	Cell E5=D5-$D$4				
28	Cells B14:C18=LINEST(E5:E12,C5:C12,TRUE,TRUE)				
29	Cell B20=-2*B14/0.0592				
30	Cell B21=2*B15/0.0592				
31	Cell B22=-2*C14/0.0592				
32	Cell B23=2*C15/0.0592				
33	Cell B24=10^B22				

The graph shows a plot of $\Delta E_{1/2}$ versus log c_L with the fitted line equation $y = -0.1176x - 0.2068$.

Figure 11-11 Determination of formula and formation constant of a complex.

Amperometric Titrations

In an amperometric titration, we measure the voltammetric current at a constant potential,

usually on the plateau of the voltammetric wave, as a function of the volume of titrant added. In

other words we monitor the progress of the titration by measuring the limiting current. One of

the species involved in the titration reaction must be electroactive in order for us to obtain the

end point of the titration. Amperometric titrations can be performed with one polarizable

microelectrode or with dual polarizable electrodes. We will consider here only the case of a

single polarizable electrode.

An example of an amperometric titration is the determination of the mass percentage of

gold in an ore sample by titration with hydroquinone, $C_6H_4(OH)_2$. The titration reaction is

$$2\,AuCl_4^- + 3\,C_6H_4(OH)_2 \rightleftharpoons 2\,Au + 3\,C_6H_4O_2 + 6\,H^+ + 8\,Cl^-$$

Hydroquinone can be oxidized at a rotating platinum electrode at 1.0 V vs. SCE to quinone. The

oxidation reaction is

$$C_6H_4(OH)_2 \rightleftharpoons C_6H_4O_2 + 2\,H^+ + 2\,e^-$$

A 100.0 g sample of the ore was decomposed with concentrated sulfuric acid and redissolved in

aqua regia. Repeated evaporations with HCl expelled the nitric acid and converted the gold

to $AuCl_4^-$. Then, 200 mL of sulfuric acid was added to the solution, and it was transferred to the

titration vessel. The electrodes were positioned in the cell and the potential was set to + 1.0 V vs.

SCE. The solution was titrated with 0.002500 M hydroquinone in 1 M sulfuric acid and the

limiting current measured as a function of titrant volume. The data shown in Figure 11-12 were

collected and plotted. Note that we have added to the spreadsheet the atomic mass of gold, the

concentration of hydroquinone c_{HQ}, and the mass of the sample. These will be needed to find the

percentage of gold in the sample.

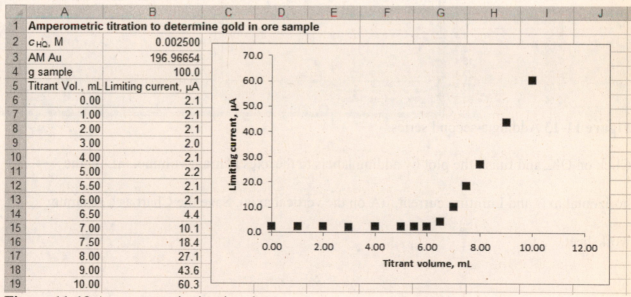

Figure 11-12 Amperometric titration data.

From the plot we can see that the titration curve consists of two linear segments. The end point is taken as the intersection of the two segments. We will treat the two segments separately in order to obtain their equations and the intersection point. Let's prepare a new plot by selecting the first 8 points and clicking on the Insert tab and selecting an XY (Scatter) Chart. Under the Chart Tools, select Design and Select Data. This brings up the Select Data Source window shown in Figure 11-13. Click the <u>A</u>dd button. In the Edit Series window, name the second series (last 6 points) Series 2. With your cursor in the Series <u>X</u> Values: blank, select cells A14:A19. In the Series <u>Y</u> Values: blank, erase anything that Excel deposits automatically, and then select cells B14:B19.

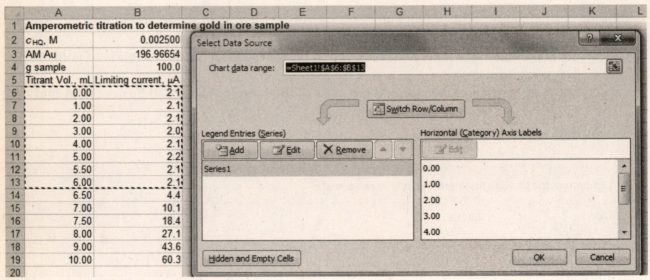

Figure 11-13 Adding a second series.

Click on OK, and finish the plot by adding labels to the axes (Titrant volume, mL on the horizontal axis and Limiting current, µA on the vertical axis). Save the Chart as a separate worksheet.

For each segment, add a linear Trendline and display the equation on the plot as shown in Figure 11-14. Note that we could also try this plot without the point at 6.50 mL which seems to be in the nonlinear portion of the second segment. The inclusion of this point, however, has only a small effect on the end point.

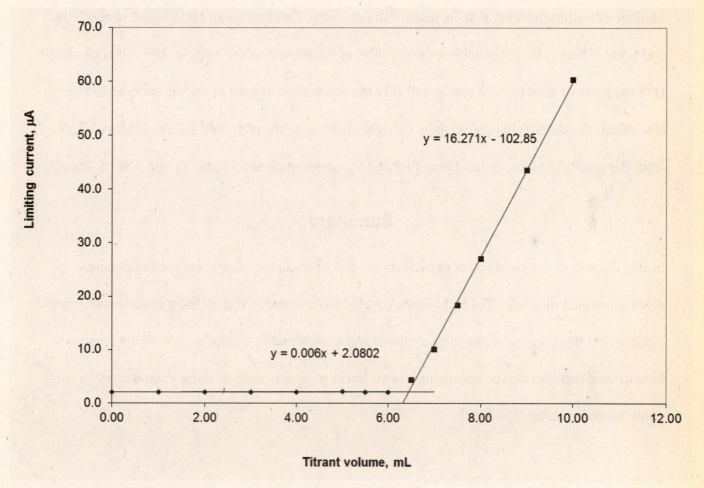

Figure 11-14 Amperometric titration curve showing two linear segments and their equations.

Now we must find the end point from the equations for the two segments. The end point will occur when the two y values are equal or when

$$0.006x + 2.0802 = 16.271x - 102.85$$

Solving for x, the titrant volume, gives

$$x = \frac{102.85 + 2.0802}{16.271 - 0.006} = 6.45 \text{ mL}$$

In the final spreadsheet shown in Figure 11-15, we've calculated the slopes and intercepts separately, and solved for the end point volume from these values. We've also found the amount of hydroquinone (mmoles) needed to reach the endpoint and the amount of $AuCl_4^-$ present (millimoles). Since 3 moles of HQ are consumed for every 2 moles of $AuCl_4^-$ present, the number of millimoles of $AuCl_4^-$ is just 2/3 the number of millimoles of HQ needed to reach the end point. This is also the number of millimoles of Au present in the sample. We calculate in cell B28 the mass of gold present and in cell B29 the percentage of gold in the ore sample. We've also added the chart to the spreadsheet by right clicking on the chart and selecting Move Chart… from the resulting menu. Select Object in: and the appropriate sheet name to add it to the sheet.

Summary

In this chapter we've used Excel to perform several of the calculations involved in dynamic electrochemical methods. The techniques covered have included controlled-potential coulometry, coulometric titrations, voltammetric methods and amperometric titrations. We've also learned how to treat titration curves consisting of two linear segments such as those encountered in some amperometric titrations.

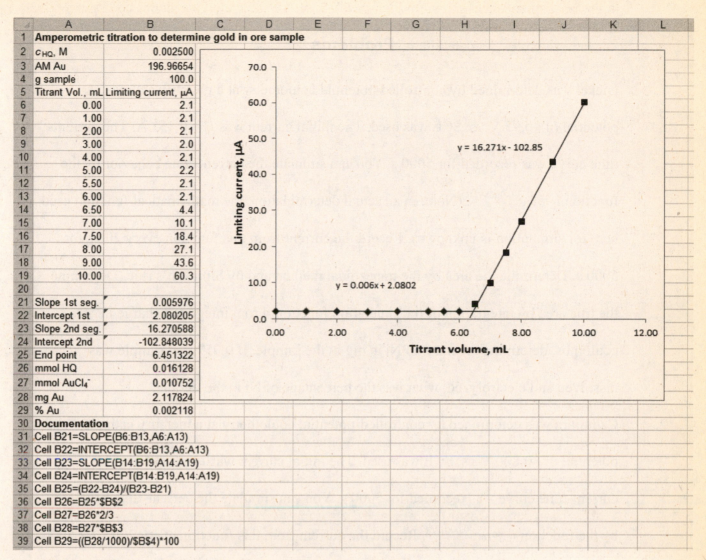

	A	B
1	**Amperometric titration to determine gold in ore sample**	
2	c_{HQ}, M	0.002500
3	AM Au	196.96654
4	g sample	100.0
5	Titrant Vol., mL	Limiting current, µA
6	0.00	2.1
7	1.00	2.1
8	2.00	2.1
9	3.00	2.0
10	4.00	2.1
11	5.00	2.2
12	5.50	2.1
13	6.00	2.1
14	6.50	4.4
15	7.00	10.1
16	7.50	18.4
17	8.00	27.1
18	9.00	43.6
19	10.00	60.3
20		
21	Slope 1st seg.	0.005976
22	Intercept 1st	2.080205
23	Slope 2nd seg.	16.270588
24	Intercept 2nd	-102.848039
25	End point	6.451322
26	mmol HQ	0.016128
27	mmol $AuCl_4^-$	0.010752
28	mg Au	2.117824
29	% Au	0.002118
30	**Documentation**	
31	Cell B21=SLOPE(B6:B13,A6:A13)	
32	Cell B22=INTERCEPT(B6:B13,A6:A13)	
33	Cell B23=SLOPE(B14:B19,A14:A19)	
34	Cell B24=INTERCEPT(B14:B19,A14:A19)	
35	Cell B25=(B22-B24)/(B23-B21)	
36	Cell B26=B25*$B$2	
37	Cell B27=B26*2/3	
38	Cell B28=B27*$B$3	
39	Cell B29=((B28/1000)/$B$4)*100	

Figure 11-15 Final spreadsheet for amperometric titration.

Problems

1. Nickel was determined by controlled-potential coulometry at a mercury cathode. A potential of -0.95 V vs. SCE was used. The initial current was $i_0 = 0.155$ A. The current-time curve was recorded for 5000 s. You can simulate the current-time behavior by the function $i_t = i_0 e^{-0.000895t}$. (Note: in an actual determination, the mathematical function used here for simulation is unknown). Record the current from this function every 200 s for 5000 s. Determine the area by the trapezoidal method and by Simpson's rule. Determine the true area by integration of the simulation function. From integrated charge in coulombs, determine the mass of Ni in mg in the sample. If 0.957 g of sample was dissolved and electrolyzed, what was the percentage of Ni in the sample.

2. Cadmium was determined by controlled-potential coulometry at a mercury cathode. A potential of -0.800 V vs. SCE was used. The initial current was $i_0 = 0.195$ A. The current-time curve was recorded for 5000 s. You can simulate the current-time behavior by the function $i_t = i_0 e^{-0.000125t}$. Record the current from this function every 200 s for 5000 s. Determine the area by the trapezoidal method and by Simpson's rule. Determine the true area by integration of the simulation function. From the charge in coulombs, determine the number mass of Cd (mg) in the sample. If 1.250 g of sample were dissolved and used in the electrolysis, what was the percentage of cadmium in the sample.

3. Generate the coulometric titration curve of Figure 11-6 by the stoichiometric method. Use 0.0815 mmol as the initial amount of Fe^{2+} present. Prior to the equivalence point, for each time given in Figure 11-6, calculate the amount of Fe^{3+} produced and the amount of Fe^{2+} remaining. Use the Nernst equation to find the system potential. Find the

equivalence point potential in the usual manner for a redox titration. After the

equivalence point, calculate the amount of Ce^{4+} produced from the electrolysis and the

amount of Ce^{3+} remaining.

4.   Use the inverse master equation approach to generate a titration curve for the coulometric

titration of Fe^{2+} with electrogenerated Ce^{4+} similar to that of Figure 11-6. Use 0.0815

mmol as the initial amount of Fe^{2+} present. Define the fraction titrated ϕ in terms of the

electrolysis time t. Calculate ϕ for various values of the system potential E.

5.   Lead was determined polarographically at the dropping mercury electrode by

measurements in 1 M HNO_3. The limiting current on the Pb(II) wave was measured at

−0.600 V vs. SCE. At this potential, the residual current was 0.12 μA. The method of

external standards was used and the following results obtained:

Concentration of Pb(II), mM	Limiting current, μA
0.50	4.37
1.00	8.67
2.00	17.49
3.00	25.75
4.00	34.35
5.50	47.10
6.50	55.70
Unknown	12.35

Determine the concentration of lead in the unknown solution and its standard deviation.

6.   Cadmium was determined by polarography at the DME in solutions that were 1 M in

      HCl. The limiting current was measured at –0.750 V vs. SCE. The residual current at this

      potential was 0.21 µA. The method of external standards was used and the following

      results obtained:

Concentration of Cd(II), mM	Limiting current, µA
1.00	4.37
2.00	8.67
3.00	12.87
5.00	21.54
8.00	34.35
12.00	51.25
Unknown	28.53

Determine the concentration of cadmium in the unknown solution and its standard

deviation.

7.   Measurements were made on a polarographic wave to determine if the couple

      $O + ne^- \rightleftharpoons R$ is reversible and, if so, the number of electrons n and the half-wave

      potential, $E_{1/2}$. The following data were obtained:

E vs. SCE, V	i, mA
–0.395	0.49
–0.406	0.96
–0.415	1.48
–0.422	1.95
–0.431	2.42
–0.445	2.95

Determine if the couple behaves reversibly. Find n and $E_{1/2}$.

8.   In a polarographic study, the Cu(II) complexes of diethylenetriamine (DT),

$H_2NCH_2CH_2NHCH_2CH_2NH_2$, were investigated in 0.1 M KNO_3 solutions. The following

results were obtained:

Concentration of free DT, M	$E_{1/2}$, V vs. SCE
0.000	+0.016
0.020	−0.504
0.040	−0.520
0.100	−0.542
0.200	−0.559
0.400	−0.575
1.00	−0.602

Determine the formula of the Cu(II)/DT complex and its formation constant.

9.   An amperometric titration of an analyte A with a titrant T yielded the following data:

Titrant volume, mL	Limiting current, μA
0.00	18.000
1.00	13.960
2.00	9.660
3.00	5.520
3.50	4.400
6.00	1.195
7.00	1.220
8.00	1.210
10.00	1.200

Determine the volume of titrant needed to reach the endpoint of the titration. Which of

the species is electroactive? the analyte? the titrant? or both?

10.   Sulfate ion can be determined by an amperometric titration procedure using Pb^{2+} as the

titrant. If the potential of a Hg electrode is adjusted to −1.00 V vs. SCE, the current can

be used to monitor the Pb^{2+} concentration during the titration. In a calibration experiment,

the limiting current, after correction for background and residual currents, was found to

be related to the Pb^{2+} concentration by $i_l = 10\,c_{Pb^{2+}}$, where i_l is the limiting current in μA

and $c_{Pb^{2+}}$ is the Pb^{2+} concentration in mM. The titration reaction is

$$SO_4^{2-} + Pb^{2+} \rightleftharpoons PbSO_4(s) \qquad K_{sp} = 1.6 \times 10^{-8}$$

If 25 mL of 0.025 M Na_2SO_4 is titrated with 0.04 M $Pb(NO_3)_2$, develop the titration curve in spreadsheet format, and plot the limiting current vs. volume of titrant.

Chapter 12

Spectrochemical Methods

In this chapter, we'll use Excel to perform calculations used in molecular spectroscopy. These topics are covered in Chapters 24-27 of FAC9 and 21-23 of AC7. Although molecular spectroscopy is stressed here, many of the calibration considerations, such as the standard additions and external standards procedures, apply as well to atomic spectroscopic methods.

Beer's Law

The Beer-Lambert law, commonly known as Beer's law, is the basis of quantitative absorption spectroscopy. According to Beer's law, the absorbance A is related linearly to the concentration of the absorbing species c and the pathlength b of the absorbing medium as

$$A = abc$$

where a is a proportionality constant known as the *absorptivity*. Because the absorbance A is unitless, absorptivity must have units that cancel the units of b and c. If c is in grams per liter $(g\ L^{-1})$, for example, and b is in centimeters (cm), a has the units of liters per gram centimeter $(L\ g^{-1}\ cm^{-1})$. If c is in ppm $(mg\ L^{-1})$ and b is in centimeters, a has the units of $ppm^{-1}\ cm^{-1}$ or $L\ mg^{-1}\ cm^{-1}$.

When we express the concentration in moles per liter and b is in centimeters, the proportionality constant is called the *molar absorptivity* and is given the special symbol ε.

$$A = \varepsilon bc$$

where ε has the units of liters per mole centimeter $(L\ mol^{-1}\ cm^{-1})$.

Absorptivity

There are several ways to determine the absorptivity or molar absorptivity of a compound. It is not good practice to rely on a single measurement of absorbance at a known concentration. Instead, several standards are used and a calibration curve of A vs. c is prepared. The slope of the curve is the absorptivity or molar absorptivity, depending on the concentration units.

As an example, consider the determination of the absorptivity and molar absorptivity of permanganate ion at a wavelength of 525 nm. For this determination, five permanganate standards were prepared at concentrations ranging from 5×10^{-5} M to 4×10^{-6} M and the absorbance values shown in Figure 12-1, cells C5:C9 were measured. Note that LINEST was used to determine the slope, the intercept, and the statistics of the linear plot. From the slope and its standard deviation, the molar absorptivity ε is 2450 ± 10 L mol^{-1} cm^{-1}.

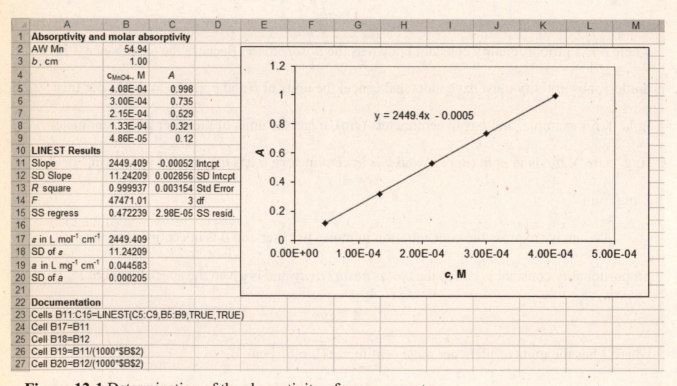

Figure 12-1 Determination of the absorptivity of permanganate.

In some experiments, we may be using permanganate standards with concentrations given in ppm. Hence, we might like to know the proportionality constant in units of $ppm^{-1}\ cm^{-1}$. We can readily convert from molar absorptivity to absorptivity as follows:

$$a = 2450\ \text{L mol}^{-1}\ \text{cm}^{-1} \times \frac{1\ \text{mol MnO}_4^-}{\text{mol Mn}} \times \frac{1\ \text{mol Mn}}{54.94\ \text{g}} \times \frac{1\ \text{g}}{1000\ \text{mg}}$$

In cell B19 (Figure 12-1), we've calculated a and in B20, the standard deviation in a from the standard deviation in ε. Thus, a is 0.0446 ± 0.0002 L mg^{-1} cm^{-1} or 0.0446 ± 0.0002 ppm^{-1} cm^{-1}.

Deviations From Beer's Law

From Beer's law, we expect a plot of A vs. c to be linear with an intercept of zero. However, calibration curves are sometimes found to be nonlinear or to have a nonzero intercept. As an example, we'll model a deviation due to a chemical factors and one due to an instrumental factor.

Deviation due to Chemical Equilibria. Sometimes the analyte can exist in several chemical forms in solution that are in chemical equilibrium. Consider, for example, an indicator whose acid form HIn is colored and whose base form In$^-$ is colorless. Since only one form absorbs, we can rewrite Beer's law as

$$A = \varepsilon b \alpha c_\text{t}$$

where c_t is the total concentration of the indicator in solution, and α is the fraction of the total indicator concentration that absorbs. We'll get a linear relationship between A and c_t only if α is independent of the total concentration, c_t.

Let's now take an even more complex situation, a monomer/dimer equilibrium in which both forms absorb at the wavelength chosen. The reaction is

$$2M \rightleftharpoons D \qquad\qquad K_{dimer} = \frac{c_D}{c_M^2}$$

The total absorbance due to both species can be written from the additive form of Beer's law as

$$A = b(\varepsilon_D c_D + \varepsilon_M c_M)$$

The total concentration can be written from the mass balance condition as

$$c_t = c_M + 2c_D = c_M + 2K_{dimer}c_M^2$$

If we solve this equation for c_M and c_D and substitute into the equation for A, we obtain

$$A = \frac{b}{2}\left[\varepsilon_D c_t + (2\varepsilon_M - \varepsilon_D)\frac{\sqrt{(8K_{dimer}c_t + 1)} - 1}{4K_{dimer}}\right]$$

We can consider three cases here. First, if $\varepsilon_D = 2\varepsilon_M$, the second term vanishes, and we predict a linear calibration curve. In other cases, we can get positive or negative deviations depending on the sign of $(2\varepsilon_M - \varepsilon_D)$. If the dimer molar absorptivity exceeds that of the monomer, $\varepsilon_D > \varepsilon_M$, a positive deviation should be observed, because the decrease in absorbance due to the loss in monomer is more than compensated by the increase due to the formation of the dimer. On the other hand, if $\varepsilon_D < 2\varepsilon_M$, a negative deviation should occur because the increase in dimer absorbance does not compensate for the loss in monomer absorbance.

We'll take a case where $K_{dimer} = 500$, $b = 1$ cm, and $\varepsilon_M = 500$ L mol^{-1} cm^{-1}. We'll calculate calibration curves for three values of ε_D, 100, 1000, and 2000 using total concentrations ranging from 1×10^{-4} M to 1×10^{-3} M. The spreadsheet is shown in Figure 12-2. Note that here we've used the equation above to calculate the absorbances for the three cases.

	A	B	C	D	E	F	G
1	Beer's law deviations due to equilibria						
2	Reaction is 2M $\rightleftharpoons$ D						
3	K_{dimer}	500					
4	b	1					
5	ε_M	500					
6	c_t	ε_D	A	ε_D	A	ε_D	A
7	0	100	0.00E+00	1000	0.00E+00	2000	0.00E+00
8	1.00E-04	100	4.62E-02	1000	5.00E-02	2000	5.42E-02
9	2.00E-04	100	8.69E-02	1000	1.00E-01	2000	1.15E-01
10	4.00E-04	100	1.58E-01	1000	2.00E-01	2000	2.47E-01
11	8.00E-04	100	2.76E-01	1000	4.00E-01	2000	5.38E-01
12	1.00E-03	100	3.28E-01	1000	5.00E-01	2000	6.91E-01
13	Documentation						
14	Cell B7=($B$4/2)*(B7*A7+(2*$B$5-B7)*(SQRT(8*$B$3*A7+1)-1)/(4*$B$3))						
15	Cell E7=($B$4/2)*(D7*A7+(2*$B$5-D7)*(SQRT(8*$B$3*A7+1)-1)/(4*$B$3))						
16	Cell G7=($B$4/2)*(F7*A7+(2*$B$5-F7)*(SQRT(8*$B$3*A7+1)-1)/(4*$B$3))						

Figure 12-2 Spreadsheet for Beer's law deviations due to dimer formation.

The three calibration curves are plotted in Figure 12-3. Here, we selected the first series, cells A7:A12 and C7:C12 and produced a scatter plot. Then, we right clicked on a member of the series and chose Select Data…. We then added Series 2 and 3 in the Select Data Source window as we did in the amperometric titration plot in Chapter 11. Note that positive deviations occur in the case where $\varepsilon_D = 2000$, and negative deviations occur for $\varepsilon_D = 100$. The curve in the center for $\varepsilon_D = 1000$ shows excellent linearity as shown by the equation near the Trendline and the R^2 value. Try different values of the molar absorptivities as well as different values for the dimerization constant K and see their effects on the deviations from linearity.

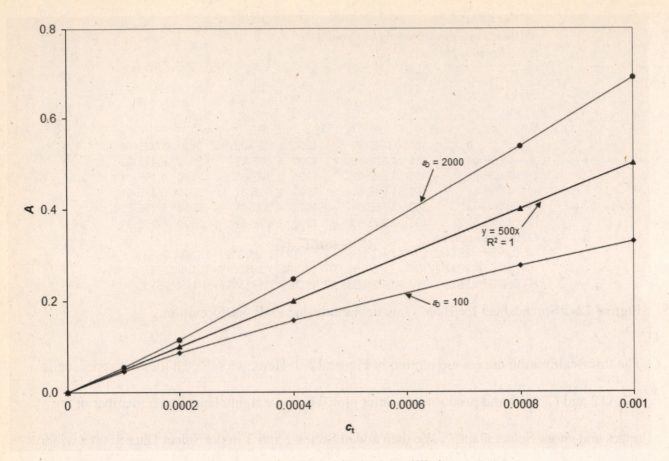

Figure 12-3 Deviations from Beer's law due to chemical equilibria.

Deviations Due to Stray Light. The radiation used for absorption measurements is usually

contaminated with a small amount of radiation outside the nominal wavelength band chosen for

the determination. This stray radiation is usually called *stray light*. When measurements are

made in the presence of stray light, the observed absorbance A' is given by

$$A' = -\log \frac{P + P_s}{P_0 + P_s}$$

where P_0 is the incident radiant power at the analysis wavelength, P is the transmitted radiant

power at the analysis wavelength, and P_s is the stray light radiant power. The true transmittance

$T = P/P_0$, and the true absorbance A is

$$A = -\log T = -\log\frac{P}{P_0}$$

If we call $P_s/P = f$, the fractional stray light power, we can write the equation for A'

$$A' = -\log\frac{\dfrac{P}{P_0}+\dfrac{P_s}{P_0}}{1+\dfrac{P_s}{P_0}} = -\log\frac{T+f}{1+f}$$

Since f is always a small fraction, usually much less than 1, we can approximate A' as

$$A' \approx -\log(T+f)$$

Consider an example of a compound with a molar absorptivity of 1.00×10^5 L mol^{-1} cm^{-1} measured in a 1.00 cm cell. Let's chose standards with concentrations ranging from 0.00 M to 3.00×10^{-5} M. These should lead to true absorbances ranging from 0.00 to 3.00 as shown in cells B7:B13 of Figure 12-4. In cells C7:C13, we calculate the corresponding true transmittances for these measurements. In columns D through G, we calculate the observed absorbances for varying degrees of stray light. At a concentration of 0.00 M, all read an absorbance of zero. The calculations for the other entries are based on the model above shown in the documentation.

	A	B	C	D	E	F	G
1	Beer's law deviations due to stray light						
2	Model uses $A' = (T + f)$						
3	b	1					
4	ε	1.00E+05					
5	f	0%		0.01%	0.10%	1%	5%
6	c, M	A	T	A'	A'	A'	A'
7	0	0	1.00000	0.00000	0.00000	0.00000	0.00000
8	5.00E-06	0.50	0.31623	0.499863	0.498629	0.486479	0.436249
9	1.00E-05	1.00	0.10000	0.999566	0.995679	0.958607	0.823909
10	1.50E-05	1.50	0.03162	1.498629	1.486479	1.380669	1.088189
11	2.00E-05	2.00	0.01000	1.995679	1.958607	1.69897	1.221849
12	2.50E-05	2.50	0.00316	2.486479	2.380669	1.880669	1.274396
13	3.00E-05	3.00	0.00100	2.958607	2.69897	1.958607	1.29243
14	Documentation						
15	Cell C7=10^-B8						
16	Cell D8=-LOG(C8+0.0001)						
17	Cell E8=-LOG(C8+0.001)						
18	Cell F8=-LOG(C8+0.01)						
19	Cell G8=-LOG(C8+0.05)						

Figure 12-4 Spreadsheet for stray light calculations.

We can then prepare the chart to display these five calibration curves. We first select cells

A7:A13 and B7:B13 and produce a scatter (XY) plot. Next, we right click on a data point, chose

S<u>e</u>lect Data… and add the four additional series in the Select Data Source window. In all cases

the <u>X</u> Values: are A7:A13, while the <u>Y</u> Values: are the A' values for the differing levels of stray

light. Your final plot should look similar to Figure 12-5, after labeling. Note that the absorbance

begins to level off with concentration at high stray light levels. Stray light always limits the

maximum absorbance that can be obtained because when the absorbance is high, the radiant

power transmitted through the sample can become comparable or lower than the stray light level.

You can try different stray light levels, as well as different values of molar absorptivity and see

their effects on the plots.

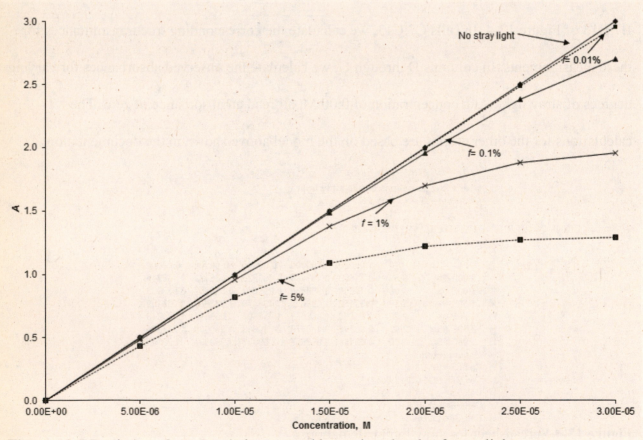

Figure 12-5 Deviations from Beer's law caused by various levels of stray light.

Precision of Absorption Measurements

The accuracy and precision of spectrophotometric measurements are often limited by indeterminate error or *noise*. As discussed in FAC9, Chapter 26 and AC7 Chapter 23, such errors can be modeled or measured experimentally. We consider here three categories of instrumental errors that lead to imprecision in spectrophotometry. In the first category, typical of older instruments, the precision is limited by the resolution of a meter readout or by transducer thermal or dark current noise. Under this case 1 scenario, the standard deviation in transmittance, σ_T, is equal to a constant value k_1 and independent of the transmittance T. This leads to a relative concentration error σ_c/c given by

$$\frac{\sigma_c}{c} = \frac{0.434}{\log T} \times \frac{k_1}{T}$$

In the second category, typical of many modern digital instruments, the major source of error is shot noise in the photon transducer. In this case the standard deviation in transmittance, σ_T, is proportional to $\sqrt{T^2 + T}$, and the relative concentration error is given by

$$\frac{\sigma_c}{c} = \frac{0.434}{\log T} \times k_2 \sqrt{1 + \frac{1}{T}}$$

where k_2 is a constant for a given instrument.

In the final category, the major sources of uncertainty are light source fluctuations (flicker noise) or imprecision in positioning the cells. In this case the standard deviation in transmittance, σ_T is directly proportional to the transmittance ($\sigma_T = k_3 T$, where k_3 is a constant) and the relative concentration error is

$$\frac{\sigma_c}{c} = \frac{0.434}{\log T} \times k_3$$

We can produce plots in Excel to show us how the relative concentration error depends on transmittance or absorbance. Such plots can be used to indicate the optimum range of T or A values that minimize the uncertainty. Let's prepare a worksheet in which we list absorbance values from 0 to 3.0. For case 1, we'll use a typical k_1 value of ± 0.003. For case 2, a typical k_2 value of ± 0.003 will be used. For case 3, we'll use a k_3 value of 0.0075. Our worksheet is shown in Figure 12-6. The absorbance values are entered in cells A9 through A25 and the corresponding transmittances calculated in cells B9:B25. In cell C9, we enter the expression for the category 1 relative concentration error, which we multiply by 100% to get the percent error. Note that we calculate the absolute value of the error, since uncertainties can have either sign. We then copy this formula into cells C10:C25. Likewise in cell D9, we enter the category 2 relative error expression and copy it into the appropriate column D cells. Finally we enter the formula for the category 3 concentration error in cell E9 and copy it into E10:E25.

	A	B	C	D	E	F
1	**Precision of molecular absorption measurements**					
2	Cases: 1. $\sigma_T = k_1$ for readout resolution or thermal noise limit					
3	2. $\sigma_T = k_2(T^2+T)^{1/2}$ for shot noise limit					
4	3. $\sigma_T = k_3T$ for flicker noise or cell positioning limit					
5	$k_1=k_2$	0.003				
6	k_3	0.0075				
7			Case 1	Case 2	Case 3	
8	A	T	σ_c/c	σ_c/c	σ_c/c	
9	0.03	0.933254	4.650394	6.246464	10.85	
10	0.05	0.891251	2.921736	3.793286	6.51	
11	0.2	0.630957	1.031765	1.046652	1.6275	
12	0.4	0.398107	0.817619	0.609988	0.81375	
13	0.6	0.251189	0.863893	0.484307	0.5425	
14	0.8	0.158489	1.026883	0.440015	0.406875	
15	1	0.1	1.302	0.431825	0.3255	
16	1.2	0.063096	1.719609	0.445365	0.27125	
17	1.4	0.039811	2.336054	0.475292	0.2325	
18	1.6	0.025119	3.239597	0.51985	0.203438	
19	1.8	0.015849	4.563925	0.579099	0.180833	
20	2	0.01	6.51	0.654247	0.16275	
21	2.2	0.00631	9.379686	0.747402	0.147955	
22	2.4	0.003981	13.62698	0.861514	0.135625	
23	2.6	0.002512	19.93598	1.00042	0.125192	
24	2.8	0.001585	29.33952	1.168952	0.11625	
25	3	0.001	43.4	1.373115	0.1085	
26	**Documentation**					
27	Cell B9=10^-A9					
28	Cell C9=ABS((0.434/LOG(B9))*($B$5/B9)*100)					
29	Cell D9=ABS((0.434/LOG(B9))*$B$5*SQRT(1+(1/B9)))*100					
30	Cell E9=ABS((0.434/LOG(B9))*$B$6)*100					

Figure 12-6 Spreadsheet to calculate relative concentration error.

Next, we'll plot the three series. Because the errors can get very large in some cases, we'll use different numbers of cells for some of the plots. In the chart shown in Figure 12-7, for case 1, we use cells A9:A20 and C9:C20 to limit the plot to errors less than 10%. For case 2, we use all the cells, A9:A25 and D9:D25. Finally, for case 3, we leave out the first row and use cells A10:A25 and E10:E25 to avoid the large error shown in cell E9.

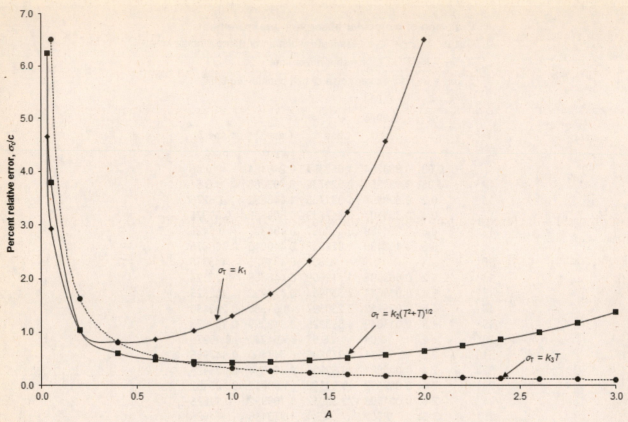

Figure 12-7 Relative concentration errors for 3 classes of instrumental uncertainties.

Calibration Methods

In most spectrophotometric methods, calibration is achieved either by the method of *external*

standards, or by the method of *standard additions*. We'll discuss both calibration methods here.

These principles of these methods were discussed in Chapter 4.

External Standard Methods

In the external standard method, a series of standard solutions of the analyte is used to prepare

the calibration curve or to produce a linear regression equation. The slope of the calibration

curve is the product of absorptivity and path length. Thus, using external standards is a way of

determining the proportionality factor between absorbance and concentration under the same conditions and with the same instrument as is used for the samples. In quantitative analysis, it is

seldom if ever safe to assume adherence to Beer's law and only use a single standard to

determine the proportionality constant. It is even less wise to base the results on a literature value

of absorptivity. The best standards are those that match the composition of the samples as closely

as is feasible.

The worksheet and calibration curve for a typical external standard calibration is shown

in Figure 12-8. Here the absorbances for standards of 5.00, 10.00, 15.00, 20.00 and 25.00 µM are

obtained and used to produce the calibration curve and regression equation. The concentration of

the unknown is found by the formula shown in the documentation and discussed in Chapter 4.

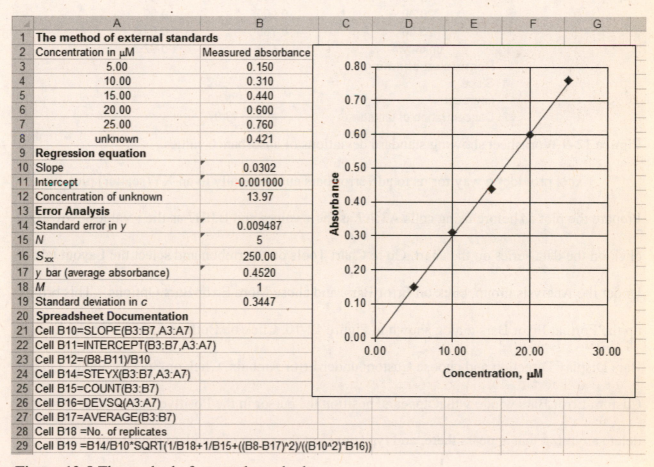

Figure 12-8 The method of external standards.

Adding Error Bars to a Chart

In many cases with experimental data, we would like a chart to have error bars indicating, for

example, $\pm 1\ s$. Let's assume that the data shown in Figure 12-8 came from triplicate

measurements on each standard, and that the absorbances were average values as shown in the

worksheet of Figure 12-9. The standard deviations are also shown in cells C3:C8.

	A	B	C
1	**The method of external standards**		
2	Concentration in µM	Measured absorbance	Std. Dev.
3	5.00	0.150	0.03
4	10.00	0.310	0.02
5	15.00	0.440	0.01
6	20.00	0.600	0.02
7	25.00	0.760	0.01
8	unknown	0.421	0.02
9	**Regression equation**		
10	Slope	0.0302	
11	Intercept	-0.0010	
12	Concentration of unknown	13.97	

Figure 12-9 Worksheet showing standard deviations of absorbance values.

Excel provides a way for us to add error bars automatically to an XY(scatter) plot.

Prepare the plot as before using cells A3:A7 as the x values and B3:B7 as the y values. Now right

click on the data series on the chart. Go to Chart Tools on the ribbon and select the Layout tab.

Under the Analysis group, click on Error Bars, and chose More Error Bars Options…This brings

up the Format Error Bars dialog shown in Figure 12-10. Click on <u>B</u>oth under the Vertical Error

Bars Display Direction and choose <u>C</u>ustom under Error Amount. Click on Specify <u>V</u>alue. On the

Custom Error Bars window that appears, position the cursor in the <u>P</u>ositive Error Value box,

delete anything that appears there, and type or select the range of standard deviations in the

worksheet (C3:C7). Repeat this for the <u>N</u>egative Error Value box as shown in Figure 12-11, and

click the OK button. Note on the plot that Excel displays horizontal error bars as well as vertical error bars. Since we want only the vertical bars, right click on any horizontal error bar on the plot and select Delete. Only the vertical error bars should remain. Add a Trendline to the plot and display it as a separate chart as shown in Figure 12-12.

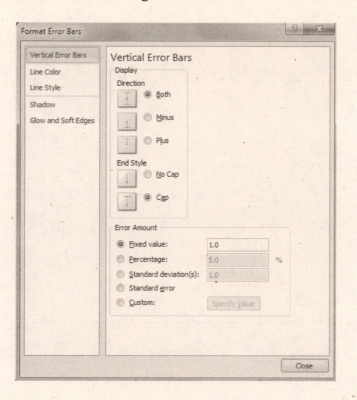

Figure 12-10 Adding error bars.

Figure 12-11 Specifying error bar values.

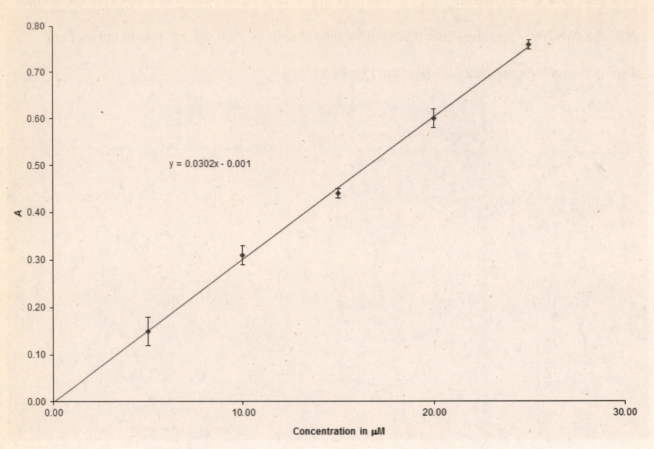

Figure 12-12 Calibration curve with error bars added to the *y* data.

Standard Additions Methods

The method of standard additions is often used when it is difficult or impossible to prepare

standards with a composition resembling that of the samples. The standard additions method

requires that Beer's law be obeyed. For this reason, the multiple standard additions method is

most often used because it provides a built-in check that Beer's law is obeyed.

As an example, let's consider the determination of Fe^{3+} in a natural water sample as

discussed in Example 26-1 of FAC9 and Example 23-2 of AC7. In this example, Fe^{3+} was

determined as the thiocyanate complex. A standard solution of 11.1 ppm Fe^{3+} was used and

volumes ranging from 0.00 to 20.00 mL were added to 10.00 mL of the sample in 50.00 ml volumetric flasks. The results are shown in Figure 12-13. Note that the error analysis is identical

to that described in conjunction with Figure 4-18 in Chapter 4.

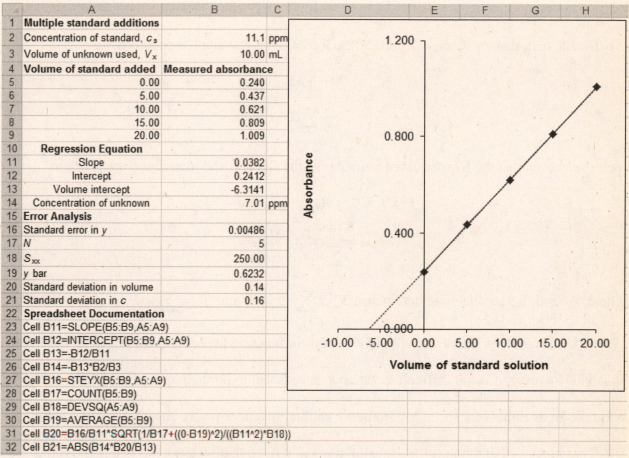

	A	B	C	D	E	F	G	H
1	Multiple standard additions							
2	Concentration of standard, c_s	11.1	ppm					
3	Volume of unknown used, V_x	10.00	mL					
4	Volume of standard added	Measured absorbance						
5	0.00	0.240						
6	5.00	0.437						
7	10.00	0.621						
8	15.00	0.809						
9	20.00	1.009						
10	Regression Equation							
11	Slope	0.0382						
12	Intercept	0.2412						
13	Volume intercept	-6.3141						
14	Concentration of unknown	7.01	ppm					
15	Error Analysis							
16	Standard error in y	0.00486						
17	N	5						
18	S_{xx}	250.00						
19	y bar	0.6232						
20	Standard deviation in volume	0.14						
21	Standard deviation in c	0.16						
22	Spreadsheet Documentation							
23	Cell B11=SLOPE(B5:B9,A5:A9)							
24	Cell B12=INTERCEPT(B5:B9,A5:A9)							
25	Cell B13=-B12/B11							
26	Cell B14=-B13*B2/B3							
27	Cell B16=STEYX(B5:B9,A5:A9)							
28	Cell B17=COUNT(B5:B9)							
29	Cell B18=DEVSQ(A5:A9)							
30	Cell B19=AVERAGE(B5:B9)							
31	Cell B20=B16/B11*SQRT(1/B17+((0-B19)^2)/((B11^2)*B18))							
32	Cell B21=ABS(B14*B20/B13)							

Figure 12-13 Multiple standard additions method.

Weighted Linear Regression

In conventional calibration methods, an implicit assumption is that the random errors in the y

values are approximately constant and do not vary with the x values (usually concentration). In

practice, however, we often find that the standard deviation of results increases with increasing

concentration. In such cases, the regression line should give more importance to the points with

best precision and less importance to those with the largest errors.

Let us assume that we collect the dependent variable y_i values at various values of the independent variable x_i. We'll assume there is negligible uncertainty in the x_i values as usual, but there is uncertainty in the y_i values that is given by σ_i. Also, suppose that each standard deviation is different such that $\sigma_1 \neq \sigma_2 \neq \sigma_3$, etc. If each data point is weighted according to

$$w_i = \frac{1/\sigma_i^2}{\frac{1}{N}\sum(1/\sigma_i^2)}$$

it can be shown that the least-squares estimates of the slope m and intercept b are given by

$$b = \bar{y}_w - b\bar{x}_w$$

$$m = \frac{\sum w_i x_i y_i - N\bar{x}_w \bar{y}_w}{\sum w_i x_i^2 - N\bar{x}_w^2}$$

where $\bar{x}_w$ and $\bar{y}_w$ are the weighted measn $\bar{x}_w = \sum w_i x_i / N$ and $\bar{y}_w = \sum w_i y_i / N$.

Unfortunately, Excel does not have a built-in function for accomplishing weighted least squares. However, it is not difficult to set up a spreadsheet to do the required calculations. We will take an example of a spectrophotometric calibration curve where absorbances of a complex NiQ^{2+} are measured at 530 nm for concentrations of 0.0, 20.0, 40.0, 60.0, 80.0 and 100.0 μM Ni^{2+}. Triplicate measurements of the corresponding absorbances along with their standard deviations in parentheses gave: 0.011 (0.001), 0.200 (0.005), 0.376 (0.012), 0.590 (0.016), 0.721 (0.022), and 0.976 (0.028). A solution of unknown Ni^{2+} concentration gave an absorbance of 0.246 (0.007). We will first set up the worksheet and do a conventional regression as shown in Figure 12-14. From this, the concentration of the unknown is 25.30 ± 2.11 μM or 25 ± 2 μM.

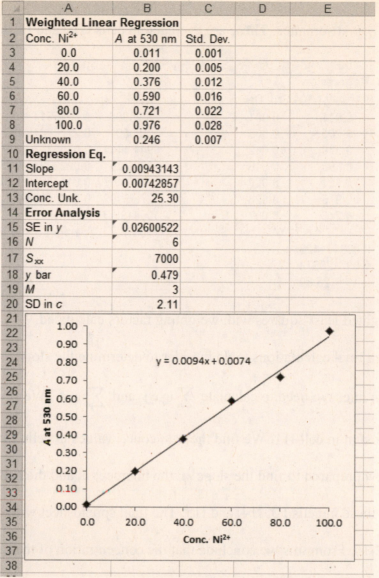

	A	B	C	D	E
1	**Weighted Linear Regression**				
2	Conc. Ni^{2+}	*A* at 530 nm	Std. Dev.		
3	0.0	0.011	0.001		
4	20.0	0.200	0.005		
5	40.0	0.376	0.012		
6	60.0	0.590	0.016		
7	80.0	0.721	0.022		
8	100.0	0.976	0.028		
9	Unknown	0.246	0.007		
10	**Regression Eq.**				
11	Slope	0.00943143			
12	Intercept	0.00742857			
13	Conc. Unk.	25.30			
14	**Error Analysis**				
15	SE in *y*	0.02600522			
16	*N*	6			
17	S_{xx}	7000			
18	*y* bar	0.479			
19	*M*	3			
20	SD in *c*	2.11			

Figure 12-14 Conventional linear regression analysis of spectrophotometric Ni^{2+} determination.

Although the regression looks good, we notice that the standard deviations increase with increasing concentration and decide to do a weighted least-squares analysis. We will first set up some addition columns to allow us to calculate the weighting factors as shown in Figure 12-15. In cells E3:E8, we calculate $1/s^2$ for each standard. In cell E9, we sum these values, and in cell E10, we divide by N to get the denominator of the weighting factor w_i. In cells F3:F8, we calculate the weighting factors, and in cell F9 we sum the weighting factors to check that they add up to N, the number of x values. In cells G3:G8, we find $w_i \times$ concentration and in cells

H3:H8, we obtain $w_i \times$ Absorbance. These are averaged in cells G10 and H10.

	A	B	C	D	E	F	G	H
1	**Weighted Linear Regression**							
2	Conc. Ni^{2+}	A at 530 nm	Std. Dev.		$1/s^2$	w_i	w_i*conc.	w_i*A
3	0.0	0.011	0.001		1000000	5.6915611	0	0.0626072
4	20.0	0.200	0.005		40000	0.2276624	4.5532489	0.0455325
5	40.0	0.376	0.012		6944.4444	0.0395247	1.5809892	0.0148613
6	60.0	0.590	0.016		3906.25	0.0222327	1.3339596	0.0131173
7	80.0	0.721	0.022		2066.1157	0.0117594	0.9407539	0.0084785
8	100.0	0.976	0.028		1275.5102	0.0072596	0.7259644	0.0070854
9	Unknown	0.246	0.007	Sum	1054192.3	6		
10	**Regression Eq.**			Sum/N	175698.72	Averages	1.522486	0.0252804
11	Slope	0.009431429						
12	Intercept	0.007428571						
13	Conc. Unk.	25.30						

Figure 12-15. Weighted least-squares with weighting factors calculated.

We have two final calculations to do in order to determine the slope and intercept of the

weighted regression line. We need to calculate $\sum w_i x_i y_i$ and $\sum w_i x_i^2$. We find $w_i \times$ conc. $\times A$ in

cells I3:I8 and their sum in cell I11. We find the $w_i \times$ conc.^{2} values in cells J3:J8 and the sum in

cell J11. We are now prepared to find the slope m, the intercept b, and the concentration of the

unknown. We find these in cells I 3, I14, and I15. The final spreadsheet with documentation is

shown in Figure 12-16. From this we conclude that the concentration of the unknown is 25.1

μM.

	A	B	C	D	E	F	G	H	I	J	K	L
1	**Weighted Linear Regression**											
2	Conc. Ni^{2+}	A at 530 nm	Std. Dev.		$1/s^2$	w_i	w_i*conc.	w_i*A	w_i*conc.*A	w_i*conc.^{2}		
3	0.0	0.011	0.001		1000000	5.6915611	0	0.0626072	0	0		
4	20.0	0.200	0.005		40000	0.2276624	4.5532489	0.0455325	0.9106498	91.06498		
5	40.0	0.376	0.012		6944.4444	0.0395247	1.5809892	0.0148613	0.5944519	63.23957		
6	60.0	0.590	0.016		3906.25	0.0222327	1.3339596	0.0131173	0.7870362	80.03758		
7	80.0	0.721	0.022		2066.1157	0.0117594	0.9407539	0.0084785	0.6782836	75.26031		
8	100.0	0.976	0.028		1275.5102	0.0072596	0.7259644	0.0070854	0.7085413	72.59644		
9	Unknown	0.246	0.007	Sum	1054192.3	6						
10	**Regression Eq.**			Sum/*N*	175698.72	Averages	1.522486	0.0252804				
11	Slope	0.009431429						Sums	3.6789627	382.1989		
12	Intercept	0.007428571										
13	Conc. Unk.	25.30						*m*	0.0093622			
14	**Error Analysis**							*b*	0.0110265			
15	SE in *y*	0.026005219						Conc. Unk.	25.098			
16	*N*	6										
17	S_{xx}	7000										
18	*y* bar	0.479										
19	*M*	3										
20	SD in *c*	2.11										

Graph: y-axis "A at 530 nm" from 0.00 to 1.00; x-axis "Conc. Ni^{2+}" from 0.0 to 100.0; equation $y = 0.0094x + 0.0074$.

38 **Documentation**		
39 Cell B11=SLOPE(B3:B8,A3:A8)	Cell E10=E9/B16	Cell J11=SUM(J3:J8)
40 Cell B12=INTERCEPT(B3:B8,A3:A8)	Cell F3=E3/$E$10	Cell I13=(I11-B16*G10*H10)/(J11-6*G10^2)
41 Cell B13=(B9-B12)/B11	Cell F9=SUM(F3:F8)	Cell I 14=H10-I13*G10
42 Cell B15=STEYX(B3:B8,A3:A8)	Cell G3=F3*A3	Cell I 15=(B9-I14)/I13
43 Cell B16=COUNT(B3:B8)	Cell G10=AVERAGE(G3:G8)	
44 Cell B17=DEVSQ(A3:A8)	Cell H3=F3*B3	
45 Cell B18=AVERAGE(B3:B8)	Cell H10=AVERAGE(H3:H8)	
46 Cell B20=B15/B11*SQRT(1/B19+1/B16+((B9-B18)^2)/((B11^2)*B17))	Cell I3=F3*A3*B3	
47 Cell E3=1/C3^2	Cell I11=SUM(I3:I10)	
48 Cell E9=SUM(E3:E8)	Cell J3=F3*A3^2	

Figure 12-16 Final spreadsheet for weighted regression analysis.

We note from the analysis in Figure 12-16 that the results are nearly identical for the unweighted and weighted regression lines. At first glance, the extra trouble to do a weighted regression, particularly since Excel has no built-in function, hardly seems worth it. However, an analysis of the residuals tells a different story. The continuation of Figure 12-16 is shown in Figure 12-17. The residual analysis for the conventional regressions is shown in cells L3:N10.

	A	B	H	I	J	K	L	M	N	O	P	Q	R
1	Weighted Linear Regression							Conventional				Weighted	
2	Conc. Ni^{2+}	A at 530 nm	w_i*A	w_i*conc.*A	w_i*conc.^{2}		A_{pred}	Resid	Resd^2		A_{pred}	Resid	Resd^2
3	0.0	0.011	0.0626072	0	0		0.007429	3.571E-03	1.276E-05		0.011026	-2.6489E-05	7.017E-10
4	20.0	0.200	0.0455325	0.9106498	91.06498		0.196057	3.943E-03	1.555E-05		0.198271	1.7288E-03	2.989E-06
5	40.0	0.376	0.0148613	0.5944519	63.23957		0.384686	-8.686E-03	7.544E-05		0.385516	-9.5160E-03	9.055E-05
6	60.0	0.590	0.0131173	0.7870362	80.03758		0.573314	0.0167	2.784E-04		0.572761	0.0172	2.972E-04
7	80.0	0.721	0.0084785	0.6782836	75.26031		0.761943	-0.0409	1.676E-03		0.760005	-0.0390	1.521E-03
8	100.0	0.976	0.0070854	0.7085413	72.59644		0.950571	0.0254	6.466E-04		0.947250	0.0287	8.265E-04
9	Unknown	0.246						Sum	2.705E-03			Sum	0.002739
10	Regression Eq.		0.0252804					SE in y	0.026005			SE in y	0.026166
11	Slope	0.00943143	Sums	3.6789627	382.1989								
12	Intercept	0.00742857						Documentaion					
13	Conc. Unk.	25.30	m	0.0093622				Cell L3=$B$11*A3+$B$12			Cell P3=$I$13*A3+$I$14		
14	Error Analysis		b	0.0110265				Cell M3=B3-L3			Cell Q3=B3-P3		
15	SE in y	0.02600522	Conc. Unk.	25.098				Cell N3=M3^2			Cell R3=Q3^2		
16	N	6						Cell N9=SUM(N3:N8)			Cell R9=SUM(R3:R8)		
17	S_{xx}	7000						Cell N10=SQRT(N9/4)			Cell R10=SQRT(R9/4)		
18	y bar	0.479											
19	M	3											
20	SD in c	2.11											

Figure 12-17 Continuation of weighted regression worksheet with columns C:G hidden.

We first calculate the absorbances predicted by the regression line in cells L3:L8. The residuals are $A - A_{pred}$. These and their squares are obtained in cells M3:M8 and N3:N8. The sum of the squares of the residuals is calculated in cell N9 and the standard error in y (standard error of the estimate) is obtained in cell N10. Note that this is identical to the value obtained previously in cell B15.

The analysis of the weighted-regression residuals is shown in cells P3:R10. While there is no improvement in the overall standard error in cell R10, note the values for the residuals for the low concentrations in cells R3:R5. These are the absorbances with the lowest standard deviations. The weighted regression has given the most precise absorbances more weight in the line and thus their residuals are smaller than the conventional (unweighted) values. An analysis of the concentration error similar to that for the unweighted case predicts that the standard deviation in concentration is 2.0 μM, which is slightly lower than that for the unweighted case. However, with attention paid to significant figures, in both cases the unknown Ni^{2+} concentration is 25 ± 2 μM. It should be noted that we use regression calculations to obtain confidence intervals and more statistics than just slope, intercept and concentration. Hence, even though the calculations for weighted regression are more complex, they can give more realistic values in many cases.

Multicomponent Methods

Mixtures can be analyzed by absorption spectrophotometry based on additivity of absorbances at a given wavelength. For a two-component mixture of species M and N measured at two wavelengths, the following equations hold:

$$A_1 = \varepsilon_{M_1} b c_M + \varepsilon_{N_1} b c_N$$

$$A_2 = \varepsilon_{M_2} b c_M + \varepsilon_{M_2} b c_N$$

where the subscript 1 indicates measurement at wavelength λ_1, and the subscript 2 indicates measurement at λ_2. With known values of molar absorptivities and pathlengths determined from separate measurements, there are two unknowns for these two equations, which can be solved by several methods.

We'll demonstrate a few of these here using Example 23-3 of AC7 for a mixture of titanium and vanadium measured as the peroxide complexes. Separate experiments on standard solutions gave the following molar absorptivities at 400 and 460 nm.

	εb at 400 nm in L mol^{-1}	εb at 460 nm in L mol^{-1}
V	145	232
Ti	644	321

A steel sample containing both metals was dissolved and treated with H_2O_2. The absorbances of the mixture were 0.172 at 400 nm and 0.116 at 460 nm in a 1.00 cm cell.

Iteration

We can solve the two absorbance equations simultaneously for the concentrations of titanium and vanadium in terms of known quantities. If we let M = Ti, N = V, $\lambda_1 = 400$ nm, and $\lambda_2 = 460$ nm, the equation for c_{Ti} is

$$c_{Ti} = \frac{A_{400} - \varepsilon_{V(400)} b c_V}{\varepsilon_{Ti(400)} b}$$

If we solve the two absorbance equations for c_V, we obtain

$$c_V = \frac{A_{460} - \varepsilon_{Ti(460)} b \left(\dfrac{A_{400} - \varepsilon_{V(400)} b c_V}{\varepsilon_{Ti(400)} b} \right)}{\varepsilon_{V(460)} b}$$

We can solve these equations algebraically to find $c_{Ti} = 0.0001895$ M and $c_V = 0.0002244$ M. We can also solve these equations by iterative methods using Solver, Goal Seek or circular references. We will illustrate the latter approach here since we've discussed Solver and Goal Seek earlier. Begin with a worksheet similar to that shown in Figure 12-18. Since the solution cells, B10 and B11, will contain circular references, we will first Enable iterative calculation in the Formulas section of Excel options. Choose 1000 iterations with a maximum step-size of 0.000001.[1]

---

[1] Note in Excel 2007, Excel Options are found in the Office button. In Excel 2010 they are part of the File tab.

	A	B	C	D
1	Spreadsheet to solve two-component absorption problem by iteration			
2	Known parameters	$\varepsilon_{(400)}b$ in L·mol^{-1}	$\varepsilon_{(460)}b$ in L·mol^{-1}	
3	Ti	644	321	
4	V	145	232	
5				
6	Data	A(400) for mixture	A(460) for mixture	
7		0.172	0.116	
8				
9	Solution	Iteration		
10	Conc. V			
11	Conc. Ti			
12				
13	Predicted A(400)			
14	Predicted A(460)			

Figure 12-18 Beginning worksheet for two-component absorption.

Now enter the formula corresponding to the equation for c_V into cell B10 and that corresponding to c_{Ti} in cell B11. It is useful at this point to check the validity of the results by using the calculated concentrations in cells B10 and B11 to predict the absorbance values A_{400} and A_{460}. Write the equation for the absorbances in cells B13 and B14 and see if they match the experimental absorbances. Your final worksheet should appear as shown in Figure 12-19 after documentation.

	A	B	C	D
1	Spreadsheet to solve two-component absorption problem by iteration			
2	Known parameters	$\varepsilon_{(400)}b$ in L mol^{-1}	$\varepsilon_{(460)}b$ in L mol^{-1}	
3	Ti	644	321	
4	V	145	232	
5				
6	Data	A(400) for mixture	A(460) for mixture	
7		0.172	0.116	
8				
9	Solution	Iteration		
10	Conc. V	0.0001895		
11	Conc. Ti	0.0002244		
12				
13	Predicted A(400)	0.172		
14	Predicted A(460)	0.116		
15				
16	Documentation			
17	Cell B10=(C7-C3*(B7-B4*B10)/B3)/C4			
18	Cell B11=(B7-B4*B10)/B3			
19	Cell B13=B3*B11+B4*B10			
20	Cell B14=C3*B11+C4*B10			

Figure 12-19 Final worksheet for iterative solution to multicomponent absorption.

Method of Determinants

We'll now take a more complicated example and use the method of determinants discussed in

Chapter 6. A mixture of Co^{2+}, Ni^{2+} and Cu^{2+} was analyzed by making absorption measurements

at three wavelengths, 392 nm, 510 nm, and 800 nm. By measurements on standard solutions of

individual metal ions, the molar absorptivities given in cells C4:E6 of Figure 12-20 were

obtained. The mixture had the absorbances at the three wavelengths given in cells I4:I6.

	A	B	C	D	E	F	G	H	I
1	Three component mixture by determinants								
2			Molar absorptivity, L M^{-1} cm^{-1}						
3		λ, nm	Co^{2+}	Ni^{2+}	Cu^{2+}			λ, nm	A
4	Known values	392	0.992	6.805	0.179		Mixture	392	0.94
5		510	6.452	0.21	0.2			510	0.523
6		800	0.459	1.175	14.99			800	1.065

Figure 12-20 Initial worksheet for three-component mixture determination.

By applying the same Cramer's rule methodology we discussed in Chapter 6, we can solve for

the concentrations of the three metals in the mixture. The final spreadsheet is shown in Figure

12-21. Note that we are showing here more figures in the concentrations than are significant.

Since we began with 3 significant figures, we would expect the calculations to degrade the

precision so that even fewer figures will be significant at the end. We show the results to 5

figures after the decimal here only for comparison purposes.

	A	B	C	D	E	F	G	H	I
1	**Three component mixture by determinants**								
2			**Molar absorptivity, L M^{-1} cm^{-1}**						
3		λ, nm	Co^{2+}	Ni^{2+}	Cu^{2+}			λ, nm	A
4	*Known values*	392	0.992	6.805	0.179		Mixture	392	0.94
5		510	6.452	0.21	0.2			510	0.523
6		800	0.459	1.175	14.99			800	1.065
7	D	-653.295							
8									
9		0.94	6.805	0.179					
10		0.523	0.21	0.2					
11		1.065	1.175	14.99					
12	D_1	-49.0921							
13									
14		0.992	0.94	0.179					
15		6.452	0.523	0.2					
16		0.459	1.065	14.99					
17	D_2	-82.0735							
18									
19		0.992	6.805	0.94					
20		6.452	0.21	0.523					
21		0.459	1.175	1.065					
22	D_3	-38.4783							
23	**Concentrations**								
24	c_{Co2+}	0.07515							
25	c_{Ni2+}	0.12563							
26	c_{Cu2+}	0.05890							
27	**Documentation**								
28	Cell B7=MDETERM(C4:E6)								
29	Cell B12=MDETERM(B9:D11)								
30	Cell B17=MDETERM(B14:D16)								
31	Cell B22=MDETERM(B19:D21)								
32	Cell B24=B12/B7								
33	Cell B25=B17/B7								
34	Cell B26=B22/B7								

Figure 12-21 Final results for three-component mixture.

Matrix Methods

We can also solve the three-component mixture problems by the matrix inversion procedures

described in Chapter 6. Figure 12-22 shows the worksheet and its documentation for

accomplishing this. Note that the concentrations are determined by first calculating the inverse

matrix and then the solution matrix in cells E8:E10. In cells B12:B14, this same calculation is

done in a single step by the formula **=MMULT(MINVERSE(A3:C5),E3:E5)**. The simplicity

of the matrix approach can be seen in Figure 12-18.

	A	B	C	D	E	F
1	Three component mixture by matrix methods					
2	Coefficient Matrix				Constant Matrix	
3	0.992	6.805	0.179		0.94	
4	6.452	0.21	0.2		0.523	
5	0.459	1.175	14.99		1.065	
6						
7	Inverse Matrix				Solution Matrix	
8	-0.00446	0.15582	-0.00203		0.07515	
9	0.147902	-0.02264	-0.00146		0.12563	
10	-0.01146	-0.003	0.066888		0.05890	
11						
12	c_{Co2+}	0.07515				
13	c_{Ni2+}	0.12563				
14	c_{Cu2+}	0.05890				
15	Documentation					
16	Cells A8:C10=MINVERSE(A3:C5)					
17	Cells E8:E10=MMULT(A8:C10,E3:E5)					
18	Cells B12:B14=MMULT(MINVERSE(A3:C5),E3:E5)					

Figure 12-22 Determination of three-component mixture using matrix methods.

Spectrophotometric Titrations

Spectrophotometric measurements are very useful in monitoring the progress of titrations. The

specific shapes of titration curves depend on which species absorb radiation, the analyte, the

titrant, the products, or a mixture of these. Most titration curves contain linear segments that can

be extrapolated to obtain the end point. In order to obtain linear portions, Beer's law must be obeyed and the absorbances must be corrected for volume changes by multiplying them by

$(V + v)/V$, where V is the initial volume, and v is the volume of titrant added.

We'll consider here an example of a spectrophotometric titration used to standardize a

titrant. The standard solution is a reagent X which reacts with the titrant T to produce product P

according to $X + T \rightarrow P$. A 50.00 mL aliquot of 0.0045 M standard is titrated with an unknown

concentration of titrant. The raw absorbances are measured as a function of volume as shown in

Figure 12-23.

	A	B
1	**Spectrophotometric titration**	
2	Titration reaction: X + T → P	
3	$c_{standard}$	0.0045
4	Vol. standard, mL	50.00
5	Vol. titrant, mL	A_{meas}
6	0.00	0.234
7	0.35	0.202
8	0.70	0.169
9	1.05	0.132
10	1.40	0.102
11	1.75	0.070
12	2.10	0.070
13	2.45	0.157
14	2.80	0.242
15	3.15	0.333
16	3.50	0.420
17	3.85	0.508
18	4.20	0.595

Figure 12-23 Data for spectrophotometric titration.

Before constructing the titration curve and finding the end point, we must correct the

measured absorbances, A_{meas}, for volume changes. Thus, we enter a formula for this correction in

cell C6 and copy it into cells C7:C18. We next do a preliminary plot of the corrected

absorbances, A_{corr}, vs. titrant volume and find there are two linear segments with the end point

somewhere between 1.75 and 2.10 mL. We thus prepare scatter chart with two series: one

containing x values in cells A6:A11 and y values in cells C6:C11, and the other containing the

remaining x and y values. We then find the intersection point of the two lines as we did for amperometric titrations in Chapter 11. The results, shown in Figure 12-24, indicate the end point

occurs at 2.0195 mL of titrant and that the titrant concentration is 0.1114 M.

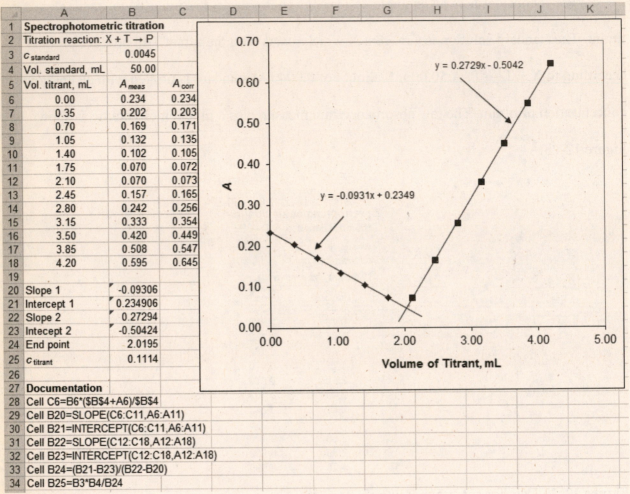

Figure 12-24 Spectrophotometric titration results.

An example of the calculations needed to determine a typical spectrophotometric titration

curve is given in Figure 12-25. Here all species absorb at the analysis wavelength. The titration

reaction is A + T ¾→ P. A 50.00-mL aliquot of 0.0013 M analyte is titrated with 0.00325 M

titrant, and the absorbance measured in a 3.00-cm cell. Since there are $50.00 \times 0.0013 = 0.065$

mmoles of analyte present, the equivalence point should occur at $0.065/0.00325 = 20.00$ mL of

titrant. Before the equivalence point, there is no unreacted titrant, and the analyte and product

concentrations are given by

$$c_{analyte} = \frac{\text{no. of mmoles analyte originally present} - \text{no. of mmoles titrant added}}{\text{total solution volume, mL}}$$

$$c_{product} = \frac{\text{no. of mmoles titrant added}}{\text{total solution volume, mL}}$$

After the equivalence point, there is no analyte remaining, and the titrant and product concentrations are given by

$$c_{titrant} = \frac{\text{no. of mmoles titrant added} - \text{no. of mmoles analyte originally present}}{\text{total solution volume, mL}}$$

$$c_{product} = \frac{\text{no. of mmoles analyte originally present}}{\text{total solution volume, mL}}$$

The measured absorbance is found from

$$A_{meas} = \varepsilon_{analyte} b c_{analyte} + \varepsilon_{titrant} b c_{titrant} + \varepsilon_{product} b c_{product}$$

Note that the absorbances are corrected for volume changes as in the previous example. The end point is found by the same method as the previous example. The small increase in absorbance beyond the equivalence point is due to the small absorbance contributed by the titrant.

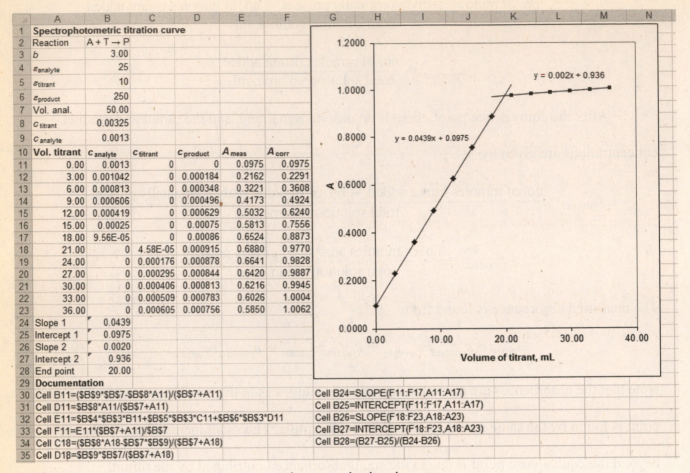

	A	B	C	D	E	F
1	Spectrophotometric titration curve					
2	Reaction	A + T → P				
3	b	3.00				
4	$\varepsilon_{analyte}$	25				
5	$\varepsilon_{titrant}$	10				
6	$\varepsilon_{product}$	250				
7	Vol. anal.	50.00				
8	$c_{titrant}$	0.00325				
9	$c_{analyte}$	0.0013				
10	Vol. titrant	$c_{analyte}$	$c_{titrant}$	$c_{product}$	A_{meas}	A_{corr}
11	0.00	0.0013	0	0	0.0975	0.0975
12	3.00	0.001042	0	0.000184	0.2162	0.2291
13	6.00	0.000813	0	0.000348	0.3221	0.3608
14	9.00	0.000606	0	0.000496	0.4173	0.4924
15	12.00	0.000419	0	0.000629	0.5032	0.6240
16	15.00	0.00025	0	0.00075	0.5813	0.7556
17	18.00	9.56E-05	0	0.00086	0.6524	0.8873
18	21.00	0	4.58E-05	0.000915	0.6880	0.9770
19	24.00	0	0.000176	0.000878	0.6641	0.9828
20	27.00	0	0.000295	0.000844	0.6420	0.9887
21	30.00	0	0.000406	0.000813	0.6216	0.9945
22	33.00	0	0.000509	0.000783	0.6026	1.0004
23	36.00	0	0.000605	0.000756	0.5850	1.0062
24	Slope 1	0.0439				
25	Intercept 1	0.0975				
26	Slope 2	0.0020				
27	Intercept 2	0.936				
28	End point	20.00				
29	Documentation					
30	Cell B11=($B$9*$B$7-$B$8*A11)/($B$7+A11)					
31	Cell D11=$B$8*A11/($B$7+A11)					
32	Cell E11=$B$4*$B$3*B11+$B$5*$B$3*C11+$B$6*$B$3*D11					
33	Cell F11=E11*($B$7+A11)/$B$7					
34	Cell C18=($B$8*A18-$B$7*$B$9)/($B$7+A18)					
35	Cell D18=$B$9*$B$7/($B$7+A18)					

Chart annotations:
y = 0.002x + 0.936
y = 0.0439x + 0.0975

Right-column documentation:
Cell B24=SLOPE(F11:F17,A11:A17)
Cell B25=INTERCEPT(F11:F17,A11:A17)
Cell B26=SLOPE(F18:F23,A18:A23)
Cell B27=INTERCEPT(F18:F23,A18:A23)
Cell B28=(B27-B25)/(B24-B26)

Figure 12-25 Constructing the spectrophotometric titration curve

Spectrophotometric Study of Complex Ions

Spectrophotometry is often used for determining the composition and formation constants of complex ions in solution. There are three common methods used for complex-ion studies: the method of continuous variations, the mole-ratio method, and the slope-ratio method. Only the method of continuous variations is illustrated here. The other methods are considered in detailed in Section 26A-5 of FAC9.

Method of Continuous Variations

In this method, metal and ligand solutions with identical analytical concentrations are mixed in such a way that the total volume in each mixture is constant, but the ratio of reactants varies systematically. Since the total volume is constant, the total number of moles of reactants is constant. The absorbance of each solution is measured and corrected for any absorbance that the mixture might have if no reaction had occurred. The corrected absorbance is plotted against the volume fraction of one reactant. For example, the corrected absorbance is plotted against $V_L/(V_L+V_M)$, where V_M is the volume of the metal ion solution and V_L is the volume of the ligand solution. Typically, for complexes having reasonably large formation constants, two linear segments are obtained and extrapolated until they intersect. The intersection corresponds to the combining ratio of metal ion and ligand in the complex.

Figure 12-26 shows an example in which 1.00×10^{-4} M concentrations of metal ion and ligand solutions are used. We use a total volume of 10.00 mL, and vary the volume of the ligand solution in 1.00 mL increments from 0.00 to 10.00 mL. The volume of the metal ion solution is $10.00 - V_L$ in mL. The corrected absorbance is shown in cells D7:D17. A preliminary chart of the total curve indicated that V_L values from 0.00 to 4.00 mL give one linear segment, while volumes of 9.00 and 10.00 give a second linear segment. These two segments are then analyzed and the intersection point found to be a volume fraction of 0.75. Since, $V_L/(V_L + V_M) = 0.75$ is indicated, $V_L = 3V_M$. Hence, the formula ML_3 is suggested. Note that the data points corresponding to V_L values of 5.00 through 8.00 do not fall on the line. For these intermediate volume fractions, the formation reaction is incomplete. The deviation of the data points from the linear segments is a measure of the formation constant of the complex. For complexes with relatively small formation constants, the continuous variations plot may be essentially a continuous curve so that linear segments are not apparent. In such cases, the method of

continuous variations may not a good choice for investigating the composition.

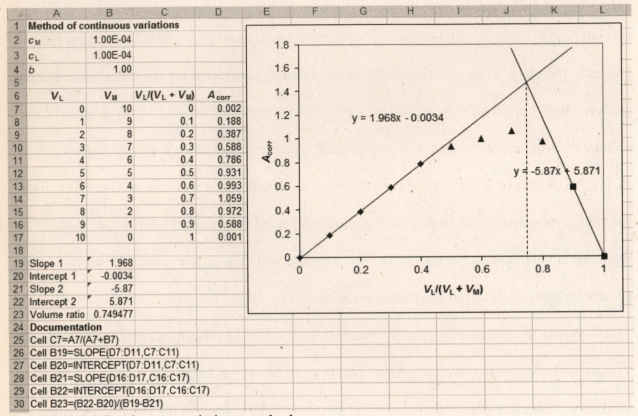

Figure 12-26 Continuous variations method.

Producing Charts with Insets

Often in making scientific plots, we would like a smaller inset plot to show a blow-up of one

region of a larger main plot. For example in Figure 12-26, we might like to produce an inset of

the region from volume fraction 0.6 to 0.8 to show the curved region more clearly. This is easy

to do in Excel. Prepare the main chart as a separate chart sheet, say Chart 1. Then create a second

chart of the inset region and right click on the new chart. Select Move Chart... and then choose

Object in Chart1 for the new location. In Figure 12-27, this was done by selecting cells C13:D15

as the points to be plotted and locating the new chart in the original chart.

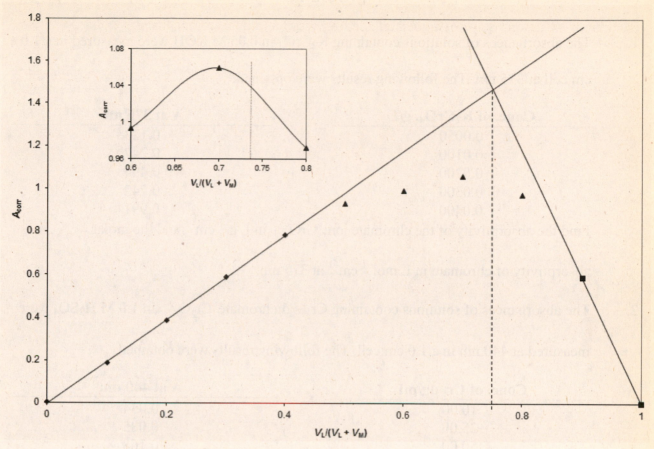

Figure 12-27 Continuous variations plot with inset showing region of curvature.

Summary

In this chapter we've explored several Excel features useful in spectrophotometric methods. We've examined Beer's law and some of its limitations and investigated some of the factors limiting precision in absorption spectroscopy. Two of the most useful calibration methods were illustrated, and multicomponent determinations were described. We also briefly looked at spectrophotometric titrations and at methods for determining the composition of complex ions. Several new Excel features, such as adding error bars to charts and producing insets, were also described. The calibration methods discussed here are applicable to atomic spectroscopy and molecular fluorescence spectroscopy as well as absorption methods.

Problems

1. The absorbances of solutions containing K_2CrO_4 in 0.05 M KOH were measured in a 1.0 cm cell at 375 nm. The following results were obtained:

Conc. of K_2CrO_4, g/L	A at 375 nm
0.0050	0.123
0.0100	0.247
0.0200	0.494
0.0300	0.742
0.0400	0.991

Find the absorptivity of the chromate ion, CrO_4^{2-} in L g^{-1} cm^{-1} and the molar absorptivity of chromate in L mol^{-1} cm^{-1} at 375 nm.

2. The absorbances of solutions containing Cr as dichromate $Cr_2O_7^{2-}$ in 1.0 M H_2SO_4 were measured at 440 nm in a 1.0 cm cell. The following results were obtained:

Conc. of Cr, µg/mL	A at 440 nm
10.00	0.034
25.00	0.085
50.00	0.168
75.00	0.252
100.00	0.335
200.00	0.669

Find the absorptivity of dichromate in L g^{-1} cm^{-1} and the molar absorptivity in L $mole^{-1}$ cm^{-1} at 440 nm.

3. Set up the spreadsheet of Figure 12-2 for the dimerization equilibrium to allow variation of the equilibrium constant for dimer formation. Vary the constant from 100 to 1000 and note the effect on the Beer's law plot of Figure 12-3. Use the same molar absorptivities and concentrations as in Figure 12-2.

4. Set up the stray light spreadsheet of Figure 12-4 to allow variation of the molar absorptivity of the absorber. Vary ε from 100 to 1.0×10^6 and note the effect. Use the

same stray light levels, but change the concentrations so as to obtain absorbances in the same range as in Figure 12-4.

5.   A compound X is to be determined by UV/Visible spectrophotometry. A calibration curve is constructed from standard solutions of X with the following results: 0.50 ppm, A = 0.24; 1.5 ppm, A = 0.36; 2.5 ppm, A = 0.44; 3.5 ppm, A = 0.59; 4.5 ppm, A = 0.70. A solution of unknown X concentration had an absorbance of A = 0.50. Find the slope and intercept of the calibration curve, the standard error in Y, the concentration of the solution of unknown X concentration and the standard deviation in the concentration of X. Construct a plot of the calibration curve and determine the unknown concentration by hand from the plot. Compare to that obtained from the regression line.

6.   One common way to determine phosphorus in urine is to treat the sample, after removing the protein, with molybdenum (VI) and then reducing the resulting 12-molybdophosphate complex with ascorbic acid to give an intense blue colored species called molybdenum blue. The absorbance of molybdenum blue can be measured at 650 nm. A 24-hour urine sample was collected and the patient produced 1122 mL in 24 hours. A 1.00 mL aliquot of the sample was treated with Mo(VI) and ascorbic acid and diluted to a volume of 50.00 mL. A calibration curve was prepared by treating 1.00 mL aliquots of phosphate standard solutions in the same manner as the urine sample. The absorbances of the standards and the urine sample were obtained at 650 nm and the following results obtained:

Solution	Absorbance at 650 nm
1.00 ppm P	0.230
2.00 ppm P	0.436
3.00 ppm P	0.638
4.00 ppm P	0.848
Urine sample	0.518

(a)   Find the slope and intercept of the calibration curve. Construct a plot of the

calibration curve. Determine the number of ppm P in the urine sample and its

standard deviation from the least-squares equation of the line. Compare the

unknown concentration to that obtained manually from a calibration curve.

(b)   What mass of phosphorus (grams) were eliminated per day by the patient?

(c)   What is the phosphate concentration in urine in mM?

7.   Nitrite is commonly determined by a colorimetric procedure using a reaction called the

Griess reaction. In this reaction, the sample containing nitrite is reacted with

sulfanilimide and N-(1-Napthyl)ethylenediamine to form a colored species that absorbs at

550 nm. Using an automated flow analysis instrument, the following results were

obtained for standard solutions of nitrite and for a sample containing an unknown

amount.

Solution	Absorbance at 550 nm
2.00 μM	0.065
6.00 μM	0.205
10.00 μM	0.338
14.00 μM	0.474
18.00 μM	0.598
Unknown	0.402

(a)   Obtain the slope, intercept and standard deviation of the calibration curve

(b)   Construct a plot of the calibration curve.

(c)   Determine the concentration of nitrite in the sample and its standard deviation.

8.   The following data were taken from a diode array spectrophotometer in an experiment to measure the spectrum of the Co(II)-EDTA complex. The column labeled $P_{solution}$ is the relative signal obtained with sample solution in the cell after subtraction of the dark signal. The column labeled $P_{solvent}$ is the reference signal obtained with only solvent in the cell after subtraction of the dark signal. Find the transmittance at each wavelength, and the absorbance at each wavelength. Plot the spectrum of the compound.

	A	B	C	D	E	F
1	Spreadsheet of data taken from diode array spectrophotometer					
2		Wavelength, nm	$P_{solvent}$	$P_{solution}$		
3		350	0.002689	0.002560		
4		375	0.006326	0.005995		
5		400	0.016975	0.015143		
6		425	0.035517	0.031648		
7		450	0.062425	0.024978		
8		475	0.095374	0.019073		
9		500	0.140567	0.023275		
10		525	0.188984	0.037448		
11		550	0.263103	0.088537		
12		575	0.318361	0.200872		
13		600	0.394600	0.278072		
14		625	0.477018	0.363525		
15		650	0.564295	0.468281		
16		675	0.655066	0.611062		
17		700	0.739180	0.704126		
18		725	0.813694	0.777466		
19		750	0.885979	0.863224		
20		775	0.945083	0.921446		
21		800	1.000000	0.977237		

9.   A standard solution was put through appropriate dilutions to give the concentrations of iron shown below. The iron(II)-1,10,phenanthroline complex was then developed in 25.0-mL aliquots of these solutions, following which each was diluted to 50.0 mL. The following absorbances (1.00-cm cells) were recorded at 510 nm:

Fe(II) concentration in original solutions, ppm	A_{510}
4.00	0.160 (0.001)
10.0	0.390 (0.002)
16.0	0.630 (0.003)
24.0	0.950 (0.008)

32.0	1.26 (0.011)
40.0	1.58 (0.050)

(a)   Plot a calibration curve from these data.

(b)   Use the method of least squares to prodice an equation relating absorbance and

the concentration of iron(II).

(c)   Calculate the standard deviation of the slope and intercept.

(d)   If the standard deviations for each absorbance value are those shown in

parentheses, find the weighted regression line and do a residual analysis. Are the

more precise points better estimated in the weighted regression?

10.   The method developed in Problem 9 above was used for the routine determination of iron

in 25.0-mL aliquots of ground water. Express the concentration (as ppm Fe) in samples

that yielded the accompanying absorbance data (1.00-cm cell). Calculate the relative

standard deviation of the result. Repeat the calculation assuming the absorbance data are

means of three measurements.

(a)   0.143   (d)   1.009

(b)   0.675   (e)   1.512

(c)   0.068   (f)   0.546

11.   The reduced form of nicotinamide adenine dinucleotide (NADH) is an important and

highly fluorescent coenzyme. It has an absorption maximum of 340 nm and an emission

maximum at 465 nm. Standard solutions of NADH gave the following fluorescence

intensities:

Concn NADH, μmol/L	Relative Intensity
0.100	2.24
0.200	4.52
0.300	6.63
0.400	9.01
0.500	10.94
0.600	13.71
0.700	15.49
0.800	17.91

(a)   Construct a spreadsheet and use it to make a calibration curve for NADH.

(b)   Find the least-squares slope and intercept for the plot in (a).

(c)   Calculate the standard deviation of the slope and about regression for the curve.

(d)   An unknown exhibits a relative fluorescence of 12.16. Use the spreadsheet to

        calculate the concentration of NADH.

(e)   Calculate the relative standard deviation for the result in part (d).

(f)   Calculate the relative standard deviation for the result in part (d) if the reading of

        12.16 was the mean of three measurements.

12.   Strontium was determined in a bone sample by atomic absorption spectrometry. The

        method of multiple standard additions was used. The following table indicates the

        volume of a 100 mg/L Sr standard added to 100 mL of the sample solution. The blank-

        corrected absorbance was recorded.

Volume of Sr added, μL	A
0.00	0.088
10.0	0.124
20.0	0.151
30.0	0.179

40.0   0.214

Find the concentration of Sr in the sample in ppm and its standard deviation.

13.   A. J. Mukhedkar and N. V. Deshpande (*Anal. Chem.*, **1963**, *35*, 47) report on a

simultaneous determination for cobalt and nickel based upon absorption by their 8-

quinolinol complexes. The molar absorptivities are ε_{Co} = 3529 and ε_{Ni} = 3228 at 365 nm

and ε_{Co} = 428.9 and ε_{Ni} = 0 at 700 nm. Use the method of determinants to find the

concentrations of nickel and cobalt in each of the following solutions (1.00-cm cells):

Solution	A_{700}	A_{365}
1	0.0235	0.617
2	0.0714	0.755
3	0.0945	0.920
4	0.0147	0.592
5	0.0540	0.685

14.   Molar absorptivity data for the cobalt and nickel complexes with 2,3-quinoxalinedithiol

are ε_{Co} = 36,400 and ε_{Ni} = 5520 at 510 nm and ε_{Co} = 1240 and ε_{Ni} = 17,500 at 656 nm. A

0.425-g sample was dissolved and diluted to 50.0 mL. A 25.0-mL aliquot was treated to

eliminate interferences; after addition of 2,3-quinoxalinedithiol, the volume was adjusted

to 50.0 mL. This solution had an absorbance of 0.446 at 510 nm and 0.326 at 656 nm in a

1.00-cm cell. Calculate the concentrations of cobalt and nickel (parts per million) in the

sample. Use the iterative method.

15.   Construct spectrophotometric titration curves for each of the following situations. Include

any dilution effects. A 1 cm cell is used in all cases.

(a)   The titration of 100.00 mL of 0.001 M A with 0.01 M T according to the reaction

$$A + T \rightleftharpoons P$$

Both the titrant T and the product P absorb at the chosen wavelength. The titrant

molar absorptivity is 500 and is twice that of P.

(b)   The same titration as in A except that the analyte A and the product P both

absorb. The molar absorptivity for A is 750, while that of P is 340.

(c)   The same titration as in (a) above except that the analyte and titrant both absorb.

The titrant molar absorptivity is 500, while that of the analyte is 250.

16.   The method of continuous variations was used to find the identity of the complex formed

between Fe(II) and 2,2'-bipyridine

The following data were obtained:

Mole fraction of Fe(II)	A at 522 nm
0.05	0.145
0.10	0.289
0.12	0.346
0.15	0.432
0.20	0.576
0.28	0.691
0.36	0.615
0.45	0.531
0.56	0.422
0.65	0.334
0.70	0.288
0.80	0.192
0.90	0.095

(a)   Plot absorbance vs. mole fraction of Fe(II).

(b)   Determine the formula of the complex.

(c)   If the sum of the concentrations of Fe(II) and the ligand remained constant at

2.75×10^{-4} M for all measurements in a 1.00 cm cell, find the molar absorptivity

of the complex.

Chapter 13

Kinetic Methods

In kinetic methods, measurements are made under *dynamic* conditions where the concentrations

of reactants and products change continuously. These methods differ significantly from

equilibrium methods in which measurements are made under *static* or *steady-state* conditions in

which concentrations are constant. Kinetic methods are the subject of Chapter 30 of FAC9 and

are briefly covered in Section 23A-2 of AC7.

Rate Laws for Simple Reactions

We'll explore some of the properties of simple first- and second-order reactions by constructing

spreadsheets and charts to display and observe reaction behavior.

Comparing First- and Second-order Reactions

Let's compare under nearly identical conditions, the concentration vs. time profiles of a first-

order and a second-order system. The reaction being considered is that of reactant A forming

product P by either a first-order or a second-order process. The second-order case is second order

in a single reactant. The rate laws for the two cases are

$$-\frac{d[A]}{dt} = k_1[A] \qquad\qquad \text{first order}$$

$$-\frac{d[A]}{dt} = k_2[A]^2 \qquad\qquad \text{second order}$$

where k_1 is the first-order rate constant in s^{-1} and k_2 is the second-order rate constant in $M^{-1}\,s^{-1}$.

Integration of the first-order rate law gives the concentration-time behavior as:

$$[A]_t = [A]_0 e^{-k_2 t}$$

Integration of the second-order rate law gives:

$$[A]_t = \frac{[A]_0}{1+[A]_0 kt}$$

Let's start with a worksheet with entries as shown in Figure 13-1. We'll begin with an initial

concentration, $[A]_0$ of 1 M and rate constants of $k_1 = 0.01$ s^{-1} for the first-order case and $k_2 = 0.01$

M^{-1} s^{-1} for the second order case. Enter times from 10 to 1000 s as shown. In cells B8 and C8,

we enter the formulas for the concentration of A corresponding to the integrated forms of the rate

laws shown above and copy these formulas into cells B9:B26 and C9:C26.

	A	B	C	D
1	First- and second-order reactions			
2	1st $d[A]/dt = k[A]$		2nd $d[A]/dt = k[A]^2$	
3	$[A]_t = [A]_0 e^{-kt}$		$[A]_t = [A]_0/(1+[A]_0 kt)$	
4	k_1	0.01	$[A]_0$ 1	
5	k_2	0.01		
6		1st order	2nd order	
7	t, s	[A], M	[A], M	
8	10	0.904837	0.909091	
9	20	0.818731	0.833333	
10	30	0.740818	0.769231	
11	40	0.67032	0.714286	
12	50	0.606531	0.666667	
13	60	0.548812	0.625	
14	70	0.496585	0.588235	
15	80	0.449329	0.555556	
16	90	0.40657	0.526316	
17	100	0.367879	0.5	
18	200	0.135335	0.333333	
19	300	0.049787	0.25	
20	400	0.018316	0.2	
21	500	0.006738	0.166667	
22	600	0.002479	0.142857	
23	700	0.000912	0.125	
24	800	0.000335	0.111111	
25	900	0.000123	0.1	
26	1000	4.54E-05	0.090909	
27	Documentation			
28	Cell B8=$D$4*EXP(-$B$4*A8)			
29	Cell C8=$D$4/(1+$D$4*$B$5*A8)			

Figure 13-1 Worksheet for comparing first- and second-order systems.

Now we'll prepare a two-series plot showing both the first- and second-order decreases in the A concentration. Select cells A8:C26, and prepare a Scatter (XY) plot. After adding labels and formatting, your chart should look similar to Figure 13-2.

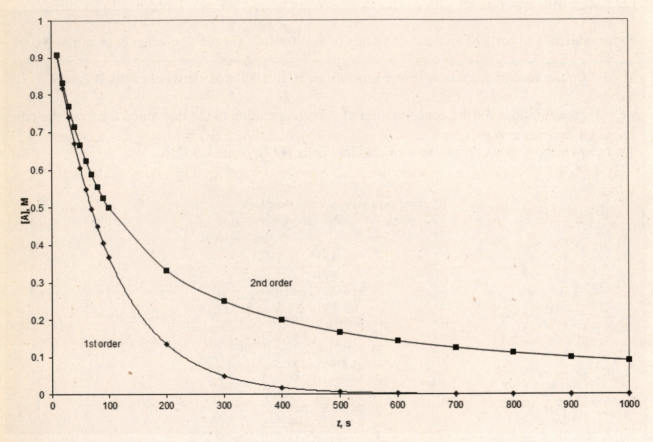

Figure 13-2 Concentration-time plots of first-and second-order systems.

Note that the very early time behavior of the two systems is quite similar. Differences in the concentration of A of as much as 10% appear only after 50 s. Later times reveal the very different courses of the two systems. For the first-order system, the reaction is essentially complete after about 500 s.

As an extension, try different values of k_1 and k_2 and different initial concentrations, $[A]_0$. You will have to change the time range for these different conditions.

Determining Reaction Orders

Although plots similar to those shown in Figure 13-2 can distinguish first-order systems from second-order systems, we often seek better diagnostic plots for finding the reaction order. There are two common methods. In the nonlinear regression approach, we try various integrated rate laws (first order, second order, second order in two reactants, etc.), and determine how closely our data fit the presumed rate laws. In the linear plotting approach, we use a linearized plotting form and determine whether our data fit a linear equation. We'll use the latter approach here.

For the first-order case, we can produce a linear plotting form by taking the natural logarithms of both sides of the exponential equation.

$$\ln[A]_t = \ln[A]_0 - kt$$

or,

$$-\ln[A]_t = -\ln[A]_0 + kt$$

A plot of $-\ln[A]_t$ vs. t should be linear with a slope of k and an intercept of $-\ln[A]_0$. To see how the first- and second-order systems differ using this type of plot, we can prepare two additional columns in our worksheet of Figure 13-1 in which we calculate $-\ln[A]_t$ for the first-order system and $-\ln[A]_t$ for the second order system. To do this we enter into cell D8 the formula **=-LN(B8)** and into cell E8 the formula **=-LN(C8)**. Use the fill handle to fill in the remaining D and E cells with these formulas. You can now prepare a two-series chart and see how these data fit the linear equation as shown in Figure 13-3. Note that the second-order data do not fall on a straight line at all. Note also, that it is much easier to distinguish the two systems with this type of plot than with the [A] vs. t plot shown previously. Our first-order system shows a slope of 0.01 and an intercept of 0.00. Since $[A]_0 = 1$, $-\ln[A]_0 = 0$.

As an extension, try this logarithmic plotting form for different first- and second-order rate constants, and different values of the initial concentration. Can you find any circumstances under which a first-order reaction appears to be second order, or vice versa?

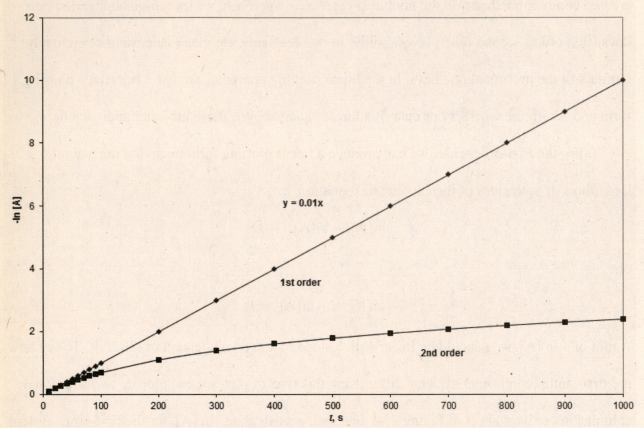

Figure 13-3 Comparison of first- and second-order systems with a logarithmic plot.

The ln [A] vs t plot is seen to be diagnostic for a first-order system. Let's now determine the appropriate linear plot for a second-order system. If we take the equation for $[A]_t$ vs. t for a second order system and take the reciprocal, we find

$$\frac{1}{[A]_t} = \frac{1+[A]_0 kt}{[A]_0} = \frac{1}{[A]_0} + kt$$

Here, a plot of $1/[A]_t$ vs. t should be linear with a slope of k and an intercept of $1/[A]_0$. We can

prepare two additional columns in our worksheet and find $1/[A]$ for the first-order system in cells

F8:F26 and $1/[A]$ for the second-order system in cells G8:G26. The final worksheet for

comparing these types of reactions should appear like that shown in Figure 13-4. We can then

prepare a chart showing the behavior of these two systems in a reciprocal plot as shown in Figure

13-5.

	A	B	C	D	E	F	G
1	First- and second-order reactions						
2	1st $d[A]/dt = k[A]$		2nd $d[A]/dt = k[A]^2$				
3	$[A]_t = [A]_0 e^{-kt}$		$[A]_t = [A]_0/(1+[A]_0 kt)$				
4	k_1	0.01	$[A]_0$ 1				
5	k_2	0.01					
6		1st order	2nd order	1st order	2nd order	1st order	2nd order
7	t, s	[A], M	[A], M	-ln[A]	-ln[A]	1/[A]	1/[A]
8	10	0.904837	0.909091	0.1	0.09531	1.105171	1.1
9	20	0.818731	0.833333	0.2	0.182322	1.221403	1.2
10	30	0.740818	0.769231	0.3	0.262364	1.349859	1.3
11	40	0.67032	0.714286	0.4	0.336472	1.491825	1.4
12	50	0.606531	0.666667	0.5	0.405465	1.648721	1.5
13	60	0.548812	0.625	0.6	0.470004	1.822119	1.6
14	70	0.496585	0.588235	0.7	0.530628	2.013753	1.7
15	80	0.449329	0.555556	0.8	0.587787	2.225541	1.8
16	90	0.40657	0.526316	0.9	0.641854	2.459603	1.9
17	100	0.367879	0.5	1	0.693147	2.718282	2
18	200	0.135335	0.333333	2	1.098612	7.389056	3
19	300	0.049787	0.25	3	1.386294	20.08554	4
20	400	0.018316	0.2	4	1.609438	54.59815	5
21	500	0.006738	0.166667	5	1.791759	148.4132	6
22	600	0.002479	0.142857	6	1.94591	403.4288	7
23	700	0.000912	0.125	7	2.079442	1096.633	8
24	800	0.000335	0.111111	8	2.197225	2980.958	9
25	900	0.000123	0.1	9	2.302585	8103.084	10
26	1000	4.54E-05	0.090909	10	2.397895	22026.47	11
27	Documentation						
28	Cell B8=$D$4*EXP(-$B$4*A8)						
29	Cell C8=$D$4/(1+$D$4*$B$5*A8)						
30	Cell D8=-LN(B8)						
31	Cell E8=-LN(C8)						
32	Cell F8=1/B8						
33	Cell G8=1/C8						

Figure 13-4 Final worksheet for comparing first- and second-order reactions.

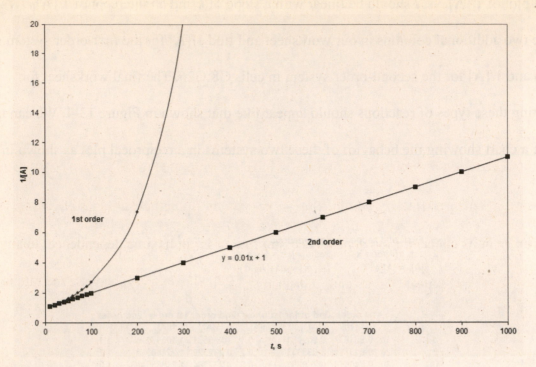

Figure 13-5 Reciprocal plot for second-order system.

Note again here that the reciprocal plot readily distinguishes the second-order reaction for the

first-order reaction. This linear plotting form is then diagnostic for a second-order system. As an

extension, you can try different values for k_1 and k_2 and different initial concentrations. You will

again have to adjust the time scale when you change these values.

Pseudo First-Order Systems

Many second-order reactions can be made to appear first order by adding a large excess of one of

the reagents. For example, consider the following reaction

$$A + B \xrightarrow{\ k\ } P$$

If the reaction occurs in a single elementary step, the rate is proportional to the concentration of

each of the reactants, and the rate law is

$$-\frac{d[A]}{dt} = k[A][B]$$

If reagent B is initially present in large excess such that $[B]_0 \gg [A]_0$, the concentration of B will change very little during the course of the reaction and we can write $k[B]_0 = \text{constant} = k'$. The rate is then given by

$$-\frac{d[A]}{dt} = k'[A]$$

which is the equation for a first-order system. This reaction is said to be *pseudo-first-order*, since the constant k' is dependent on $[B]_0$ and not truly constant. Sometimes it is said that the system is second order in its concentration dependence, but first order in its time dependence. In any case, such pseudo-first-order systems are very useful in analytical chemistry, because the rate is directly proportional to the analyte concentration, $[A]$.

One question that often arises is how much excess B is needed to make the reaction appear first order and not second order. If the reaction is truly second order in species A and B, we can write[1]

$$\frac{1}{[B]_0 - [A]_0} \ln \frac{[A]_0[B]}{[B]_0[A]} = kt$$

which holds as long as $[B]_0 \neq [A]_0$. We are interested in obtaining the time dependence of $[A]$. If we rearrange the equation and solve for $[B]/[A]$, we obtain

$$\frac{[A]_0[B]}{[B]_0[A]} = e^{([B]_0 - [A]_0)kt}$$

From which we can obtain the desired ratio

$$\frac{[B]}{[A]} = \frac{[B]_0}{[A]_0} e^{([B]_0 - [A]_0)kt}$$

From the reaction stoichiometry, $d[A] = d[B]$ and

---

[1] J. W. Moore and R. G. Pearson, *Kinetics and Mechanism*, 3rd. Ed., (Wiley; NY, 1981), p.23.

so that

$$B] = [B]_0 - [A]_0 + [A]$$

By dividing both sides of this last equation by [A], we obtain

$$\frac{[B]}{[A]} = \frac{[B]_0 - [A]_0 + [A]}{[A]}$$

We can solve this last equation for [A] in terms of the ratio [B]/[A] to obtain

$$[A] = \frac{[B]_0 - [A]_0}{\frac{[B]}{[A]} - 1}$$

If we find the ratio [B]/[A] as a function of time, we can find [A] from this last equation and

compare it to the concentration if the reaction were truly first order.

	A	B	C	D	E	F	G	H	I	J
1	First- and pseduo-first-order reactions									
2	1st $d[A]/dt = k[A]$		2nd $d[A]/dt = k[A][B]$							
3	$[A]_t = [A]_0 e^{-kt}$		$[B]/[A] = ([B]_0/[A]_0)e^{([B]_0-[A]_0)kt}$							
4			$[A] = [B]_0 - [A]_0 /([B]/[A] - 1)$							
5	$[A]_0$	0.001	$[B]_0'$	0.25						
6	k	0.01	$[B]_0''$	0.1						
7	$[B]_0$	1	$[B]_0'''$	0.01						
8		1st order	1000-fold excess		250-fold excess		100-fold excess		10-fold excess	
9	t, s	[A], M	B/A	[A]	B/A	[A]	B/A	[A]	B/A	[A]
10	10	0.000905	1105.06	0.000905	256.3031	0.000975	100.9949	0.00099	10.009	0.000999
11	20	0.000819	1221.159	0.000819	262.7652	0.000951	101.9997	0.00098	10.01802	0.000998
12	30	0.000741	1349.454	0.000741	269.3902	0.000928	103.0145	0.00097	10.02704	0.000997
13	40	0.00067	1491.228	0.00067	276.1822	0.000905	104.0395	0.000961	10.03606	0.000996
14	50	0.000607	1647.897	0.000607	283.1455	0.000883	105.0746	0.000951	10.0451	0.000995
15	60	0.000549	1821.026	0.000549	290.2843	0.000861	106.12	0.000942	10.05415	0.000994
16	70	0.000497	2012.344	0.000497	297.6032	0.00084	107.1758	0.000932	10.0632	0.000993
17	80	0.000449	2223.761	0.000449	305.1065	0.000819	108.2421	0.000923	10.07226	0.000992
18	90	0.000407	2457.39	0.000407	312.799	0.000799	109.319	0.000914	10.08133	0.000991
19	100	0.000368	2715.565	0.000368	320.6855	0.000779	110.4066	0.000905	10.09041	0.00099
20	200	0.000135	7374.293	0.000135	411.3568	0.000607	121.8962	0.000819	10.18163	0.00098
21	300	4.98E-05	20025.37	4.99E-05	527.6646	0.000473	134.5815	0.000741	10.27368	0.00097
22	400	1.83E-05	54380.19	1.84E-05	676.8576	0.000368	148.5869	0.000671	10.36656	0.000961
23	500	6.74E-06	147672.9	6.76E-06	868.2337	0.000287	164.0498	0.000607	10.46028	0.000951
24	600	2.48E-06	401015.5	2.49E-06	1113.72	0.000224	181.1219	0.00055	10.55485	0.000942
25	700	9.12E-07	1088984	9.17E-07	1428.615	0.000174	199.9706	0.000498	10.65027	0.000933
26	800	3.35E-07	2957205	3.38E-07	1832.545	0.000136	220.7808	0.00045	10.74655	0.000923
27	900	1.23E-07	8030483	1.24E-07	2350.682	0.000106	243.7566	0.000408	10.84371	0.000914
28	1000	4.54E-08	21807299	4.58E-08	3015.319	8.26E-05	269.1234	0.000369	10.94174	0.000905
29	Documentation									
30	Cell B10=$B$5*EXP(-$B$6*A10)									
31	Cell C10=($B$7/$B$5)*EXP(($B$7-$B$5)*$B$6*A10)									
32	Cell D10=($B$7-$B$5)/(C10-1)									
33	Cell E10=($D$5/$B$5)*EXP(($D$5-$B$5)*$B$6*A10)									
34	Cell F10=($D$5-$B$5)/(E10-1)									
35	Cell G10=($D$6/$B$5)*EXP(($D$6-$B$5)*$B$6*A10)									
36	Cell H10=($D$6-$B$5)/(G10-1)									
37	Cell I10=($D$7/$B$5)*EXP(($D$7-$B$5)*$B$6*A10)									
38	Cell J10=($D$7-$B$5)/(I10-1)									

Figure 13-6 Worksheet for comparing first- and pseudo-first-order systems.

The worksheet for doing this comparison of first-order and pseudo-first-order reactions is shown in Figure 13-6. Here we have entered in cells B7, D5, D6 and D7, values for $[B]_0$ that represent an excess of 1000-fold, 250-fold, 100-fold and 10-fold over the 0.001 M initial concentration of [A]. In cells B10:B28, we calculate the first order decrease in [A] with the rate constant shown in cell B6. In columns C, E, G, and I, we find the ratio [B]/[A] using the full second-order expression and the different initial concentrations of B. The A concentrations are found in columns D, F, H, and J. A plot of the reaction progress curves for the various cases are

shown in Figure 13-7. Note that with a 1000-fold excess, the second-order data points (dots) fall

on top of the first order decay. Excesses of B less than 1000-fold, however, show large

deviations from first-order behavior.

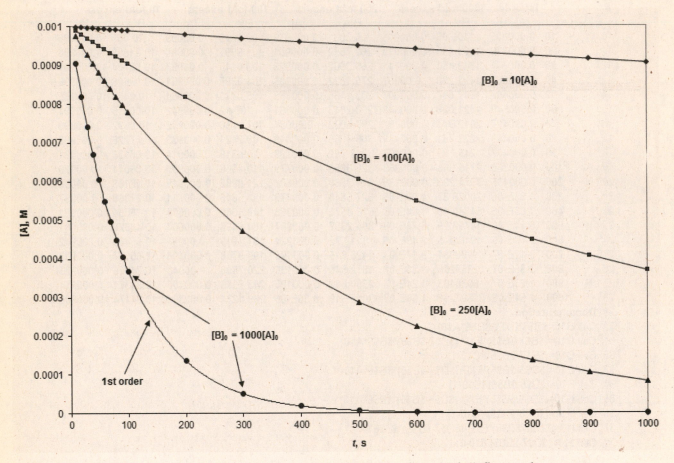

Figure 13-7 Comparison of a first-order system with several "pseudo"-first-order systems.

As an extension, try excesses of B of 500-fold and 750-fold. Find the smallest excess for which

the concentration of A after 100 s deviates by less than 1 % from its first-order value.

Enzyme-Catalyzed Reaction

Many enzyme-catalyzed reactions follow a Michaelis-Menten mechanism

$$E + S \underset{k_{-1}}{\overset{k_1}{\rightleftharpoons}} ES \xrightarrow{k_2} P + E$$

where E is the enzyme, S the substrate, P the product, and ES the enzyme-substrate complex.

Under steady-state conditions, the rate law can be expressed as

$$\frac{d[P]}{dt} = v = \frac{k_2[E]_0[S]}{K_m + [S]} = \frac{v_{max}[S]}{K_m + [S]}$$

where the rate v is $d[P]/dt$, the maximum rate, v_{max} is $k_2[E]_0$, K_m is the Michaelis constant, and

$[E]_0$ is the initial concentration of the enzyme.

For determinations of the substrate, it is desirable to operate under conditions where

$[S] \ll K_m$ so that the rate is directly proportional to the substrate concentration. Conversely, for

determinations of enzyme activity, it is best to operate under conditions of saturation where

$[S] \gg K_m$. In either case, it the value of the Michaelis constant, K_m is very important. There are

two common methods to determine K_m. Since the relationship between rate and substrate

concentration is nonlinear, we can either linearize the equation and use linear regression

methods, or use nonlinear regression methods on the untransformed equation.

Linear Transformation

We can linearize the Michaelis-Menten equation by taking the reciprocal of both sides to yield

$$\frac{1}{v} = \frac{K_m + [S]}{v_{max}[S]} = \frac{K_m}{v_{max}[S]} + \frac{1}{v_{max}}$$

A plot of the reciprocal of the rate, $1/v$, vs. the reciprocal of the substrate concentration, $1/[S]$, should give a straight line of slope K_m/v_{max} and intercept $1/v_{max}$. From these parameters, both K_m and v_{max} can be determined. Such a double reciprocal plot is called a *Lineweaver-Burk* plot.

Let's take an example of the hydrolysis of the substrate fumarate to produce malate, catalyzed by the enzyme fumarase. The rate of this reaction was measured as a function of the fumarate concentration, and the results shown in Figure 13-8 obtained. To accomplish the double reciprocal plot, we add columns for $1/v$ and $1/[S]$. We'll use LINEST for the linear regression.

	A	B	C	D
1	**Michaelis-Menten kinetics, linearized treatment**			
2	Rxn: Fumarate + $H_2O \rightarrow$ malate			
3	Enzyme: fumarase			
4				
5	Conc fumarate	d[P]/dt		
6	5.00E-05	1.90		
7	1.00E-04	2.86		
8	1.50E-04	3.52		
9	2.00E-04	4.00		
10	3.00E-04	4.46		
11	4.00E-04	4.81		
12	5.00E-04	5.00		

Figure 13-8 Data for enzymatic conversion of fumarate to malate.

From the slope and intercept, we can calculate v_{max} and K_m. The equation is of the form $y = mx + b$, where $y = 1/v$, $m = K_m/v_{max}$, $x = 1/[S]$, and $b = 1/v_{max}$. The maximum rate, $v_{max} = 1/b$, while $K_m = v_{max} \times m$. We can also determine the standard deviations in v_{max} and K_m from the standard deviations in the intercept, s_b and slope, s_m. From propagation of error mathematics, the standard deviation in v_{max} is given by

$$s_{v_{max}} = v_{max}\frac{s_b}{b}$$

and the standard deviation in K_m can be expressed as

$$s_{K_m} = K_m\sqrt{\left(\frac{s_{v_{max}}}{v_{max}}\right)^2 + \left(\frac{s_m}{m}\right)^2}$$

Figure 13-9 shows the completed spreadsheet with an embedded chart showing the linear plot. We can see from the plot one of the disadvantages of the double reciprocal plot. Unless care is taken to spread the data points evenly in reciprocal space, we end up with a cluster of points near the y axis intercept. Before doing the experiment, one can correct this problem by preparing substrate concentrations so as to obtain evenly spaced points on the 1/[S] axis. Another potential problem with linearization is that while the linear least squares procedure is optimal for the linear equation, it may not be optimal for the original nonlinear equation. For our case we estimate v_{max} = 6.15 ± 0.06 and $K_m = (1.12 \pm 0.03) \times 10^{-4}$.

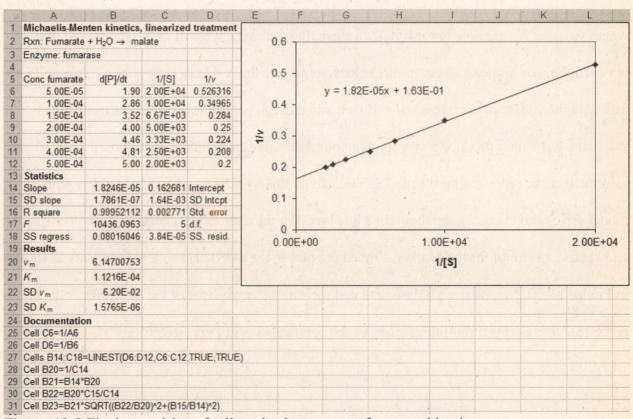

Figure 13-9 Final spreadsheet for linearized treatment of enzyme kinetics.

Nonlinear Regression Approach

A second approach is to use a nonlinear regression method to solve for the parameters of the

original nonlinear equation. With Excel, we can do that using the Solver function. Let's use the

same data as in Figure 13-9, but instead of computing $1/v$ and $1/[S]$, let us put in a model and

compute the values of v from the model. To start let us assume from the data that $v_{max} = 5.00$ and

$K_m = 5 \times 10^{-5}$ as our initial estimates. We'll add formulas in column C to calculate the rate from

our model

$$v_{model} = \frac{v_{max}[S]}{K_m + [S]}$$

where v_{max} and K_m are now our estimated values. In column D, we will calculate the differences

between the data values and our model values, which are the residuals. In column E, we will

calculate the square of each residual and then add these to get the sum of the squares of the

residuals. Let's next produce a plot showing our data values as points superimposed on a

smoothed curve representing the model. You can do this by selecting none for Marker with the

model series and checking the Smoothed line box. For the data series, check None for the line

and choose a symbol for the Marker. Embed the plot in the worksheet so that it appears as shown

in Figure 13-10. Note that we have calculated the sum of the squares of the residuals, SSR, in

cell E13. Note also that our initial guess of the parameters is not very good.

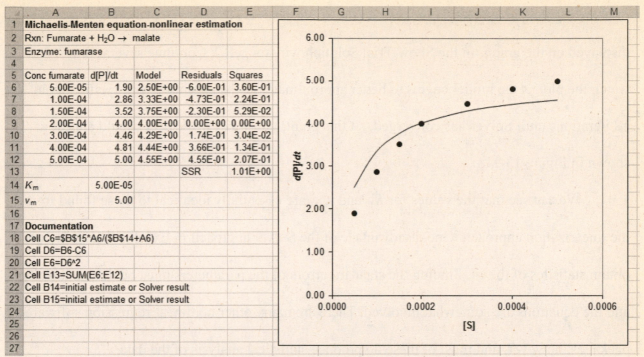

The worksheet contains the following cells:

	A	B	C	D	E
1	Michaelis-Menten equation-nonlinear estimation				
2	Rxn: Fumarate + H_2O → malate				
3	Enzyme: fumarase				
4					
5	Conc fumarate	d[P]/dt	Model	Residuals	Squares
6	5.00E-05	1.90	2.50E+00	-6.00E-01	3.60E-01
7	1.00E-04	2.86	3.33E+00	-4.73E-01	2.24E-01
8	1.50E-04	3.52	3.75E+00	-2.30E-01	5.29E-02
9	2.00E-04	4.00	4.00E+00	0.00E+00	0.00E+00
10	3.00E-04	4.46	4.29E+00	1.74E-01	3.04E-02
11	4.00E-04	4.81	4.44E+00	3.66E-01	1.34E-01
12	5.00E-04	5.00	4.55E+00	4.55E-01	2.07E-01
13				SSR	1.01E+00
14	K_m	5.00E-05			
15	v_m	5.00			
16					
17	Documentation				
18	Cell C6=$B$15*A6/($B$14+A6)				
19	Cell D6=B6-C6				
20	Cell E6=D6^2				
21	Cell E13=SUM(E6:E12)				
22	Cell B14=initial estimate or Solver result				
23	Cell B15=initial estimate or Solver result				

Figure 13-10 Worksheet with chart embedded for nonlinear estimation.

Now we'll use Solver to minimize the sum of the squares of the residuals by changing the K_m and v_{max} values in cells B14 and B15, respectively. Activate Solver to obtain the Solver Parameters window. Enter or select cell E13 in the Set Objective: box (In Excel 2007, this is called Set Target Cell). Choose Min in the To: box (Equal To in Excel 2007) and B14:B15 in the By Changing Variable Cells box (By Changing Cells in Excel 2007). No constraints will be used. Choose GRG Nonlinear for the Solving method in Excel 2010. Click on the Options button. In Excel 2010, click on the All Methods tab and set the Constraint Precision to 0.000001. Choose Use Automatic Scaling and Show Iteration results. Choose a Maximum Time of 100 seconds and an Iterations number of 100. Click on the GRG Nonlinear tab, set the Convergence to 0.000001, and choose Forward for the derivatives. Click the OK button. In Excel 2007, set the Tolerance: to 0.1%. Set the Precision: and Convergence: to 0.000001. Check the box labeled Show Iteration Results and click on OK. In both versions of Excel, choosing to show iterations causes the program to pause after each iteration and update the plot so that you can see the effect of changing the parameters. Now click on the Solve button. The first calculation results will be

displayed on the graph. In the Show Trial Solution window, click Continue several times, and

watch the plot as the model begins to better approximate the data points. It will require about 10-

12 iterations until Solver has converged. At this point, your final worksheet should appear as

shown in Figure 13-11.

We can see that the values for K_m and v_{max} are essentially identical to those found from

the linearization approach. One disadvantage of the Solver approach in Excel is that we do not

obtain statistics of the fit. Finding the standard errors of the parameters must be done manually,

and the equations are somewhat involved. There are many other nonlinear regression software

packages, however, and most of these do produce statistical analysis of the data.

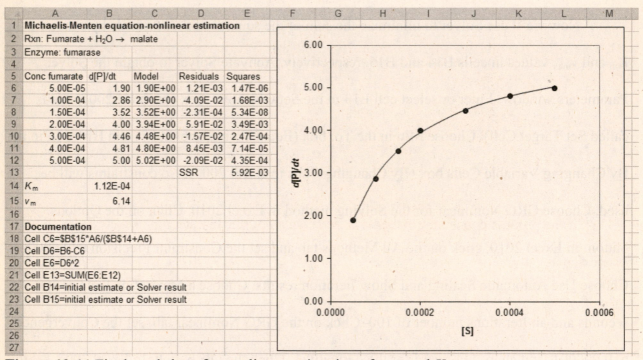

	A	B	C	D	E
1	Michaelis-Menten equation-nonlinear estimation				
2	Rxn: Fumarate + H_2O → malate				
3	Enzyme: fumarase				
4					
5	Conc fumarate	d[P]/dt	Model	Residuals	Squares
6	5.00E-05	1.90	1.90E+00	1.21E-03	1.47E-06
7	1.00E-04	2.86	2.90E+00	-4.09E-02	1.68E-03
8	1.50E-04	3.52	3.52E+00	-2.31E-04	5.34E-08
9	2.00E-04	4.00	3.94E+00	5.91E-02	3.49E-03
10	3.00E-04	4.46	4.48E+00	-1.57E-02	2.47E-04
11	4.00E-04	4.81	4.80E+00	8.45E-03	7.14E-05
12	5.00E-04	5.00	5.02E+00	-2.09E-02	4.35E-04
13			SSR		5.92E-03
14	K_m	1.12E-04			
15	v_m	6.14			
16					
17	Documentation				
18	Cell C6=$B$15*A6/($B$14+A6)				
19	Cell D6=B6-C6				
20	Cell E6=D6^2				
21	Cell E13=SUM(E6:E12)				
22	Cell B14=initial estimate or Solver result				
23	Cell B15=initial estimate or Solver result				
24					
25					
26					
27					

Figure 13-11 Final worksheet for nonlinear estimation of v_{max} and K_m

Kinetic Methods of Analysis

There are several ways in which kinetics can be used to determine the concentrations of various analytes. Section 30B-2 of FAC9 discusses the most popular methods. The differential method in which the rate of the reaction is related to concentration is one of most often used method and is illustrated here. In particular, we consider measurements of the initial rate of a reaction before reactant concentrations have changed significantly from their initial values. We'll take a reaction of the form

$$A + R \rightarrow P$$

If reagent R is in large excess, the reaction is pseudo-first-order and we can write

$$\text{rate} = -\frac{d[P]}{dt} = k'[A] = k'[A]_0 e^{-k't}$$

where k' is the pseudo-first-order rate constant and is equal to $k[R]$, and $[A]_0$ is the initial concentration of [A]. If measurements are made very early in the course of the reaction, the rate is the *initial rate*, $-(d[A]/dt)_i$, which is given by

$$\text{initial rate} = -\left(\frac{d[A]}{dt}\right)_i = k'[A]_0$$

If the product concentration is measured instead of the analyte, the initial rate is expressed as

$$\text{initial rate} = \left(\frac{d[P]}{dt}\right)_i = k'[A]_0$$

There are several ways in which we can estimate the initial reaction rate. One method is to determine the slope of the [P] vs. time curve near $t = 0$. Another way is to determine the change in product concentration per unit change in time, $\Delta[P]/\Delta t$, near the start of the reaction. We'll illustrate the first of these approaches here. Usually, the constant k' is evaluated by using

standards of known $[A]_0$. A calibration curve is then used to determine the calibration equation.

Let's take the case of a pseudo-first-order reaction in which four standards containing

0.0100 mM, 0.0500 mM, 0.0750 mM and 0.1000 mM A are reacted with R and the product

concentration is measured vs. time. A solution with an unknown concentration of A is also

reacted. The data shown in Figure 13-12 are obtained and the progress curves for the standards

plotted.

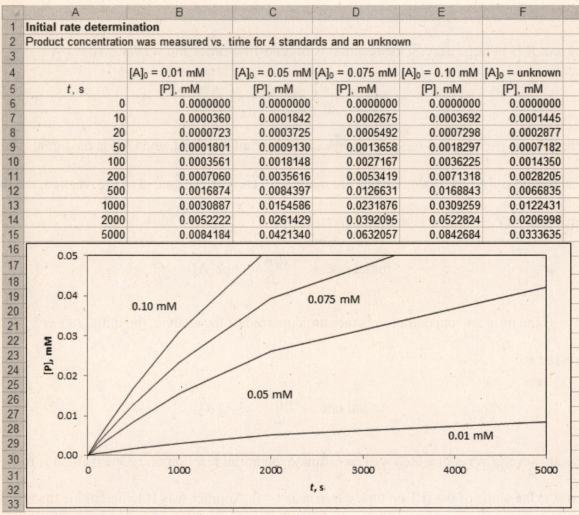

	A	B	C	D	E	F
1	Initial rate determination					
2	Product concentration was measured vs. time for 4 standards and an unknown					
3						
4		$[A]_0$ = 0.01 mM	$[A]_0$ = 0.05 mM	$[A]_0$ = 0.075 mM	$[A]_0$ = 0.10 mM	$[A]_0$ = unknown
5	t, s	[P], mM	[P], mM	[P], mM	[P], mM	[P], mM
6	0	0.0000000	0.0000000	0.0000000	0.0000000	0.0000000
7	10	0.0000360	0.0001842	0.0002675	0.0003692	0.0001445
8	20	0.0000723	0.0003725	0.0005492	0.0007298	0.0002877
9	50	0.0001801	0.0009130	0.0013658	0.0018297	0.0007182
10	100	0.0003561	0.0018148	0.0027167	0.0036225	0.0014350
11	200	0.0007060	0.0035616	0.0053419	0.0071318	0.0028205
12	500	0.0016874	0.0084397	0.0126631	0.0168843	0.0066835
13	1000	0.0030887	0.0154586	0.0231876	0.0309259	0.0122431
14	2000	0.0052222	0.0261429	0.0392095	0.0522824	0.0206998
15	5000	0.0084184	0.0421340	0.0632057	0.0842684	0.0333635

Figure 13-12 Results for initial rate determination of A.

Preliminary examination of the results indicates that in all cases the data fall on straight lines for times up to 100 s. Hence, the first 4 data points can be used to define a straight line for each standard and the unknown. Move the chart to a new sheet, Chart 1. We can use the SLOPE function to find the slope. For the 0.01 mM standard, for example, we can calculate in cell B18 `=SLOPE(B6:B10,A6:A10)` as the initial rate. For the 0.05 mM standard, we find in cell B19 `=SLOPE(C6:C10,A6:A10)`, and so on. The results and a plot showing the data points and Trendlines are shown in Figure 13-13.

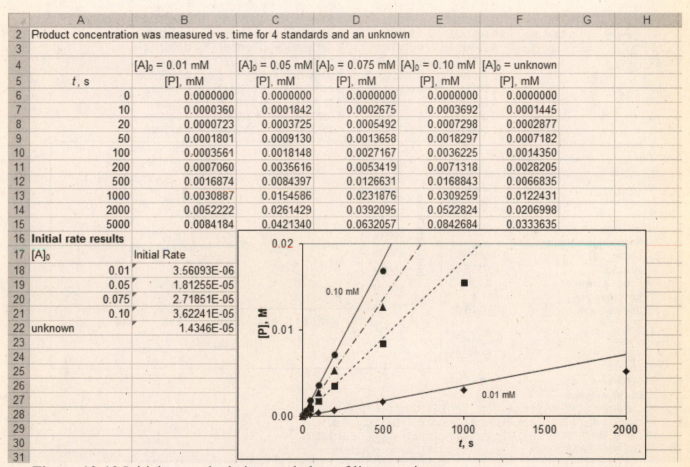

	A	B	C	D	E	F	G	H
2	Product concentration was measured vs. time for 4 standards and an unknown							
3								
4		$[A]_0$ = 0.01 mM	$[A]_0$ = 0.05 mM	$[A]_0$ = 0.075 mM	$[A]_0$ = 0.10 mM	$[A]_0$ = unknown		
5	t, s	[P], mM	[P], mM	[P], mM	[P], mM	[P], mM		
6	0	0.0000000	0.0000000	0.0000000	0.0000000	0.0000000		
7	10	0.0000360	0.0001842	0.0002675	0.0003692	0.0001445		
8	20	0.0000723	0.0003725	0.0005492	0.0007298	0.0002877		
9	50	0.0001801	0.0009130	0.0013658	0.0018297	0.0007182		
10	100	0.0003561	0.0018148	0.0027167	0.0036225	0.0014350		
11	200	0.0007060	0.0035616	0.0053419	0.0071318	0.0028205		
12	500	0.0016874	0.0084397	0.0126631	0.0168843	0.0066835		
13	1000	0.0030887	0.0154586	0.0231876	0.0309259	0.0122431		
14	2000	0.0052222	0.0261429	0.0392095	0.0522824	0.0206998		
15	5000	0.0084184	0.0421340	0.0632057	0.0842684	0.0333635		
16	Initial rate results							
17	$[A]_0$	Initial Rate						
18	0.01	3.56093E-06						
19	0.05	1.81255E-05						
20	0.075	2.71851E-05						
21	0.10	3.62241E-05						
22	unknown	1.4346E-05						
23								
24								
25								
26								
27								
28								
29								
30								
31								

Figure 13-13 Initial rate calculations and plots of linear regions.

The initial rate vs. $[A]_0$ data are next used to form a calibration curve and to calculate the unknown concentration. The calibration data and the unknown concentration are displayed along with the calibration curve in Figure 13-14. The unknown concentration would be reported as

0.0397 ± 0.0001 mM.

As an extension of this worksheet, find the initial rate by calculating $\Delta[P]/\Delta t$ for each of

the standards and the unknown. For a given analyte concentration, find $\Delta[P]/\Delta t$ for the 0 to 10 s

interval, the 10 to 20 s interval, the 20 to 50 s interval, and the 50 to 100 s interval; then average

the 4 values obtained. Do the same for the unknown and find the regression equation as above.

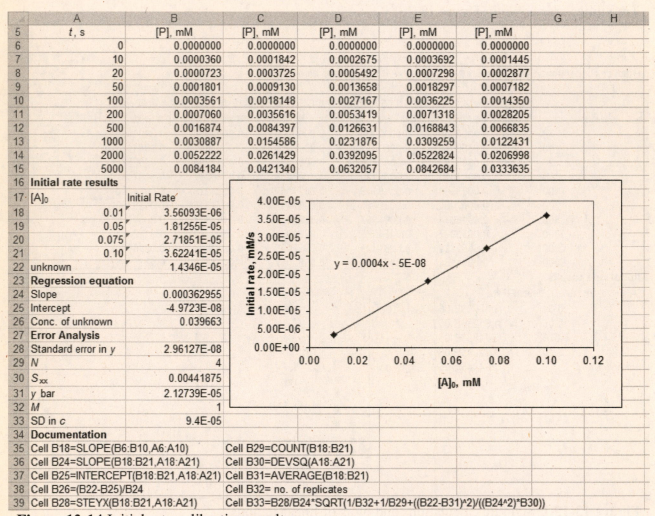

	A	B	C	D	E	F	G	H
5	t, s	[P], mM	[P], mM	[P], mM	[P], mM	[P], mM		
6	0	0.0000000	0.0000000	0.0000000	0.0000000	0.0000000		
7	10	0.0000360	0.0001842	0.0002675	0.0003692	0.0001445		
8	20	0.0000723	0.0003725	0.0005492	0.0007298	0.0002877		
9	50	0.0001801	0.0009130	0.0013658	0.0018297	0.0007182		
10	100	0.0003561	0.0018148	0.0027167	0.0036225	0.0014350		
11	200	0.0007060	0.0035616	0.0053419	0.0071318	0.0028205		
12	500	0.0016874	0.0084397	0.0126631	0.0168843	0.0066835		
13	1000	0.0030887	0.0154586	0.0231876	0.0309259	0.0122431		
14	2000	0.0052222	0.0261429	0.0392095	0.0522824	0.0206998		
15	5000	0.0084184	0.0421340	0.0632057	0.0842684	0.0333635		
16	**Initial rate results**							
17	[A]$_0$	Initial Rate						
18	0.01	3.56093E-06						
19	0.05	1.81255E-05						
20	0.075	2.71851E-05						
21	0.10	3.62241E-05						
22	unknown	1.4346E-05						
23	**Regression equation**							
24	Slope	0.000362955						
25	Intercept	-4.9723E-08						
26	Conc. of unknown	0.039663						
27	**Error Analysis**							
28	Standard error in y	2.96127E-08						
29	N	4						
30	S_{xx}	0.00441875						
31	y bar	2.12739E-05						
32	M	1						
33	SD in c	9.4E-05						
34	**Documentation**							
35	Cell B18=SLOPE(B6:B10,A6:A10)		Cell B29=COUNT(B18:B21)					
36	Cell B24=SLOPE(B18:B21,A18:A21)		Cell B30=DEVSQ(A18:A21)					
37	Cell B25=INTERCEPT(B18:B21,A18:A21)		Cell B31=AVERAGE(B18:B21)					
38	Cell B26=(B22-B25)/B24		Cell B32= no. of replicates					
39	Cell B28=STEYX(B18:B21,A18:A21)		Cell B33=B28/B24*SQRT(1/B32+1/B29+((B22-B31)^2)/((B24^2)*B30))					

Figure 13-14 Initial rate calibration results.

Summary

In this chapter, we have discussed several aspects of kinetics as applied to analytical chemistry. We've dealt with rate laws for simple first- and second-order systems and investigated the behavior of these systems. We have illustrated several applications of Excel in dealing with enzyme-catalyzed reactions. Finally, we have described the initial rate method for determining concentrations using kinetic methods

Problems

1. Hydrogen peroxide reacts with thiosulfate ion in slightly acidic solution according to the

   following stoichiometry

   $$H_2O_2 + 2S_2O_3^{2-} + 2H^+ \rightarrow 2H_2O + S_4O_6^{2-}$$

   The rate is independent of hydrogen ion in the pH range 4 to 6. The following data were

   obtained at 25°C and pH 5.0. The initial concentrations were $[H_2O_2]_0 = 0.1000$ M,

   $[S_2O_3^{2-}]_0 = 0.02040$ M.

$[S_2O_3^{2-}]$, M	t, min
0.0170	16
0.0140	36
0.0134	43
0.0125	52
0.0097	90
0.0083	120

   Determine the order of the reaction with respect to thiosulfate ion. What is the pseudo-

   order rate constant?

2. The following data were obtained on the hydrolysis of 17% sucrose in 0.100 M HCl at

   35°C:

t, min	Sucrose remaining, %
9.83	96.5
59.60	80.3
93.18	71.0
142.9	59.1
294.8	32.8
589.4	11.1

   Determine the order of the reaction and the value of the rate constant, k.

3.   The hydrolysis of an ester was studied by measuring the acid produced by titration at

measured intervals with a solution of NaOH. The data obtained were:

t, hours	Vol NaOH, mL
0	0.35
1.0	2.90
3.0	7.15
5.0	10.50
9.0	14.95
12.0	17.32
∞	21.97

Determine the order of the reaction and the value for the rate constant, k.

4.   The hydrolysis of carbobenzoxyglycyl-L-trptophan (CBT) is catalyzed by the enzyme

carboxypeptidase to give carbobenzoxyglycine (CB) plus tryptophan (T) according to

$$CBT + H_2O \rightarrow CB + T$$

The following data were obtained on the rate of formation of tryptophan at 25°C and pH

7.5.

[CBT], mM	Rate, d[T]/dt, mM /s
1.00	0.0115
1.25	0.0139
1.50	0.0159
2.50	0.0240
5.00	0.0360
10.00	0.0532
15.00	0.0600
20.00	0.0640

Prepare a Lineweaver-Burk double reciprocal plot and determine the Michaelis constant,

K_m and the maximum rate, v_{max}. Find the standard deviations in K_m and v_{max}.

5. Use Solver to perform a nonlinear regression of the data of Problem 4. Use initial

    estimates of 1.0×10^{-4} for K_m and 0.064 for v_{max}. Set up the problem as was done in

    Figure 13-10. Check the Show Iteration Results box and watch the model converge.

    Compare the results to those obtained in Problem 4. Why might these be different?

6. The decarboxylation of a β-keto acid catalyzed by a decarboxylation enzyme can be

    measured by the rate of CO_2 formation. From the initial rates given below, find the

    Michaelis constant and the maximum velocity using the Lineweaver-Burk approach. Also

    find the standard deviations of these values.

Concentration of keto acid, M	Initial rate, μmol CO_2/2 min
0.125	0.156
0.250	0.256
0.526	0.370
0.714	0.430
1.000	0.480
1.750	0.553
2.500	0.588

7. Find the Michaelis constant and the maximum velocity for the data of Problem 6 by using

    Solver and a nonlinear regression approach. Choose initial estimates of 1.0 M for K_m and

    0.600 for v_{max}. Set up the problem as was done in Figure 13-10. Check the Show Iteration

    Results box to see the results of the minimization. Compare the results to the linearization

    results of Problem 7. Why might these results differ?

8. Copper(II) forms a 1:1 complex with the organic complexing agent R in acidic medium.

    The formation of the complex can be monitored by spectrophotometry at 480 nm. Use the

    following data collected under pseudo-first-order conditions to construct a calibration

    curve of rate versus concentration of R. Find the concentration of copper(II) in an

    unknown whose rate under the same conditions was 6.2×10^{-3} A s^{-1}.

$c_{Cu^{2+}}$, ppm	Rate, A s^{-1}
3.0	3.6×10^{-3}
5.0	5.4×10^{-3}
7.0	7.9×10^{-3}
9.0	1.03×10^{-2}

9. Use Excel to calculate the product concentrations vs. time for a pseudo-first-order

    reaction with $k' = 0.015$ s^{-1} and $[A]_0 = 0.005$ M. Use times of 0.000s, 0.001 s, 0.01 s, 0.1

    s, 0.2 s 0.5 s, 1.0 s, 2.0 s, 5.0 s, 10.0 s, 20.0 s, 50.0 s, 100.0 s, 200.0 s, 500.0 s and 1000.0

    s. From the two earliest time values, find the "true" initial rate of the reaction. Determine

    approximately what percentage completion of the reaction occurs before the initial rate

    drops to (a) 99 % and (b) 95% of the true value.

10. Aluminum forms a 1:1 complex with 2-hydroxy-1-naphthaldehyde *p*-

    methoxybenzoylhydraxonal that exhibits fluorescence emission at 475 nm. In order to

    measure the initial rate of formation of the complex, the change in fluorescence emission

    readings were measured over a 50 second interval. The data below were obtained:

Concentration of Al(III) in μM	ΔF/50 s
0.25	9.5
0.50	34.5
1.00	77.2
1.50	122.0
2.00	161.8
Unknown1	47.5
Unknown2	93.8
Unknown3	25.1

    Prepare a plot of initial rate vs. concentration of Al(III) and determine the concentrations

    of the three unknowns. Show the statistics of the linear regression analysis and find the

    standard deviations of the unknowns.

Chapter 14

Chromatography

In this chapter we explore the use of Excel for the study of chromatography. We will simulate

chromatograms and study the effect of important variables on the shapes of chromatographic

peaks and on the quality of separations. We will also investigate quantitative chromatographic

methods and the use of Excel in optimizing the chromatographic separation.

Chromatographic Terminology and Basic Equations

The shape of a symmetrical chromatographic peak may be approximated by the following

equation:

$$c_t = \frac{A}{\sigma} e^{-\frac{(t-t_R)^2}{2\sigma^2}}$$

where σ is the standard deviation of the Gaussian-shaped peak, t_R is the retention time, A is an

amplitude function containing the initial concentration of the solute, and c_t is the concentration of

the solute at time t. A peak calculated from this expression is shown in Figure 14-1. As we have

defined it in Chapter 31 of FAC9 and Chapter 24 of AC7, W is the width of the base of the peak.

The base is determined by extending to the baseline, tangents drawn to the points of inflection of

the sides of the peak as shown in Figure 14-1. The tangents intersect the baseline such that $W =$

4σ or 4τ when time units are used. Here, we will use σ as is common in the chromatographic

literature. The number of theoretical plates N in a separation is given by

$$N = 16\left(\frac{t_R}{W}\right)^2$$

where t_R is the retention time as shown in Figure 31-9 of FAC9 or Figure 24-7 of AC7. By

substituting 4σ for W and solving for σ, we find that

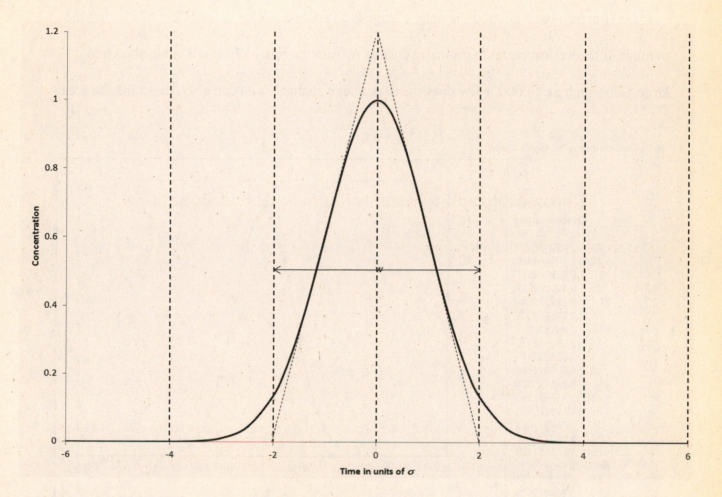

Figure 14-1 A Gaussian chromatographic peak

$$\sigma = \frac{t_R}{\sqrt{N}}$$

So, given a retention time t_R, a number of theoretical plates N, and an amplitude, A, we can

calculate the shape of a chromatographic peak as shown in the worksheet of Figure 14-2. Note

that we have entered A, t_R and N in cells B2:B4. The standard deviation σ is calculated in cell

B5. The concentrations are computed from the equation for the Gaussian peak. These are

calculated every 2.5 s in the region where there is significant change and every 10 s outside this region. The chart is an XY(Scatter) chart with a smooth line added. The chart is embedded in the

worksheet

After you have entered the data, change the variables. What is the effect of varying the

number of theoretical plates? Gradually change N from the relatively small value shown to a

large value such as 10,000. How does the peak shape change? You can also determine the area

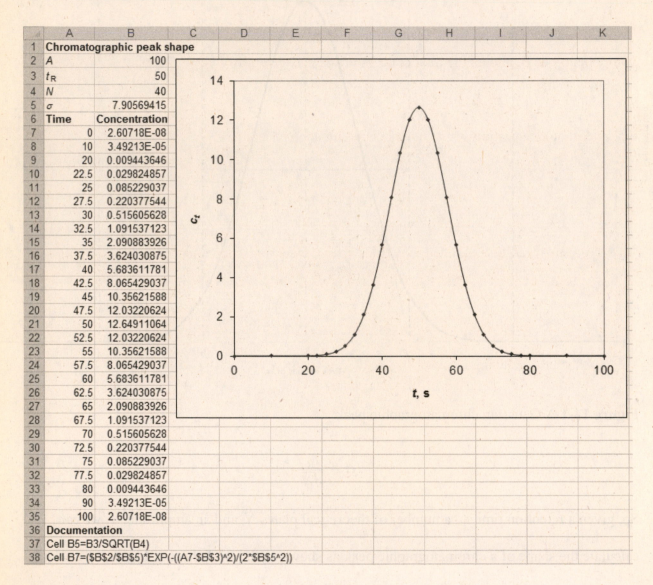

Figure 14-2 Simulating a Gaussian chromatographic peak.

under the peak by doing a Simpson's rule integration as described in Chapter 11. Does the area

change with changes in N? Change the retention time and note its effect. Change the amplitude and see what happens. Does the area change? Manipulating the variables in this way should help you visualize how these variables influence chromatographic peak shape.

Simulating a Chromatogram

In our next exercise, we will simulate complete chromatograms containing several peaks and calculate the resolution between pairs of components. If our chromatogram contains three components, A, B, and C, the total concentration, c_t is given by

$$c_t = c_A + c_B + c_C$$

where each concentration is given by a Gaussian so that

$$c_t = \sum_{i=1}^{3} \frac{A_i}{\sigma_i} e^{-\frac{(t-t_{R_i})^2}{2\sigma_i^2}}$$

This expression can be expanded to include any number of components, but three is enough for our purposes.

Another important figure of merit for chromatographic separations is the resolution R_s, which for solutes A and B is given by:

$$R_s = \frac{2\left[(t_R)_B - (t_R)_A\right]}{W_A + W_B}$$

Since for component A, $W_A = 4\sigma_A$ and for component B, $W_B = 4\sigma_B$, we can substitute these values into the above equation to arrive at the following equation for the resolution between the peaks for solutes A and B:

$$R_s = \frac{(t_R)_B - (t_R)_A}{2(\sigma_A + \sigma_B)}$$

We now have sufficient equations to calculate a chromatogram containing three components: A, B, and C, and compute the resolution for adjacent pairs of peaks. The spreadsheet shown in Figure 14-3 is designed to accomplish this task and allow you to change the various chromatographic variables to observe the effects on the resulting chromatogram.

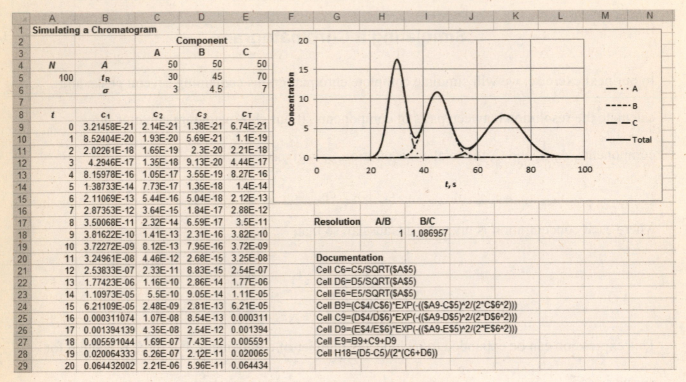

	A	B	C	D	E
1	Simulating a Chromatogram				
2				Component	
3			A	B	C
4	N	A	50	50	50
5	100	t_R	30	45	70
6		σ	3	4.5	7
7					
8	t	c_1	c_2	c_3	c_T
9	0	3.21458E-21	2.14E-21	1.38E-21	6.74E-21
10	1	8.52404E-20	1.93E-20	5.69E-21	1.1E-19
11	2	2.02261E-18	1.65E-19	2.3E-20	2.21E-18
12	3	4.2946E-17	1.35E-18	9.13E-20	4.44E-17
13	4	8.15978E-16	1.05E-17	3.55E-19	8.27E-16
14	5	1.38733E-14	7.73E-17	1.35E-18	1.4E-14
15	6	2.11069E-13	5.44E-16	5.04E-18	2.12E-13
16	7	2.87353E-12	3.64E-15	1.84E-17	2.88E-12
17	8	3.50068E-11	2.32E-14	6.59E-17	3.5E-11
18	9	3.81622E-10	1.41E-13	2.31E-16	3.82E-10
19	10	3.72272E-09	8.12E-13	7.95E-16	3.72E-09
20	11	3.24961E-08	4.46E-12	2.68E-15	3.25E-08
21	12	2.53833E-07	2.33E-11	8.83E-15	2.54E-07
22	13	1.77423E-06	1.16E-10	2.86E-14	1.77E-06
23	14	1.10973E-05	5.5E-10	9.05E-14	1.11E-05
24	15	6.21109E-05	2.48E-09	2.81E-13	6.21E-05
25	16	0.000311074	1.07E-08	8.54E-13	0.000311
26	17	0.001394139	4.35E-08	2.54E-12	0.001394
27	18	0.005591044	1.69E-07	7.43E-12	0.005591
28	19	0.020064333	6.26E-07	2.12E-11	0.020065
29	20	0.064432002	2.21E-06	5.96E-11	0.064434

Chart axes: Concentration (vertical, 0–20) vs t, s (horizontal, 0–100). Legend: A, B, C, Total.

Resolution	A/B	B/C
	1	1.086957

Documentation
Cell C6=C5/SQRT($A$5)
Cell D6=D5/SQRT($A$5)
Cell E6=E5/SQRT($A$5)
Cell B9=(C$4/C$6)*EXP(-(($A9-C$5)^2/(2*C$6^2)))
Cell C9=(D$4/D$6)*EXP(-(($A9-D$5)^2/(2*D$6^2)))
Cell D9=(E$4/E$6)*EXP(-(($A9-E$5)^2/(2*E$6^2)))
Cell E9=B9+C9+D9
Cell H18=(D5-C5)/(2*(C6+D6))

Figure 14-3 Simulation of a chromatogram with three components.

Let us begin our discussion of the spreadsheet with the chromatographic variables that you can change in cells A5 and C4 through E5. Cell A5 contains the number of theoretical plates N for the column. Cells C4, D4, and E4 contain the factors A_A, A_B, and A_C for the three peaks in our chromatogram, and cells C5, D5, and E5 contain the retention times $(t_R)_A$, $(t_R)_B$, and $(t_R)_C$. The spreadsheet calculates σ_A, σ_B, and σ_C in cells C6, D6, and E6 from the retention times and the number of theoretical plates. Note that once you have entered the correct formula in cell C6, you can simply copy the formula into cells D6 and E6.

The formulas for peaks A, B, and C begin in cells B9, C9, and D9. After you have entered the formula for peak A in cell B9, you can just copy it to cells C9 and D9. Cell E9 contains the sum of cells B9, C9, and D9, and the sequence of numbers in column E beneath cell

E9 represent the entire chromatogram. In other words, columns B through D represent individual

chromatograms for components A, B, and C, and column E is the sum of all three, which

represents a chromatogram of a mixture of the three components. Although we have shown only

the first 29 rows of the spreadsheet, the formulas in cells A9 through E9 must be copied to row

109 in order to calculate a complete chromatogram. Before you copy the cells, be sure to enter 0

in cell A9 and 1 in cell A10 so that cells A9 through A109 will contain one hundred consecutive

times at 1-second intervals beginning at zero and ending in 100. The resolution between peaks A

and B is computed by the formula in cell H18, which can be copied into cell I18 to determine the

resolution between peaks B and C. To create the graph, highlight cells A9:E109, and insert a

Scatter (XY) chart. Once the plot has been created, right click on an individual point on each

peak to view the Format Data Series menu. Click on Line Style, and select Smoothed line .

Select a different Line Color for each peak and for the overall chromatogram in column E. Under

Marker Options, click on None. It is also a good idea to double-click on the axes and manually

set the scale on the x-axis to a minimum of 0 and a maximum of 100. Set the y-axis to a

minimum of 0 and allow the maximum to scale automatically.

Begin your study of chromatographic variables with the numbers shown in the

spreadsheet. Note the resolutions calculated between pairs of peaks, and then change the number

of theoretical plates N to 200, 300, 400, 800, etc., noting the resolutions under each set of

conditions. What resolution constitutes a good separation? Reset N to 100, vary the retention

times, and note the results both visually and by inspection of the calculated resolutions. It is

instructive to choose one peak, say peak A, and systematically change its retention time $(t_R)_A$ in

cell C5 through the entire range of the time scale, noting the results; for example, 10, 20, 30, etc.

Now return the retention times to their original settings, and systematically vary the amplitudes in cells C4 to E4, and note the effect on resolution. As you can see, the number of theoretical plates has a powerful influence on the resolving ability of a particular column. By working with this spreadsheet, you can develop a feel for and a visual sense of the significance of chromatographic resolution.

Finding the Number of Plates to Achieve a Given Resolution

Another calculation we can make with Excel is to find the number of theoretical plates needed to achieve a given resolution. The equation used is Equation 31-31 of FAC9 or Equation 24-21 of AC7 which is repeated here:

$$N = 16R_s^2 \left(\frac{\alpha}{\alpha - 1} \right)^2 \left(\frac{1 + k_B}{k_B} \right)^2$$

In order to calculate N, we need the desired resolution R_s, the selectivity factor α, and the retention factor of component B, k_B. Since $\alpha = k_B/k_A$, we can calculate α given the two retention factors.

Let us begin with a desired resolution of 1.0, a k_A value of 5.5, and a k_B value of 7.5. As shown in the spreadsheet of Figure 14-4 this gives an α value of 1.364. The required number of theoretical plates is then 289. If the retention factor of component B is closer to component A, k_B = 6.5, N increases to 900 as shown. Changing k_B to 6.0, leads to a requirement that $N = 3136$, and so on. Plot N vs. k_B. Now change R_s to 1.25, 1.5, 2.0, then 2.5, and repeat the calculations. Repeat for $k_A = 6.0$. Given that the retention time of component A is 30 s, $\sigma_A = 3$ and $\sigma_B = 4.5$, simulate chromatograms for the cases where $k_A = 5.5$, $k_B = 7.5$, and $R_s = 1.0, 1.25, 1.5, 2.0$. and 2.5. These should give you more of a feel for the factors that determine resolution in chromatography.

	A	B	C	D	E	F	G	H
1	Spreadsheet to Calculate Number of Theoretical Plates to Achieve a Given Resolution							
2								
3	Desired R_s	k_A	k_B	α	N			
4	1.0	5.5	7.5	1.364	289			
5	1.0	5.5	6.5	1.182	900			
6	1.0	5.5	6.0	1.091	3136			
7	1.0	5.5	5.8	1.055	8220			
8								
9	Documentation							
10	Cell D4=C5/B5							
11	Cell F4=16*A4^2*(D4/(D4-1))^2*((1+C4)/C4)^2							

Figure 14-4 Calculation of number of theoretical plates to achieve desired resolution.

Nonideal Peak Shapes

Most chromatographic peaks do not have the ideal Gaussian shape just described. As a result of various mass transport processes, peaks exhibit the phenomenon of *tailing*.[1] Peaks with these characteristics can be represented by the following equation.[2]

$$c_t = A \frac{\sigma}{\tau\sqrt{2}} \exp\left[\frac{1}{2}\left(\frac{\sigma}{\tau}\right)^2 - \frac{t - t_R}{\tau}\right] \int_{-\infty}^{z/\sqrt{2}} \exp(-x^2)\, dx$$

where,

$$z = \frac{t - t_R}{\sigma} - \frac{\sigma}{\tau}$$

This expression, or model, is considerably more complex than the basic Gaussian model; however, most of the terms are familiar from our previous examples. The new term τ is the time constant of the exponentially tailing portion of the peak. Thus, we have two characteristics: σ is characteristic of the width of the ideal Gaussian portion of the peak, and τ is characteristic of the length of the tailing portion. The ratio τ/σ, which appears as the inverse σ/τ in the equations, is

---

[1] D. A. Skoog, F. J. Holler, and S. R. Crouch, *Principles of Instrumental Analysis*, 6th ed., p. 769. Belmont, CA: Brooks/Cole 2007.
[2] J. P. Foley and J. G. Dorsey, *Anal. Chem.*, **1983**, *55*, 730; *J. Chromatogr. Sci.*, **1984**, *22*, 40, and references found in these papers.

an important fundamental characteristic of this model. The model is usually called an

exponentially modified Gaussian, or EMG, model because it is the result of the convolution of a

normal Gaussian function and an exponential decay.

We can calculate the EMG function in Excel by modifying the normal Gaussian function.

We will first compute a Gaussian array of data points by the methods described previously. If

$G(t)$ represents the unmodified Gaussian and $EMG(t)$ represents the modified function at time t,

we can obtain $EMG(t)$ from the value at the previous time, $G(t)$, and a parameter, Z by the

following equation:

$$EMG(t) = EMG(t-1) + \left[\frac{G(t) - EMG(t-1)}{Z} \right]$$

where

$$Z = \left[\frac{1}{1 - e^{-1/\tau}} \right] WF$$

WF is a weighting factor, and τ is the time constant of the exponential.

Let us now do these calculations in a spreadsheet. We will first enter the equation for a

normal Gaussian with an amplitude $A = 5$, a width $\sigma = 5$, and a retention time $t_R = 15$ s. Then we

will calculate $EMG(t)$ for τ values of 25, 15, 10, 5, and 1. We will use a weighting factor WF of

1.13. The spreadsheet is shown in Figure 14-5. Only the first 31 rows (19 s) of the spreadsheet

are shown. The calculations were done, however, for times up to 120 s. A chart of the 6 peaks

with their different distortion levels is shown in Figure 14-6. Note the large amount of tailing

that occurs for τ values greater than 10. You can try other values for the weighting factor WF and

other values of τ. Also, you may want to try different amplitudes and values of σ.

	A	B	C	D	E	F	G	H	I	J	K	L	M
1	Exponentially modified Gaussian peaks												
2	A	5											
3	σ	5											
4	t_R	15											
5	EMG parameters												
6	τ	25	15	10	5	1							
7	x_1	1.13											
8	Z	28.81877	17.52128	11.87442	6.233821	1.787634							
9	*Peak calculations*												
10		Gaussian	$\tau = 25$	$\tau = 15$	$\tau = 10$	$\tau = 5$	$\tau = 1$						
11	t	c_t	c_t	c_t	c_t	c_t	c_t		Documentation				
12	0	0.011109	0.011109	0.011109	0.011109	0.011109	0.011109		Cell B8=$B$7*(1/(1-EXP(-1/B6)))				
13	1	0.019841	0.011412	0.011607	0.011844	0.01251	0.015994		Cell C8=$B$7*(1/(1-EXP(-1/C6)))				
14	2	0.034047	0.012197	0.012888	0.013714	0.015965	0.026093		Cell D8=$B$7*(1/(1-EXP(-1/D6)))				
15	3	0.056135	0.013722	0.015356	0.017287	0.022409	0.042898		Cell E8=$B$7*(1/(1-EXP(-1/E6)))				
16	4	0.088922	0.016331	0.019555	0.023319	0.033078	0.068644		Cell F8=$B$7*(1/(1-EXP(-1/F6)))				
17	5	0.135335	0.020461	0.026163	0.032753	0.049482	0.105951		Cell B12=($B$2/$B$3)*EXP(-((A12-$B$4)^2/(2*$B$3^2)))				
18	6	0.197899	0.026618	0.035965	0.04666	0.07329	0.157386		Cell C12:G12=B12				
19	7	0.278037	0.035342	0.04978	0.066146	0.106135	0.224878		Cell C13=C12+(B13-C12)/$B$8				
20	8	0.375311	0.047139	0.06836	0.092182	0.149315	0.30903		Cell D13=D12+(B13-D12)/$C$8				
21	9	0.486752	0.062393	0.092239	0.125411	0.203445	0.408448		Cell E13=E12+(B13-E12)/$D$8				
22	10	0.606531	0.081275	0.121591	0.165928	0.268106	0.519255		Cell F13=F12+(B13-F12)/$E$8				
23	11	0.726149	0.103651	0.156095	0.213107	0.341583	0.634991		Cell G13=G12+(B13-G12)/$F$8				
24	12	0.83527	0.129038	0.194858	0.265502	0.420778	0.747027						
25	13	0.923116	0.156593	0.236422	0.320883	0.501361	0.845531						
26	14	0.980199	0.185171	0.278872	0.376407	0.578174	0.920864						
27	15	1	0.213446	0.32003	0.428923	0.645841	0.965133						
28	16	0.980199	0.240052	0.357708	0.475348	0.699477	0.973561						
29	17	0.923116	0.263754	0.389978	0.513057	0.735352	0.945342						
30	18	0.83527	0.283585	0.415392	0.540192	0.751381	0.883768						
31	19	0.726149	0.298942	0.433128	0.555852	0.747333	0.795596						

Figure 14-5 Worksheet for calculating peak shapes of exponentially modified Gaussian peaks.

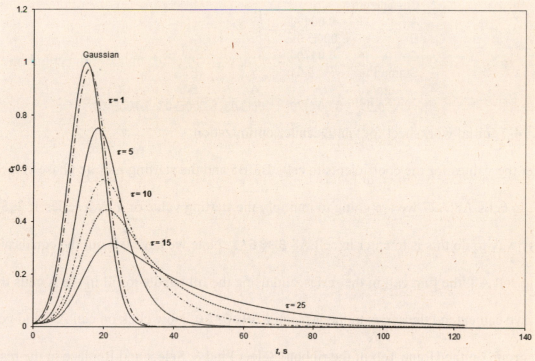

Figure 14-6 Chart showing exponentially modified Gaussian peaks.

Optimizing Chromatographic Methods: van Deemter Equation

The mobile phase flow rate in chromatographic methods is often optimized by plotting plate height H vs. mobile phase velocity u in order to minimize H. One important model of how H varies with u is called the van Deemter equation, which can be written as:

$$H = A + B/u + Cu$$

where the constants A, B, and C are coefficients of eddy diffusion, longitudinal diffusion, and mass transfer, respectively.[3] We will use Excel here to plot H vs. u and also to plot the three components A, B/u, and Cu.

van Deemter Plots

Let us begin by setting up a worksheet as shown in Figure 14-7.

	A	B	C	D	E
1	van Deemter optimization				
2	Coefficients				
3	A	0.00100			
4	B	6.000E-05			
5	C	0.01700			
6	Starting u	0.01			
7	u, cm/s	H	A	B/u	Cu
8	0.01	0.007170	1.000E-03	6.000E-03	1.700E-04

Figure 14-7 Initial worksheet for van Deemter optimization.

We enter the values for the coefficients in cells B3:B5 and the starting mobile phase velocity in cell B6. In cells A8:A27 we are going to multiply the starting value of u by a series 1, 2, 3, 4, etc. The easy way to do this is to enter in cell A8 **$B$6*1**. Note we have left out the equals sign. By selecting cell A8, the first cell in the series, and using the fill handle to fill in the A cells through row 27, we can extend the series to the last entry **$B$6*20**. Now click on cell A7, and go to the Editing group on the Home Tab of the ribbon. Select Find & Select and Replace in the resulting

---

[3] The mass transfer coefficient is often broken into a coefficient for the mobile phase C_m and the stationary phase, C_s

menu. In the Find what: box, enter **$B$6** and in the Replace with: box, enter **=$B$6** as shown in Figure 14-8. Click on Options >>. In the Search: box, choose By Rows. Click on the Find Next button and replace all the values until you have replaced cell A27. Then click on Close.

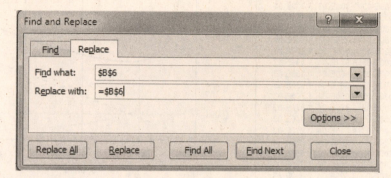

Figure 14-8 Replacing part of the contents of cells.

Now enter in cell B8, the formula for calculating H. In cell C8, enter the formula for A, in D8 the formula for B/u, and in E8 the formula for Cu. Fill in the values up to and including row 27. Make a chart showing H, A, B/u, and Cu vs. u. The final worksheet should appear as shown in Figure 14-9. Find the optimum value of u.

The fun part of this exercise comes in varying the coefficients and flow velocities. Keeping the same flow velocity, vary A from 0 to 0.01 and observe the effect. Now vary u and note the effect. You will have to change the scale for widely different values of u. Vary the B/u term and the Cu term. Try to find values for the parameters that will allow you to duplicate closely the GC and HPLC plots shown in Figure 31-13 of FAC9 or Figure 24-11 of AC7. Why are plots for HPLC and GC so different?

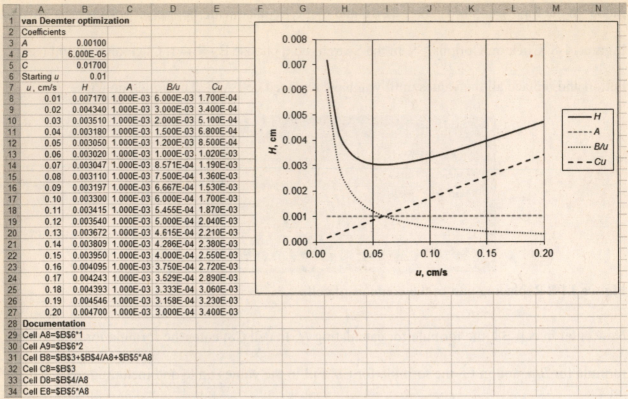

Figure 14-9 Final worksheet and chart for van Deemter optimization.

Using Solver to Find van Deemter Coefficients

We will now use results from an HPLC experiment that obtained H values as a function of u to

find the coefficients, A, B, and C. The data that were collected are shown in Figure 14-10.

	A	B	C	D
1	van Deemter Calculations			
2	Coefficients		u, cm/s	H, cm
3	A		0.03037	0.004182
4	B		0.04521	0.003352
5	C		0.06527	0.002731
6			0.10170	0.002293
7			0.12950	0.002246
8			0.16590	0.00211
9			0.24960	0.002305
10			0.30820	0.002407
11			0.48390	0.002778
12			0.60420	0.003098

Figure 14-10 Experimental results for plate height H vs. u for an HPLC experiment.

We will use Solver to obtain the van Deemter coefficients by calculating a value of H and

then minimizing the sums of the squares of the deviations between the experimental H and the calculated H. Looking at the values in Figure 14-9, we pick our initial estimates of 1.00×10^{-3}

for A and C, and 1.00×10^{-6} for B. A plot of these values is shown in Figure 14-11.

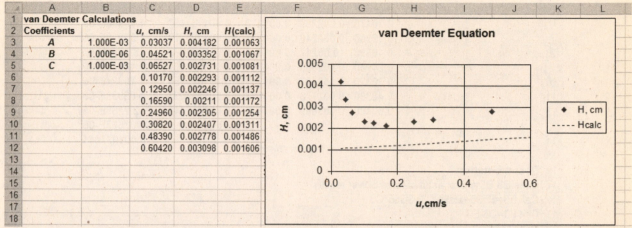

Figure 14-11 van Deemter plots with initial estimates.

The upper values are the experimental data points, while the dashed line is H(calc). It can

be seen that these initial values do not fit the experimental curve very well at all. Now move the

chart out of the way, and calculate in column F, the deviations (residuals) $(H - H(\text{calc}))$. In

column G, find the deviations squared. In cell G 13 find the sum of the squares (SSR) and in cell

G14 find the standard error $= \sqrt{\text{SSR} / N - 3}$ as shown in Figure 14-12.

	A	B	C	D	E	F	G
1	van Deemter Calculations						
2	Coefficients		u, cm/s	H, cm	H(calc)	dev	dev^2
3	A	1.000E-03	0.03037	0.004182	0.001063	0.003118703	9.72631E-06
4	B	1.000E-06	0.04521	0.003352	0.001067	0.002284671	5.21972E-06
5	C	1.000E-03	0.06527	0.002731	0.001081	0.001650409	2.72385E-06
6			0.10170	0.002293	0.001112	0.001181467	1.39586E-06
7			0.12950	0.002246	0.001137	0.001108778	1.22939E-06
8			0.16590	0.00211	0.001172	0.000938072	8.7998E-07
9			0.24960	0.002305	0.001254	0.001051394	1.10543E-06
10			0.30820	0.002407	0.001311	0.001095555	1.20024E-06
11			0.48390	0.002778	0.001486	0.001292033	1.66935E-06
12			0.60420	0.003098	0.001606	0.001492145	2.2265E-06
13						SSR	2.737663E-05
14						Std. Error	0.001977611
15							
16	Documentation						
17	Cells B3:B5=initial estimates or Solver results						
18	Cell E3=$B$3+$B$4/C3+$B$5*C3						
19	Cell F3=D3-E3						
20	Cell G3=F3^2						
21	Cell G13=SUM(G3:G12)						
22	Cell G14=SQRT(G13/7)						

Figure 14-12 Worksheet after calculating deviations (residuals) and deviations squared.

Now we will invoke Solver to minimize the standard error shown in cell G14. Bring up Solver and set the Set Objective as G14 (Target cell in Excel 2007) By Changing Variable Cells B3:B5. We will use no constraints. Select GRG Nonlinear as the Solving Method. In the Solver Options, enter 0.1% for optimality, 100 seconds for the Max Time and 500 for Iterations. Have Solver obtain a solution. The values should appear similar to Figure 14-13. Note how much better the fit is after this optimization.

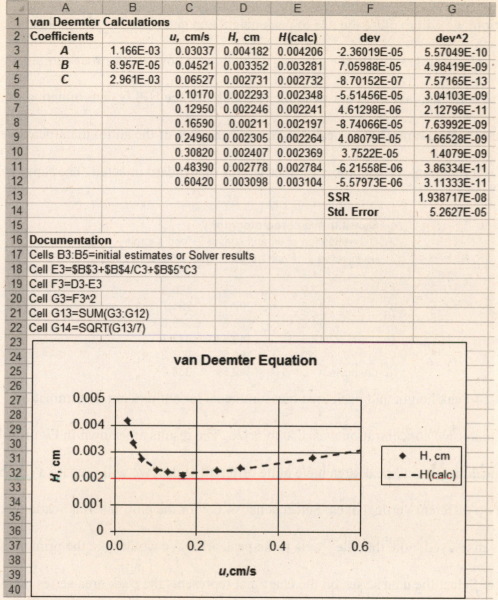

	A	B	C	D	E	F	G
1	van Deemter Calculations						
2	Coefficients		u, cm/s	H, cm	H(calc)	dev	dev^2
3	A	1.166E-03	0.03037	0.004182	0.004206	-2.36019E-05	5.57049E-10
4	B	8.957E-05	0.04521	0.003352	0.003281	7.05988E-05	4.98419E-09
5	C	2.961E-03	0.06527	0.002731	0.002732	-8.70152E-07	7.57165E-13
6			0.10170	0.002293	0.002348	-5.51456E-05	3.04103E-09
7			0.12950	0.002246	0.002241	4.61298E-06	2.12796E-11
8			0.16590	0.00211	0.002197	-8.74066E-05	7.63992E-09
9			0.24960	0.002305	0.002264	4.08079E-05	1.66528E-09
10			0.30820	0.002407	0.002369	3.7522E-05	1.4079E-09
11			0.48390	0.002778	0.002784	-6.21558E-06	3.86334E-11
12			0.60420	0.003098	0.003104	-5.57973E-06	3.11333E-11
13						SSR	1.938717E-08
14						Std. Error	5.2627E-05
15							
16	Documentation						
17	Cells B3:B5=initial estimates or Solver results						
18	Cell E3=$B$3+$B$4/C3+$B$5*C3						
19	Cell F3=D3-E3						
20	Cell G3=F3^2						
21	Cell G13=SUM(G3:G12)						
22	Cell G14=SQRT(G13/7)						

Figure 14-13 Final Solver results for van Deemter coefficients.

Quantitative Chromatography

The most straightforward method for quantitative chromatographic analyses involves the preparation of a series of standard solutions that approximate the composition of the unknown (external standard method). Either peak height or peak area can be used as the dependent variable.

Method of External Standards

In this section, we will explore the use of peak height and peak area for the method of external

standards in chromatography. In the process, we will learn how to construct combination charts

that have two different axes. As our example, we will take the HPLC determination of an

antioxidant in a food sample. The peak height and peak area of the antioxidant peak were

measured as a function of the number of micrograms of standard injected. The same volume of a

	A	B	C	D
1	**Quantitative chromatography**			
2	Determination of an antioxidant in food sample			
3	µg injected	Peak height	Peak area	
4	4.5	38.7	114.6	
5	7.5	64	190.9	
6	11.5	97.7	291.7	
7	14.8	125.6	375.6	
8	17.5	148.5	443.7	
9	20	169.2	507.2	
10	unknown	112.5	336.2	

Figure 14-14 Peak height and peak area measurements for antioxidant determination.

sample of unknown concentration was also injected. The results are shown in Figure 14-14.

Since the peak height and peak area have quite different scales, we will prepare a combination

plot with two different vertical axes. Select cells A4:C9 for the plot. Insert a Scatter chart. When

the chart is displayed, note that the y axis is the peak area. We would like the primary axis to be

peak height. Select the data series on the chart that represents the peak area series. Right click

and select Format Data Series…. Under Series Options, choose Secondary axis for plotting the

series. Note that the peak height series is now plotted on the left (primary) axis, and the peak area

series on the right (secondary) axis.

The completed spreadsheet with calculations of the unknown concentration and statistical

analysis is shown in Figure 14-15. From the peak height data, we would report the unknown as

13.25 ± 0.02 μg. From the peak area data, we would report 13.249 ± 0.009 μg. Note that there is no statistically significant difference between the two methods of treating the data.

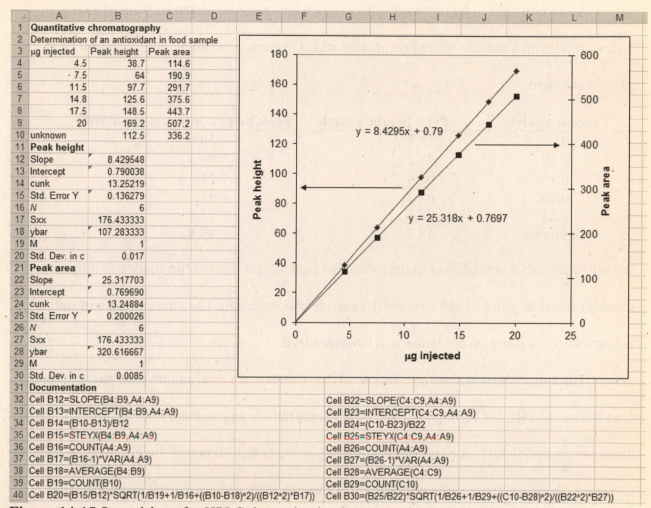

Figure 14-15 Spreadsheet for HPLC determination by external standard method.

Internal Standard Method

The addition of an *internal standard* can be used to compensate for various uncertainties that occur in chromatography. We can often compensate for variations in sample injection, flow rate, and column conditions by using the internal standard method. Here, we add the same amount of internal standard to mixtures containing known amounts of the analyte and to the samples of unknown analyte concentration. We then ratio the peak height or area for the analyte to that of the internal standard as discussed in Chapter 4.

As an illustration, let us take the data shown below for the determination of a C_7 hydrocarbon with a closely related compound added to each standard and to the unknown as an internal standard.

Percent analyte	Peak height analyte	Peak height, internal standard
0.05	18.8	50.0
0.10	48.1	64.1
0.15	63.4	55.1
0.20	63.2	42.7
0.25	93.6	53.8
Unknown	58.9	49.4

We will construct a spreadsheet to determine the peak height ratio of the analyte to internal standard and then plot this ratio versus the analyte concentration. The concentration of the unknown and its standard deviation will be determined.

The spreadsheet is shown in Figure 14-16. The data are entered into columns A through C as shown. In cells D4 through D9, the peak height ratio is calculated by the formula shown in documentation cell A22. A plot of the calibration curve is also shown in the Figure. The linear regression statistics are calculated in cells B11 through B20 using our usual approach. The statistics are calculated by the formulas in documentation cells A23through A31. The percentage of the analyte in the unknown is found to be 0.163 ± 0.008.

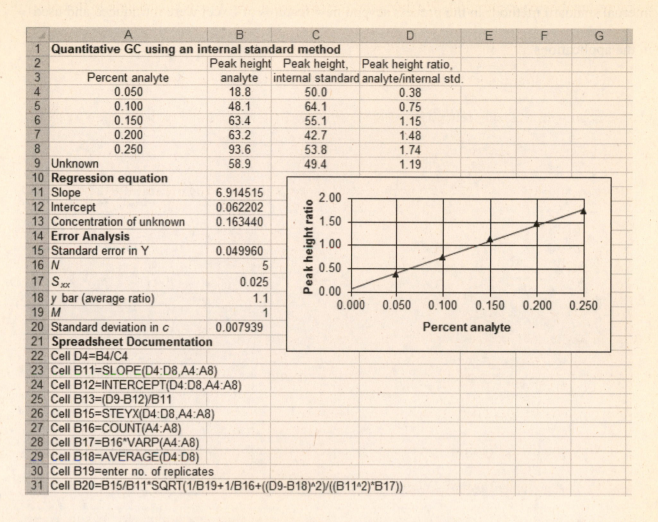

	A	B	C	D	E	F	G
1	**Quantitative GC using an internal standard method**						
2		Peak height	Peak height,	Peak height ratio,			
3	Percent analyte	analyte	internal standard	analyte/internal std.			
4	0.050	18.8	50.0	0.38			
5	0.100	48.1	64.1	0.75			
6	0.150	63.4	55.1	1.15			
7	0.200	63.2	42.7	1.48			
8	0.250	93.6	53.8	1.74			
9	Unknown	58.9	49.4	1.19			
10	**Regression equation**						
11	Slope	6.914515					
12	Intercept	0.062202					
13	Concentration of unknown	0.163440					
14	**Error Analysis**						
15	Standard error in Y	0.049960					
16	N	5					
17	S_{xx}	0.025					
18	y bar (average ratio)	1.1					
19	M	1					
20	Standard deviation in c	0.007939					
21	**Spreadsheet Documentation**						
22	Cell D4=B4/C4						
23	Cell B11=SLOPE(D4:D8,A4:A8)						
24	Cell B12=INTERCEPT(D4:D8,A4:A8)						
25	Cell B13=(D9-B12)/B11						
26	Cell B15=STEYX(D4:D8,A4:A8)						
27	Cell B16=COUNT(A4:A8)						
28	Cell B17=B16*VARP(A4:A8)						
29	Cell B18=AVERAGE(D4:D8)						
30	Cell B19=enter no. of replicates						
31	Cell B20=B15/B11*SQRT(1/B19+1/B16+((D9-B18)^2)/((B11^2)*B17))						

Figure 14-16 The internal standard method for GC determinations.

Summary

In this chapter, we have explored the use of Excel for several chromatographic applications. We have modeled chromatographic peaks and simulated chromatograms containing several components. We have seen how to calculate the number of plates needed to achieve a desired resolution and how to incorporate exponential peak distortions. We have investigated the optimization of chromatographic methods and the use of Solver for obtaining the van Deemter coefficients. We have also studied quantitative applications of chromatography, including the internal standard method. In the process several new features of Excel were introduced and used in the applications.

Problems

1. Prepare a spreadsheet for simulating a three-component chromatogram similar to that shown in Figure 14-3. Change the number of theoretical plates from 100 to 1000 and note the effect on resolution in cell H18. Change N back to 100 and focus on peak A. Vary the retention time systematically for a from 5 to 40 s and note the effect on resolution. Now return the retention time to its original setting and vary the amplitude of peak A. What variables have the largest influence on resolution?

2. Set up a spreadsheet similar to Figure 14-4 to find the number of theoretical plates needed to achieve a given resolution. Use with a retention factor for species A of $k_A = 4.0$ and a resolution of 1.0. Vary k_B over a range from 4.2 to 7.0 and note the values of α and N. Now change the resolution to 1.25 and vary k_B over the same range. Do the same for R_s values of 1.50, 1.75 and 2.0. Note the effect on selectivity and number of theoretical plates.

4. Two components in an HPLC separation have retention times that differ by 15 s. The first peak elutes in 9.0 min and the peak widths are approximately equal. Find the minimum number of theoretical plates needed to achieve the following resolution, R_s values: 0.50, 0.75, 0.90, 1.0, 1.10, 1.25, 1.50, 1.75, 2.0, 2.5. How would the results change if peak 2 were twice as broad as peak 1?

5. Construct a spreadsheet to study peak distortion similar to Figure 14-5. Vary the weighting factor, WF from 1.0 to 3.0 and note its effect. Vary σ from 1.0 to 15 and note the effect of peak width. Vary the amplitude A and the retention time t_R. Which variables have the largest effect on peak distortion?

6. Construct the van Deemter optimization spreadsheet of Figure 14-9. Find values for the coefficients that allow you to duplicate closely the plots of Figure 31-13 of FAC9 or 24-11 of AC7 for HPLC and GC. Why are the plots so different in terms of flow rates and sharpness of the minima?

7. One method for quantitative determination of the concentration of constituents in a sample analyzed by gas chromatography is the area normalization method. Here, complete elution of all of the sample constituents is necessary. The area of each peak is then measured and corrected for differences in detector response to the different eluates. This correction involves dividing the area by an empirically determined correction factor. The concentration of the analyte is found from the ratio of its corrected area to the total corrected area of all peaks. For a chromatogram containing three peaks, the relative areas were found to be 16.4, 45.2, and 30.2 in the order of increasing retention time. Calculate the percentage of each compound if the relative detector responses were 0.60, 0.78, and 0.88, respectively.

8. Peak areas and relative detector responses are to be used to determine the concentration of the five species in a sample. The area-normalization method described in Problem 7 is to be used. The relative areas for the five gas chromatographic peaks are given below. Also shown are the relative responses of the detector. Calculate the percentage of each component in the mixture.

Compound	Relative Peak Area	Relative Detector Response
A	32.5	0.70
B	20.7	0.72
C	60.1	0.75
D	30.2	0.73
E	18.3	0.78

9.   An HPLC method was developed for the separation and determination of ibuprofen in rat

plasma samples as part of a study of the time course of the drug in laboratory animals.

Several standards were chromatographed and the results below obtained:

Ibuprofen concentration, µg/mL	Relative peak area
0.5	5.0
1.0	10.1
2.0	17.2
3.0	19.8
6.0	39.7
8.0	57.3
10.0	66.9
15.0	95.3

Next a 10mg/kg sample of ibuprofen was administered orally to a laboratory rat. Blood

samples were drawn at various times after administration of the drug and subjected to

HPLC analysis. The following results were obtained.

Time, hr.	Peak Area
0	0
0.5	91.3
1.0	80.2
1.5	52.1
2.0	38.5
3.0	24.2
4.0	21.2
6.0	18.5
8.0	15.2

Find the concentration of ibuprofen in the blood plasma for each of the times given above

and plot the concentration versus time. On a percentage basis, during what half-hour

period (1st, 2nd, 3rd, etc.) is most of the ibuprofen lost?

10.   For the data below, given in Example 32-1 of FAC9 or 25-1 of AC7, compare the method

of external standards to the internal standard method. Plot the analyte peak height versus

percent analyte and determine the unknown without using the internal standard results.

Are your results any more precise using the internal standard method? If so, give some

possible reasons.

Percent analyte	Peak height analyte	Peak height, internal standard
0.05	18.8	50.0
0.10	48.1	64.1
0.15	63.4	55.1
0.20	63.2	42.7
0.25	93.6	53.8
Unknown	58.9	49.4

Chapter 15

Electrophoresis and Other Separation Methods

In this chapter, we'll explore the use of Excel for problems involving electrophoresis, and other separation methods. We begin by using Excel to resolve two severely overlapped Gaussian peaks as commonly encountered in separations. Then we explore the use of Excel for problems in capillary electrophoresis and micellar electrokinetic capillary chromatography. These methods are the subject of Chapter 34 of FAC9 and Chapter 26 of AC7.

Resolution of Overlapped Gaussian Peaks

Overlapped peaks occur often in chromatography and electrophoresis. In many cases, these peaks are approximated by Gaussian curves as discussed in Chapter 14. If the Gaussian approximation holds, we can model the resulting response as a summation of Gaussian curves and use Solver to obtain the parameters even though the peaks are severely overlapped. The overall process of separating overlapped peaks is often called *deconvolution*.

Let's first develop a simulated chromatogram as we did in Chapter 14. This chromatogram will contain two components, A and B. The equation used to generate the chromatogram is the sum of Gaussians equations for the total concentration, c_t.

$$c_t = \sum_{i=1}^{2} \frac{A_i}{\sigma_i} e^{-\frac{(t-t_{R_i})^2}{2\sigma_i^2}}$$

The simulated two-peak chromatogram (or electropherogram) is given in Figure 15-1. Note that we've chosen the retention times, peak heights and peak widths to give severely overlapped peaks. Only the first 29 rows of the spreadsheet are shown, but you'll need to carry out the simulation up to row 69 (60 s). Show only the curve, not the data points on the chart.

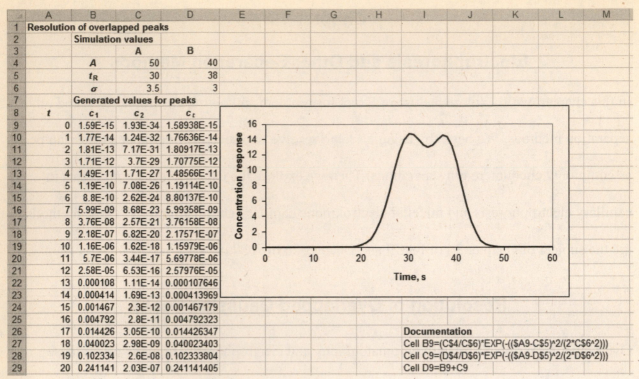

Figure 15-1 Simulated two-peak chromatogram.

We'll now use Solver to resolve the two overlapped peaks so that we can determine the individual peak heights, retention times, and peak widths. To do that we need to add columns for the Solver model, the residuals, the sum of the squares of the residuals, and the total sum of the squares as shown in Figure 15-2. Note that we've put initial guesses for the peak heights (*A* values), retention times (*t_R* values) and peak widths (*σ* values) in cells F4 through G6. We've also added a new series to the chart to display the model values as points. Our initial guesses for the model parameters are not very good as you can see. Be sure to copy the formulas for the model, the residuals and the sum of the squares of the residuals through row 69.

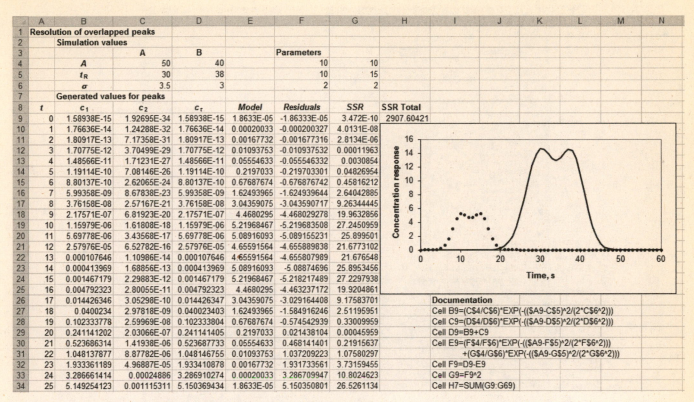

	A	B	C	D	E	F	G	H	I	J	K	L	M	N
1	Resolution of overlapped peaks													
2		Simulation values												
3			A		B		Parameters							
4		A	50		40		10	10						
5		t_R	30		38		10	15						
6		σ	3.5		3		2	2						
7		Generated values for peaks												
8	t	c_1	c_2		c_τ	Model	Residuals	SSR	SSR Total					
9	0	1.58938E-15	1.92695E-34		1.58938E-15	1.8633E-05	-1.86333E-05	3.472E-10	2907.60421					
10	1	1.76636E-14	1.24288E-32		1.76636E-14	0.00020033	-0.000200327	4.0131E-08						
11	2	1.80917E-13	7.17358E-31		1.80917E-13	0.00167732	-0.001677316	2.8134E-06						
12	3	1.70775E-12	3.70499E-29		1.70775E-12	0.01093753	-0.010937532	0.00011963						
13	4	1.48566E-11	1.71231E-27		1.48566E-11	0.05554633	-0.055546332	0.0030854						
14	5	1.19114E-10	7.08146E-26		1.19114E-10	0.2197033	-0.219703301	0.04826954						
15	6	8.80137E-10	2.62065E-24		8.80137E-10	0.67687674	-0.676876742	0.45816212						
16	7	5.99358E-09	8.67838E-23		5.99358E-09	1.62493965	-1.624939644	2.64042885						
17	8	3.76158E-08	2.57167E-21		3.76158E-08	3.04359075	-3.043590717	9.26344445						
18	9	2.17571E-07	6.81923E-20		2.17571E-07	4.4680295	-4.468029278	19.9632856						
19	10	1.15979E-06	1.61808E-18		1.15979E-06	5.21968467	-5.219683508	27.2450959						
20	11	5.69778E-06	3.43568E-17		5.69778E-06	5.08916093	-5.089155231	25.899501						
21	12	2.57976E-05	6.52782E-16		2.57976E-05	4.65591564	-4.655889838	21.6773102						
22	13	0.000107646	1.10986E-14		0.000107646	4.65591564	-4.655807989	21.676548						
23	14	0.000413969	1.68856E-13		0.000413969	5.08916093	-5.08874696	25.8953456						
24	15	0.001467179	2.29883E-12		0.001467179	5.21968467	-5.218217489	27.2297938						
25	16	0.004792323	2.80055E-11		0.004792323	4.4680295	-4.463237172	19.9204861						
26	17	0.014426346	3.05298E-10		0.014426347	3.04359075	-3.029164408	9.17583701			Documentation			
27	18	0.0400234	2.97818E-09		0.040023403	1.62493965	-1.584916246	2.51195951			Cell B9=(C$4/C$6)*EXP(-(($A9-C$5)^2/(2*C$6^2)))			
28	19	0.102333778	2.59969E-08		0.102333804	0.67687674	-0.574542939	0.33009959			Cell C9=(D$4/D$6)*EXP(-(($A9-D$5)^2/(2*D$6^2)))			
29	20	0.241141202	2.03066E-07		0.241141405	0.2197033	0.021438104	0.00045959			Cell D9=B9+C9			
30	21	0.523686314	1.41938E-06		0.523687733	0.05554633	0.468141401	0.21915637			Cell E9=(F$4/F$6)*EXP(-(($A9-F$5)^2/(2*F$6^2)))			
31	22	1.048137877	8.87782E-06		1.048146755	0.01093753	1.037209223	1.07580297			+(G$4/G$6)*EXP(-(($A9-G$5)^2/(2*G$6^2)))			
32	23	1.933361189	4.96887E-05		1.933410878	0.00167732	1.931733561	3.73159455			Cell F9=D9-E9			
33	24	3.286661414	0.00024886		3.286710274	0.00020033	3.286709947	10.8024623			Cell G9=F9^2			
34	25	5.149254123	0.001115311		5.150369434	1.8633E-05	5.150350801	26.5261134			Cell H7=SUM(G9:G69)			

Figure 15-2 Spreadsheet after adding Gaussian model.

We'll now invoke Solver and ask it to minimize the sum of the squares (value in cell H9).

We'll also add some constraints so that Solver's search is more efficient. Call up Solver from the

Tools menu and ask it to minimize the value in cell H9 as shown in Figure 15-3. Add the

constraints shown. These constrain cells F4 and G4 (peak heights) to be between 10 and 60, cells

F5 and G5 (retention times) to be between 10 and 60 s, and cells F6 and G6 (peak widths) to be

between 1 and 10. The Parameters window is slightly different for Excel 2007. However, choose

the same constraints and Objective (Target Cell) as shown in Figure 15-3. Next, click the

Options button and select a Constraint Precision and a Convergence of 1.0×10^{-10}. For Excel

2007, use a Tolerance of 0.1%. Choose to show the iteration results with a maximum of 100

iterations and a maximum time of 100 s. Use the GRG nonlinear solving method for Excel 2010.

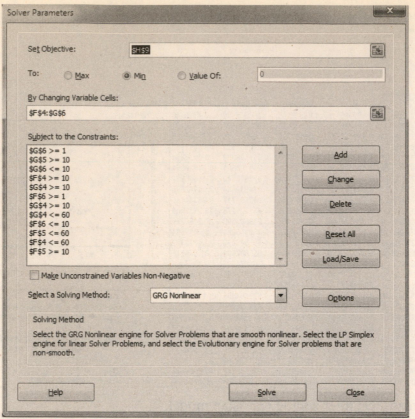

Figure 15-3 Solver parameters window in Excel 2010.

Click the Solve button. After the first result is displayed, click on the Continue button and

observe the **SSR Total** value (cell H9) and the chart. Keep on clicking the Continue button until

convergence is reached. This may require about 60 iterations. When Solver has minimized the

sum of the squares, the results shown in Figure 15-4 should be displayed. The results shown are

for Excel 2010. The results for Excel 2007 are very slightly different because a different solving

method is used. However, the results are extremely close to those in Figure 15-4. The Parameters

of the peak (height, retention time and width) are given in Cells F4:F6 for component B and in

Cells G4:G6 for component A. Note how closely the predicted values of the Parameters match

the original data set.

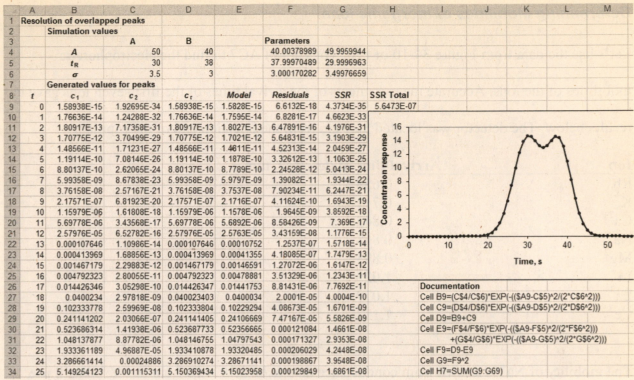

Figure 15-4 Final spreadsheet for resolving two component peaks.

You can extend this exercise by changing the retention time for component B in the simulation values so that it is closer to that of component A. Determine when Solver is no longer able to resolve the two components. You can also vary the constraints or the initial parameter values and investigate conditions under which Solver fails to resolve the components. In dealing with actual data, Solver may not be as successful as we've seen with generated data because of nonideal peak shapes and the presence of noise.

Electrophoresis

Electrophoresis, particularly capillary electrophoresis, has applicability to a wide array of chemical problems. We'll take two examples here. First, we'll find the electrophoretic mobilities of some inorganic ions by capillary electrohoresis. Then, we'll see how capillary electrophoresis can be used to determine the pK_a values of weak organic acids.

Determining Mobilities of Inorganic Ions

Let's use the data from Figure 34-10 of FAC9 or 26-9 of AC7 to find the electrophoretic

mobilities of the ions. Accurate measurements of the arrival times of the first seven ions gave the

results below. The detector was located 30.0 cm from the injection end of the capillary.

Ion	Arrival time at detector, min
Rb^+	0.658
K^+	0.711
Ca^{2+}	0.842
Na^+	0.868
Mg^{2+}	0.895
Li^+	1.026
La^{3+}	1.079
Ce^{3+}	1.132

The mobility of Na^+ is known from conductance data to be 5.19×10^{-4} cm^2 s^{-1} V^{-1}. We'll use

this value to calculate the electroosmotic flow velocity (cm/s) for the data in Figure 34-10 of

FAC9 or 26-9 of AC7, and then use the velocity to calculate the electrophoretic mobilities of the

other ions. We'll then compare the values obtained to the limiting ionic mobilities obtained from

conductance data shown in the table that follows.

Ion	Limiting ionic mobility, cm^2 s^{-1} $V^{-1} \times 10^{-4}$ at 25°C
Rb^+	8.06×10^{-4}
K^+	7.62×10^{-4}
Ca^{2+}	6.17×10^{-4}
Na^+	5.19×10^{-4}
Mg^{2+}	5.50×10^{-4}
Li^+	4.01×10^{-4}
La^{3+}	7.21×10^{-4}
Ce^{3+}	7.26×10^{-4}

Source: J. A. Dean, *Analytical Chemistry Handbook*, (New York: McGraw-Hill, 1995), p. 14.41

Electroosmotic Flow Velocity. The arrival time of an ion at the detector t_D is given by the

distance to the detector l_D divided by the velocity with which the ion migrates. The migration rate is the sum of the ions electrophoretic velocity v_e and the velocity of the electroosmotic flow v_{eo}. Hence we can write

$$t_D = \frac{l_D}{v_e + v_{eo}}$$

From this equation, we can relate the electrophoretic velocity to the electrophoretic mobility, the applied voltage and the length of the capillary. Thus, we can write the arrival time as

$$t_D = \frac{l_D}{\dfrac{\mu_e V}{L} + v_{eo}}$$

We can solve this latter equation for v_{eo} and obtain

$$v_{eo} = \frac{l_D}{t_D} - \frac{\mu_e V}{L}$$

For Na^+ we can write

$$v_{eo} = \frac{30.0 \text{ cm}}{0.868 \text{ min} \times 60.0 \text{ s/min}} - \frac{5.19 \times 10^{-4} \text{ cm}^2 \text{ s}^{-1} \text{ V}^{-1} \times 30,000 \text{ V}}{36.5 \text{ cm}} = 0.14946 \text{ cm s}^{-1}$$

Mobilities of the Other Ions. Before entering values into the spreadsheet for calculating mobilities, let's use the equation for t_D and solve for μ_e. The result is

$$\mu_e = \frac{l_D L}{t_D V} - \frac{v_{eo} L}{V}$$

Now we can make entries into the spreadsheet as shown in Figure 15-5. We first enter the known values in the final equation above into cells B2 through B5. The table of arrival times given is entered into cells B8 through B14 after putting the appropriate labels in A8 through A14 and titles in A7 and B7. The formula to calculate μ_e from above is then entered into cell C8 as shown in the documentation cell. The fill handle is then used to apply this equation to the remaining values in the table.

	A	B	C	D
1	Calculation of electrophoretic mobilities of ions			
2	Capillary length, cm	36.5		
3	Distance to detector, cm	30.0		
4	Applied voltage, V	30000.0		
5	Electroosmotic flow, cm/s	0.14946		
6				
7	Ion	Arrival time, s	Electrophoretic mobility, $cm^2\ s^{-1}\ V^{-1}$	Ionic mobility, $cm^2\ s^{-1}\ V^{-1}$
8	Rb^+	0.658	7.43E-04	8.06E-04
9	K^+	0.711	6.74E-04	7.62E-04
10	Ca^{2+}	0.842	5.41E-04	6.17E-04
11	Na^+	0.868	5.19E-04	5.19E-04*
12	Mg^{2+}	0.895	4.98E-04	5.50E-04
13	Li^+	1.026	4.11E-04	4.01E-04
14	La^{3+}	1.079	3.82E-04	7.21E-04
15	Ce^{3+}	1.132	3.56E-04	7.26E-04
16				
17	*Na^+ used to calibrate			
18	Spreadsheet documentation			
19	Cell C8=$B$3*$B$2/(B8*60*$B$4)-$B$5*$B$2/$B$4			

Figure 15-5 Spreadsheet to calculate electrophoretic mobilities.

If we then add the values from the table of ionic mobilities, we can compare the capillary electrophoresis values directly to the conductance values as shown in Figure 15-5. The limiting ionic mobilities obtained from conductance values are called *limiting* because they are infinite dilution values found by extrapolating conductances at finite concentration to infinite dilution. Because the capillary electrophoresis results were obtained in a buffer solution, we would not expect very good correlations. Note that for the univalent ions, the values compare fairly well with the trends in both electrophoretic and ionic mobilities being the same ($Rb^+ > K^+ > Na^+ > Li^+$). In all cases except for Li^+, the electrophoretic mobilities are smaller probably because the ionic atmosphere surrounding the migrating ion exhibits a drag on the ion. The electrophoretic mobilities of the divalent and trivalent ions do not correlate well with the limiting ionic mobilities obtained from conductance. The electrophoresis values are lower due not only to the ionic atmosphere, but also because of the tendency of the rare earth ions to form complexes with anions from the buffer.

Determining the pKₐ Values of Weak Acids

Capillary electrophoresis can also be used to determine pK_a values of weak acids and bases.

We'll first find an appropriate equation to use to calculate the pK_a value.

Expression for pK_a. Let's take a general weak acid HA and start with the expression for the

dissociation constant, K_a:

$$K_a = \frac{[H_3O^+][A^-]}{[HA]}$$

If we take the logarithm of both sides of this equation, we obtain

$$\log K_a = \log [H_3O^+] + \log\{[A^-]/[HA]\}$$

Multiplying both sides by minus one and converting to p functions gives

$$pK_a = pH - \log\{[A^-]/[HA]\} = pH + \log\{[HA]/[A^-]\}$$

Now let's convert the ratio in the logarithmic term to the fraction dissociated, by using

the definition of α

$$\alpha = \frac{[A^-]}{[HA] + [A^-]}$$

and

$$\frac{1}{\alpha} = \frac{[HA] + [A^-]}{[A^-]} = \frac{[HA]}{[A^-]} + 1$$

Or,

$$\frac{[HA]}{[A^-]} = \frac{1}{\alpha} - 1$$

Substituting this latter expression into the equation for pK_a, gives

$$pK_a = pH + \log\left[\frac{1}{\alpha} - 1\right]$$

Now we'll use capillary electrophoresis to find the fraction ionized and the equation

above to find the pK_a. We can rearrange the pK_a equation to obtain

$$pH = pK_a - \log\left[\frac{1}{\alpha} - 1\right]$$

If we calculate α values as a function of pH, we can plot pH vs. $\log\{1/\alpha - 1\}$ and find the pK_a

value from the intercept. This allows us to use several data points to determine the pK_a value.

Finding α by capillary electrophoresis. To find the fraction dissociated, α, we first find the

mobility of the totally deprotonated species, μ_{A^-}. To do this, we add enough base to cause HA

to dissociate completely and collect an electropherogram. From the arrival time at the detector

and the capillary length, we can caclulate the velocity v. If we know the electric field strength E,

the mobility of A^- is found from

$$\mu_{A^-} = \frac{v}{E} = \frac{l_D/t_D}{E}$$

where l_D is the length of capillary from the point of injection to the detector, and t_D is the arrival

time at the detector.

At pH values where HA is only partially dissociated, the electrophoretic mobility μ_e is

related to the mobility of A^- by

$$\mu_e = \alpha\mu_{A^-}$$

Or, solving for $1/\alpha$,

$$\frac{1}{\alpha} = \frac{\mu_{A^-}}{\mu_e}$$

For all pH values, we'll maintain the length of capillary and the field strength the same.

Therefore, we can write

$$\frac{1}{\alpha} = \frac{(1/t_D)_{A^-}}{(1/t_D)_{pH}} = \frac{(t_D)_{pH}}{(t_D)_{A^-}}$$

where $(t_D)_{pH}$ represents the arrival times at the pH values where HA is only partially dissociated, and $(t_D)_{A^-}$ is the arrival time in very basic solution where HA is totally dissociated.

We can now substitute this expression into the equation for pH as a function of pK_a that we derived previously.

$$pH = pK_a - \log\left[\frac{1}{\alpha} - 1\right] = pK_a - \log\left[\frac{(t_D)_{pH}}{(t_D)_{A^-}} - 1\right]$$

If we measure the arrival times for various pH values and plot pH vs. the logarithmic term, we can obtain pK_a as the intercept of the resulting straight line.

The following electrophoresis results were obtained for a weak acid, HA. The arrival times were measured at pH values from 3.00 to 5.00 and in very basic solution where A^- is the only species present. The capillary length was 57.0 cm and the length to the detector was 50.0 cm. An applied voltage of 20 kV was used in all cases.

pH	Arrival time at detector, min
3.00	14.62
3.50	7.83
4.00	5.84
4.50	4.99
5.00	4.89
Very basic	4.78

The spreadsheet to find the pK_a value is shown in Figure 15-6. Note that we have used LINEST to find the slope, the intercept, and their standard deviations. From the results, we can say that $pK_a = 3.3 \pm 0.3$.

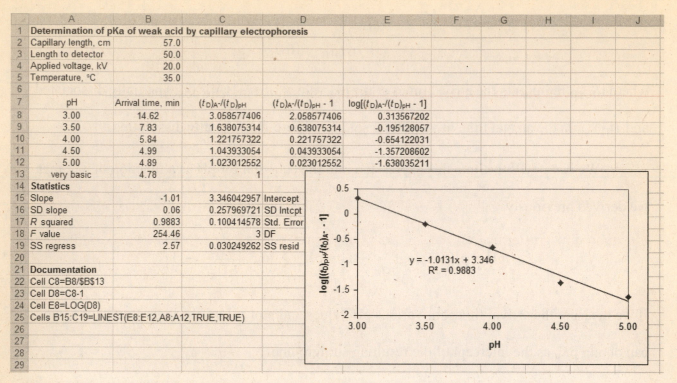

Figure 15-6 Spreadsheet for determining pK_a value of a weak acid by capillary electrophoresis.

Micellar Electrokinetic Capillary Chromatography

In micellar electrokinetic capillary chromatography (MECC), the retention factor k_A for solute A

is directly related to the surfactant concentration [Surf] by the following equation

$$k_A = \left([\text{Surf}] - \text{cmc}\right) K_A V$$

where cmc is the *critical micelle concentration*, K_A is the distribution constant for solute A

between water and the micelle, and V is the partial molar volume of the solute. Solving this

equation for the surfactant concentration gives:

$$[\text{Surf}] = \frac{k_A}{K_A V} + \text{cmc}$$

This equation implies that we can determine the critical micelle concentration by measuring the retention factor for various surfactant concentrations. A plot of k_A vs. [Surf] should be linear and the intercept on the x axis where $k_A = 0$ should be the critical micelle concentration.

An important quantity in MECC is the elution range, t_{mc}/t_M where t_{mc} is the time required for the micelles to travel the length of the column and t_M is the time for the solvent or an unretained solute to travel the same distance. Neutral solutes to be separated will always have a retention time greater than t_M and less than t_{mc}. The retention factor in MECC can be defined as

$$k_A = \frac{t_R - t_M}{t_M\left(1 - \dfrac{t_R}{t_{mc}}\right)}$$

This is analogous to the equation for normal chromatography [$k_A = (t_R - t_M)/t_M$], except for the extra term in parentheses in the denominator. Note that when the micelles are stationary, $t_{mc} = \infty$, and this equation reduces to the usual equation (Equation 31-18 of FAC9 or 24-12 of AC7) as it should. Now if we determine the retention times for a solute as a function of surfactant concentration, we can calculate k_A and determine the critical micelle concentration.

To determine the cmc for sodium dodecyl sulfate (SDS), the retention time of 4-methoxyphenol was measured as a function of the SDS concentration. The antimalarial drug halofantrine is totally incorporated into the micelle and was used as a micelle marker to determine t_{mc}. The results shown in Figure 15-7 were obtained using a capillary electrophoresis apparatus with a 270 mm, 50 μm inside diameter capillary. The applied voltage was 10 kV.

	A	B	C	D	E
1	Determination of cmc of sodium dodecyl sulfate				
2	Retention time of 4-methoxyphenol measured				
3	E	10 kV			
4	l	270 mm			
5	T	27°C			
6	[Surf], mM	t_{mc}, s	t_M, s	t_R, s	
7	20	200	55	60	
8	40	225	58.2	72	
9	60	240	61.9	84	
10	80	255	63.25	95	
11	100	260	72.8	115	

Figure 15-7 Data for determining cmc of SDS.

We now must calculate the retention factor k_A. Let's add a label in cell E6 for this and type the

formula for $k_A =$ `(D7-C7)/(C7*(1-(D7/B7)))` in cell E7. We then copy this into cells

E8:E11. In cells B13:C17, we'll use LINEST to find the slope and intercept and their standard

deviations. The x axis intercept will be found in cell B18 from the equation

$$x = \frac{-b}{m} \qquad \text{when } y = 0$$

We use propagation of error mathematics to find the standard deviation of the x axis intercept

from the relative standard deviations of the intercept b slope m.

$$s_x = x \sqrt{\left(\frac{s_b}{b}\right)^2 + \left(\frac{s_m}{m}\right)^2}$$

The final results are shown in Figure 15-8 along with a plot of the data. From these results we

conclude that the cmc for SDS is 9 ± 2 mM. These results are in good agreement with the

literature value of 10 mM[1].

---

[1] H. J. Chang and E. W. Kaler, *J. Phys. Chem.*, **1985**, *89*, 2996.

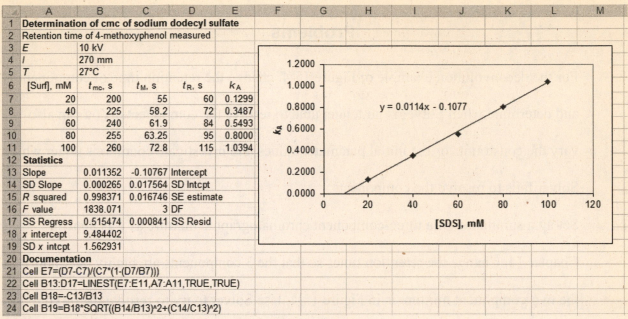

The spreadsheet (columns A–M):

	A	B	C	D	E
1	Determination of cmc of sodium dodecyl sulfate				
2	Retention time of 4-methoxyphenol measured				
3	E	10 kV			
4	l	270 mm			
5	T	27°C			
6	[Surf], mM	t_{mc}, s	t_M, s	t_R, s	k_A
7	20	200	55	60	0.1299
8	40	225	58.2	72	0.3487
9	60	240	61.9	84	0.5493
10	80	255	63.25	95	0.8000
11	100	260	72.8	115	1.0394
12	Statistics				
13	Slope	0.011352	-0.10767	Intercept	
14	SD Slope	0.000265	0.017564	SD Intcpt	
15	R squared	0.998371	0.016746	SE estimate	
16	F value	1838.071	3	DF	
17	SS Regress	0.515474	0.000841	SS Resid	
18	x intercept	9.484402			
19	SD x intcpt	1.562931			
20	Documentation				
21	Cell E7=(D7-C7)/(C7*(1-(D7/B7)))				
22	Cell B13:D17=LINEST(E7:E11,A7:A11,TRUE,TRUE)				
23	Cell B18=-C13/B13				
24	Cell B19=B18*SQRT((B14/B13)^2+(C14/C13)^2)				

Chart: k_A vs [SDS], mM with fit line $y = 0.0114x - 0.1077$

Figure 15-8 Final spreadsheet for cmc determination.

Summary

In this chapter we have shown how Excel can be used to resolve overlapping chromatographic bands and to solve problems in capillary electrophoresis or micellar electrokinetic capillary chromatography. The resolution of overlapping bands is a general technique called deconvolution and is useful in spectroscopy as well as in separation methods. The next chapter discusses the use of Excel in processing data obtained directly from analytical instruments.

Problems

1. For the deconvolution example of Figure 15-4, change the retention time of component B and determine when Solver is no longer able to resolve the components. You can also vary the constraints or the initial parameter values and investigate conditions under which Solver fails to resolve the components.

2. Set up a simulation of a three-component chromatographic mixture as was done in Chapter 14. Change the retention times so that the 3 components are highly overlapped as the two-component mixture is in Figure 15-4. Use Solver to try to resolve the mixture by inputting a model, calculating the residuals and the residuals squared. Investigate the limitations of Solver for this type of calculation.

3. The following electrophoresis results were obtained for a weak acid, HA. The arrival times were measured at pH values from 4.50 to 6.50 and in very basic solution where A^- is the only species present. The capillary length was 60.0 cm and the length to the detector was 54.0 cm. An applied voltage of 20 kV was used in all cases.

pH	Arrival time, min
4.50	20.00
5.00	9.83
5.50	6.75
6.00	5.47
6.50	5.00
very basic	4.67

Find the pK_a of the weak acid and its standard deviation.

4. The following electrophoresis results were obtained for a weak acid, HA. The arrival times were measured at pH values from 4.00 to 6.00 and in very basic solution where A^- is the only species present. The capillary length was 50.0 cm and the length to the detector was 42.0 cm. An applied voltage of 15.0 kV was used in all cases.

pH	Arrival time, min
4.00	18.70
4.50	10.05
5.00	6.35
5.50	5.36
6.00	4.95
very basic	4.75

Find the pK_a of the weak acid and its standard deviation.

5.   To determine the critical micelle concentration of a new surfactant (compound X), the

retention time of 4-methoxyphenol was measured as a function of the SDS concentration.

A micelle marker was incorporated into the micelle and used to determine t_{mc}. The

following data were collected:

[X], mM	t_{mc}, s	t_M, s	t_R, s
25	150	80	90
35	175	95	110
50	190	120	140
75	200	145	169
100	220	170	196

Determine the cmc for compound X and its standard deviation.

Chapter 16

Data Processing with Excel

In this final chapter we'll consider some data processing techniques that can be implemented with Excel. These techniques are useful when data from various laboratory instruments are imported into an Excel workbook. First, we'll consider several methods for smoothing raw instrumental data to improve the signal-to-noise ratio. Next, we'll look at techniques, such as differentiation, that can be used to enhances or amplify certain features of a data set. We'll conclude by considering some methods that can be used to process data as the data points are received rather than processing them after the entire set has been collected.

Smoothing of Data

Smoothing is widely used to improve the signal-to-noise ratio and hence the precision of data. Let's consider a Gaussian chromatographic peak recorded from an HPLC apparatus. We'll generate synthetic data as we did in Chapters 14 and 15. The generated data are found in Figure 16-1. Note from the formula in the Documentation that we have added a baseline offset of 0.2 units so that when we add random noise, the signal will not go below zero. We can use a scaling factor of 0.3 shown in cell B7 when we add noise to the data. Although we show the data only up to row 31 (100 s), the calculations should be done to row 131 (600 s). We also show the chart of the Gaussian peak in Figure 16-1.

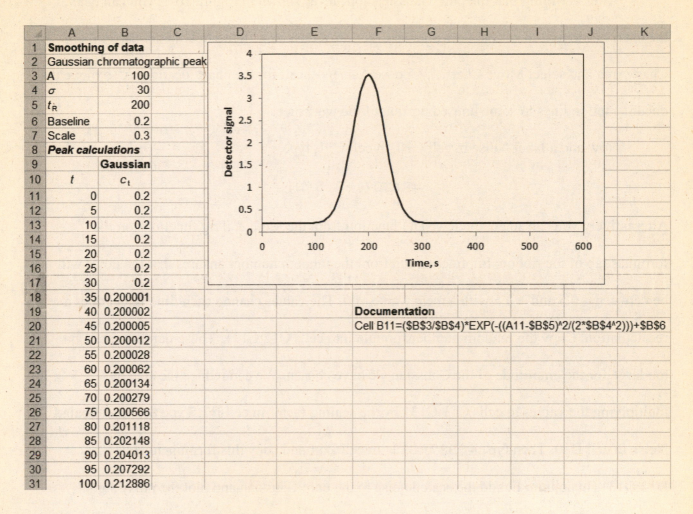

The spreadsheet shown contains the following data:

	A	B
1	**Smoothing of data**	
2	Gaussian chromatographic peak	
3	A	100
4	σ	30
5	t_R	200
6	Baseline	0.2
7	Scale	0.3
8	**Peak calculations**	
9		Gaussian
10	t	c_t
11	0	0.2
12	5	0.2
13	10	0.2
14	15	0.2
15	20	0.2
16	25	0.2
17	30	0.2
18	35	0.200001
19	40	0.200002
20	45	0.200005
21	50	0.200012
22	55	0.200028
23	60	0.200062
24	65	0.200134
25	70	0.200279
26	75	0.200566
27	80	0.201118
28	85	0.202148
29	90	0.204013
30	95	0.207292
31	100	0.212886

Documentation
Cell B11=($B$3/$B$4)*EXP(-((A11-$B$5)^2/(2*$B$4^2)))+$B$6

Figure 16-1 Simulated chromatographic peak without noise.

Generating Noise

We'll generate random noise in column C by using the Excel function RAND(), which returns a random number between 0 and 1. Noise added in this way is said to have a **uniform distribution**. We could also add noise with a **normal distribution**, but this a little more difficult with Excel. For the purposes of our exercise, the difference between these distributions is not important.

After verifying that the pure Gaussian appears as shown in Figure 16-1, you can delete

the chart or move it to a separate worksheet. To move the chart, right click the mouse in the

Chart Area and select <u>M</u>ove Chart…. Choose As new <u>s</u>heet: for the chart location. Also move the

documentation lines to some unused portion of the worksheet.

Now add a label **Noise** in cell C10. In cell C11, type

$$\texttt{=RAND()-0.5[\dashv]}$$

Alternatively you can select the RAND() function from the Math & Trig functions on the

formulas tab of the ribbon. Subtracting 0.5 from the random number ensures that the noise will

be between -0.5 and $+0.5$ with a mean value of 0. The values change each time Excel does a

recalculation. Copy the random number formula into cells C12:C131. These numbers are a bit

too large for the amplitude of the Gaussian we have chosen. We can reduce the size by

multiplying the values in cells C11:C131 by the scaling factor in cell B7. Type the label **Scaled**

Noise in cell D10. Then type **=$B$7*C11** in cell D11 and copy this formula to cells

D12:D131. In column E, add the scaled noise to the pure Gaussian and plot the noisy signal

versus time as shown in Figure 16-2. Note that the noisy baseline makes the determination of

retention time, peak height or peak area imprecise. The purpose of smoothing, or filtering, the

data is to improve the signal-to-noise ratio and enhance the precision.

We'll explore two methods for smoothing the chromatographic data. First, we'll

investigate moving average smoothing of the data. Next, we'll look at the Savitzky-Golay

method for smoothing the data set.

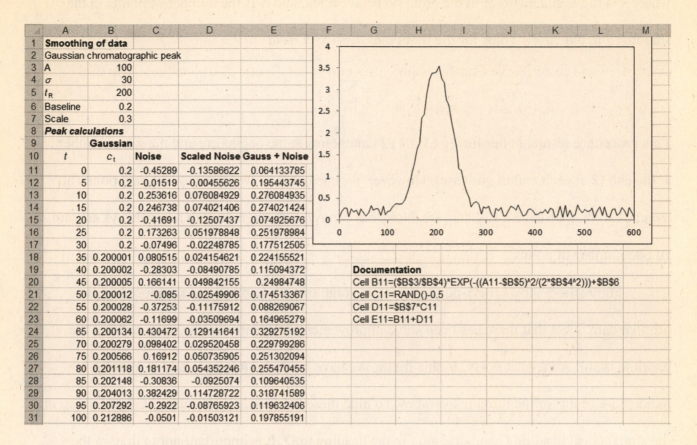

Figure 16-2 Gaussian chromatographic peak after adding scaled noise.

Moving Average Smoothing

Moving average smoothing is easy to implement with Excel. The procedure assumes that the

data points are equally spaced on the *x*-axis (time). We'll illustrate first with a 5 point smooth. In

this method, we begin with the 3rd data point in the set since there are two points before it and

two points after it. We replace this point with the average of points 1 through 5. We next replace

the 4th data point with the average of points 2 through 6, and so on. The equation for this

procedure is

$$\bar{x}_{(k+\frac{N-1}{2})} = \sum_{i=k}^{k+N-1} \frac{x_i}{N}$$

where k is the first data point in the group to be averaged, and N is the number of points in the smooth (5 in our case). Thus, for the first point, $k = 1$, we have

$$\bar{x}_3 = \sum_{i=1}^{5} \frac{x_i}{5}$$

This procedure results in our losing $(N - 1)/2$ data points at the beginning and the same number at the end (2 at each end in our case). However, we can use the raw, unsmoothed data points at the ends so that we have a full data set. Note that a moving average smooth must always contain an odd number of points.

Figure 16-3 shows the 5-point moving average smooth (circles) superimposed on the original data. Note that there has been some improvement in signal-to-noise ratio, but that the baseline is still somewhat noisy. In this figure, we have hidden columns C, and D in order to make more room for the moving averages. To hide these columns, select column C by clicking in the alphabetical heading C and dragging in the heading to D. It is important not to drag in the data below, but only in the heading. Then right click in one of the selected alphabetical headings (C or D). Choose Hide from the menu that appears. When you want the columns to reappear in the worksheet, select the column before the hidden columns and that after the hidden columns and again right click in one of the alphabetical headings. Choose Unhide from the menu.

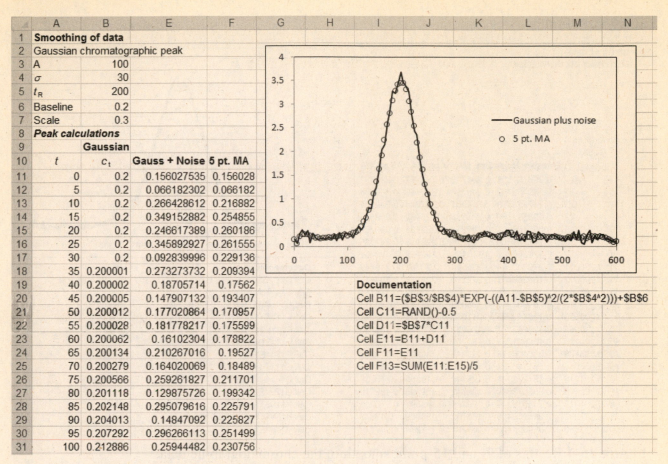

	A	B	E	F
1	**Smoothing of data**			
2	Gaussian chromatographic peak			
3	A	100		
4	σ	30		
5	t_R	200		
6	Baseline	0.2		
7	Scale	0.3		
8	**Peak calculations**			
9		**Gaussian**		
10	t	c_t	**Gauss + Noise**	**5 pt. MA**
11	0	0.2	0.156027535	0.156028
12	5	0.2	0.066182302	0.066182
13	10	0.2	0.266428612	0.216882
14	15	0.2	0.349152882	0.254855
15	20	0.2	0.246617389	0.260186
16	25	0.2	0.345892927	0.261555
17	30	0.2	0.092839996	0.229136
18	35	0.200001	0.273273732	0.209394
19	40	0.200002	0.18705714	0.17562
20	45	0.200005	0.147907132	0.193407
21	50	0.200012	0.177020864	0.170957
22	55	0.200028	0.181778217	0.175599
23	60	0.200062	0.16102304	0.178822
24	65	0.200134	0.210267016	0.19527
25	70	0.200279	0.164020069	0.18489
26	75	0.200566	0.259261827	0.211701
27	80	0.201118	0.129875726	0.199342
28	85	0.202148	0.295079616	0.225791
29	90	0.204013	0.14847092	0.225827
30	95	0.207292	0.296266113	0.251499
31	100	0.212886	0.25944482	0.230756

Documentation

Cell B11=($B$3/$B$4)*EXP(-((A11-$B$5)^2/(2*$B$4^2)))+$B$6

Cell C11=RAND()-0.5

Cell D11=$B$7*C11

Cell E11=B11+D11

Cell F11=E11

Cell F13=SUM(E11:E15)/5

Figure 16-3 Five point moving average smooth of chromatographic data.

We'll now try a 7- and a 15-point smooth to see if we can improve the signal-to-noise ratio. The 7-point smooth begins with data point 4, while the 15-point smooth begins with data point 8. The results are shown in Figure 16-4. Note that we have dropped the 5-point smooth from the chart and added the 7- and 15-point smooths. Also, we have changed the horizontal axis for a maximum time of 400 s.

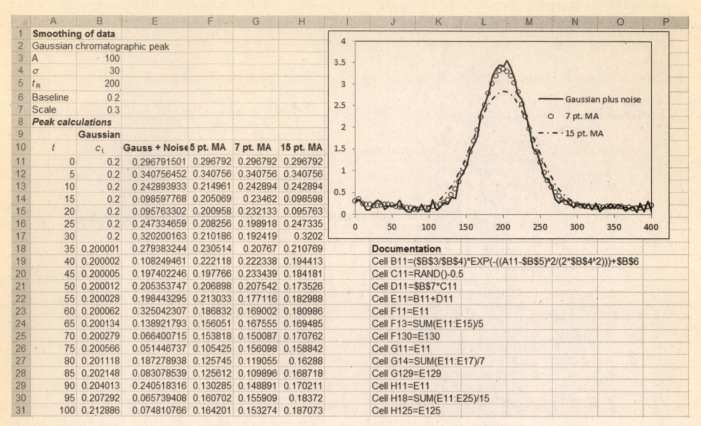

Figure 16-4 Results of 7- and 15-point smooth of the chromatographic data.

Note that the 15-point smooth seriously distorts the data. The peak height is reduced and the

peak is broadened. The 7-point smooth improves the signal-to-noise ratio with less distortion.

You can see that the more points in the smooth, the less noise, but the more distortion. The

lesson to be learned from this example is that you should apply smoothing only when needed and

you should be aware of what smoothing may do to your data.

Exponential Moving Averages

The moving averages discussed above are unweighted averages. That is, the points surrounding

the central point are given equal weighting in determining the average. Another approach to

moving average smoothing is to calculate a weighted moving average. The most applicable in

scientific data processing is the exponentially weighted moving average. This type of filtering weights the current data point highest while previous data points are given exponentially decreasing weights. The exponential moving average is similar to an instrument time constant and has the advantage that it can be applied as the data are received. This method of finding the average is often called **recursive estimation** as opposed to a nonrecursive or batch calculation where all the data points are first collected and averaged. The equation for calculating the exponential moving average (EMA) is

$$\bar{x}_k = s\,x_k + \left[(1-s)\bar{x}_{k-1}\right]$$

where $\bar{x}_k$ is the EMA of the current data point x_k, $\bar{x}_{k-1}$ is the EMA of the previous data point, and s is a smoothing coefficient given by

$$s = \frac{2}{1+N}$$

where N is the number of points in the smooth. For a 5-point smooth, $s = 0.3333$ and $(1 - s) = 0.6666$.

A 5-point EMA is shown in Figure 16-5 superimposed on the raw data. To implement the 5-point EMA, we enter the formula shown in the documentation for cell I15 after using the data points themselves for the first 4 points. We then copy this formula into the remaining column I cells. Note that the EMA also gives significant distortion of the peak, particularly the peak position. Although exponential moving averages can distort peak-shaped signals, they are often beneficial for other signal shapes. As an extension of this exercise, you can superimpose the 5-point unweighted moving average smooth and the 5-point EMA smooth and determine which gives the least distortion and which gives the best signal-to-noise ratio. The signal-to-noise ratio for a peak-shaped signal can be estimated by finding the ratio of the peak height to the standard deviation of the baseline well away from the peak. For our example, you can find the standard deviation of the points from 400 to 600 s and use this as an estimate of the noise. Note these points are not shown in Figure 16-5.

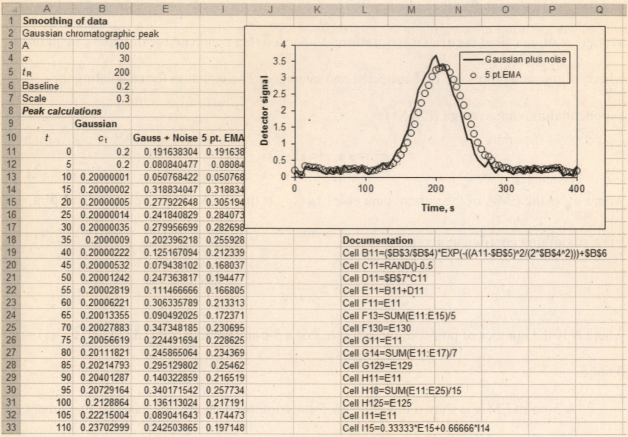

	A	B	E	I	Documentation
1	Smoothing of data				
2	Gaussian chromatographic peak				
3	A	100			
4	σ	30			
5	t_R	200			
6	Baseline	0.2			
7	Scale	0.3			
8	Peak calculations				
9		Gaussian			
10	t	c_t	Gauss + Noise	5 pt. EMA	
11	0	0.2	0.191638304	0.191638	
12	5	0.2	0.080840477	0.08084	
13	10	0.20000001	0.050768422	0.050768	
14	15	0.20000002	0.318834047	0.318834	
15	20	0.20000005	0.277922648	0.305194	
16	25	0.20000014	0.241840829	0.284073	
17	30	0.20000035	0.279956699	0.282698	
18	35	0.2000009	0.202396218	0.255928	Documentation
19	40	0.20000222	0.125167094	0.212339	Cell B11=($B$3/$B$4)*EXP(-((A11-$B$5)^2/(2*$B$4^2)))+$B$6
20	45	0.20000532	0.079438102	0.168037	Cell C11=RAND()-0.5
21	50	0.20001242	0.247363817	0.194477	Cell D11=$B$7*C11
22	55	0.20002819	0.111466666	0.166805	Cell E11=B11+D11
23	60	0.20006221	0.306335789	0.213313	Cell F11=E11
24	65	0.20013355	0.090492025	0.172371	Cell F13=SUM(E11:E15)/5
25	70	0.20027883	0.347348185	0.230695	Cell F130=E130
26	75	0.20056619	0.224491694	0.228625	Cell G11=E11
27	80	0.20111821	0.245865064	0.234369	Cell G14=SUM(E11:E17)/7
28	85	0.20214793	0.295129802	0.25462	Cell G129=E129
29	90	0.20401287	0.140322859	0.216519	Cell H11=E11
30	95	0.20729164	0.340171542	0.257734	Cell H18=SUM(E11:E25)/15
31	100	0.2128864	0.136113024	0.217191	Cell H125=E125
32	105	0.22215004	0.089041643	0.174473	Cell I11=E11
33	110	0.23702999	0.242503865	0.197148	Cell I15=0.33333*E15+0.66666*I14

Figure 16-5 A 5-point exponential moving average smooth.

Savitzky-Golay Smoothing

The Savitzky-Golay algorithm[1] applies a least-squares polynomial fit to the data points surrounding the central point to be smoothed. In practice, this is done by calculating a weighted average using appropriate weighting coefficients.[2] For a seven point smooth, the weighting coefficients for the first central point (point 4) are –2, 3, 6, 7, 6, 3, –2. The equation for computing the smoothed value of the first central point $\bar{x}_4$ is

---

[1] A. Savitzky and M. J. Golay, *Anal. Chem.* **1964**, *36*, 1627-1639.
[2] D. A. Skoog, F. J. Holler, and S. R. Crouch, *Principles of Instrumental Analysis*, 6th ed. Belmont, CA: Brooks/Cole, 2007, p. 122.

$$\bar{x}_4 = \frac{-2x_1 + 3x_2 + 6x_3 + 7x_4 + 6x_5 + 4x_6 - 2x_7}{21}$$

The Savitzky-Golay filter involves losing the first and last $(N-1)/2$ points in the data set, where N is the number of points in the smooth. For a 7-point smooth, the first and last 3 data points are lost. You can use the raw, unsmoothed data points, for the missing points to complete the data set as shown in Figure 16-6. Note that the Savitzky-Golay filter introduces only a small distortion to the signal while still improving the signal-to-noise ratio. The effect of polynomial smoothing on signal-to-noise ratio and distortion has been discussed extensively by Enke and Nieman.[3]

As an extension to this exercise, you can compare on the same chart the 7-point moving average smooth to the 7-point Savitzky-Golay smooth. Compare the signal-to-noise ratios and the peak distortion introduced by the two procedures. Although we've shown only a 7-point Savitzky-Golay smooth, the coefficients or equations for calculating the coefficients are available for Savitzky-Golay smooths of various point lengths.[4]

---

[3] C. G. Enke and T. A. Nieman, *Anal. Chem.*, **1976**, *48*, 705A-712A.
[4] P. Gorry, *Anal. Chem.*, **1990**, *62*, 570-573.

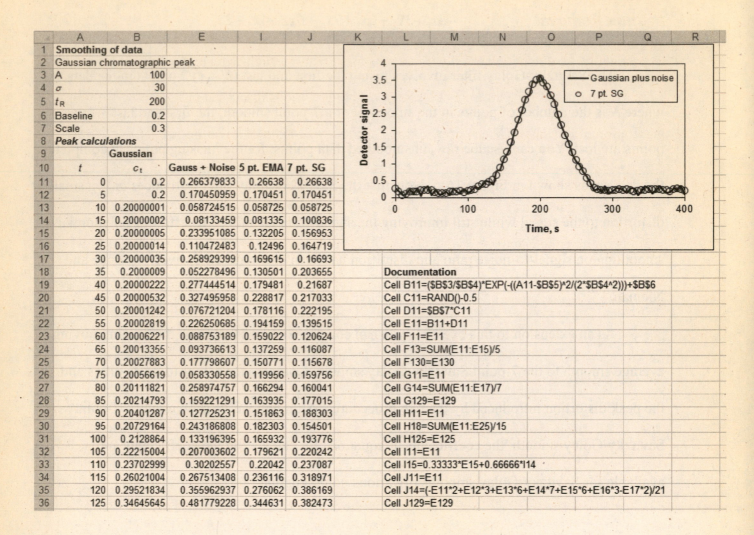

	A	B	E	I	J	K	L	M	N	O	P	Q	R
1	Smoothing of data												
2	Gaussian chromatographic peak												
3	A	100											
4	σ	30											
5	t_R	200											
6	Baseline	0.2											
7	Scale	0.3											
8	*Peak calculations*												
9		Gaussian											
10	t	c_t	Gauss + Noise	5 pt. EMA	7 pt. SG								
11	0	0.2	0.266379833	0.26638	0.26638								
12	5	0.2	0.170450959	0.170451	0.170451								
13	10	0.20000001	0.058724515	0.058725	0.058725								
14	15	0.20000002	0.08133459	0.081335	0.100836								
15	20	0.20000005	0.233951085	0.132205	0.156953								
16	25	0.20000014	0.110472483	0.12496	0.164719								
17	30	0.20000035	0.258929399	0.169615	0.16693								
18	35	0.2000009	0.052278496	0.130501	0.203655			Documentation					
19	40	0.20000222	0.277444514	0.179481	0.21687			Cell B11=($B$3/$B$4)*EXP(-((A11-$B$5)^2/(2*$B$4^2)))+$B$6					
20	45	0.20000532	0.327495958	0.228817	0.217033			Cell C11=RAND()-0.5					
21	50	0.20001242	0.076721204	0.178116	0.222195			Cell D11=$B$7*C11					
22	55	0.20002819	0.226250685	0.194159	0.139515			Cell E11=B11+D11					
23	60	0.20006221	0.088753189	0.159022	0.120624			Cell F11=E11					
24	65	0.20013355	0.093736613	0.137259	0.116087			Cell F13=SUM(E11:E15)/5					
25	70	0.20027883	0.177798607	0.150771	0.115678			Cell F130=E130					
26	75	0.20056619	0.058330558	0.119956	0.159756			Cell G11=E11					
27	80	0.20111821	0.258974757	0.166294	0.160041			Cell G14=SUM(E11:E17)/7					
28	85	0.20214793	0.159221291	0.163935	0.177015			Cell G129=E129					
29	90	0.20401287	0.127725231	0.151863	0.188303			Cell H11=E11					
30	95	0.20729164	0.243186808	0.182303	0.154501			Cell H18=SUM(E11:E25)/15					
31	100	0.2128864	0.133196395	0.165932	0.193776			Cell H125=E125					
32	105	0.22215004	0.207003602	0.179621	0.220242			Cell I11=E11					
33	110	0.23702999	0.30202557	0.22042	0.237087			Cell I15=0.33333*E15+0.66666*I14					
34	115	0.26021004	0.267513408	0.236116	0.318971			Cell J11=E11					
35	120	0.29521834	0.355962937	0.276062	0.386169			Cell J14=(-E11*2+E12*3+E13*6+E14*7+E15*6+E16*3-E17*2)/21					
36	125	0.34645645	0.481779228	0.344631	0.382473			Cell J129=E129					

Figure 16-6 Seven-point Savitzky-Golay filter applied to chromatographic data.

Feature Removal and Enhancement

We'll examine here the use of Excel to remove unwanted features from a data set or to enhance

hidden features. Feature removal involves integration or averaging of the data, while feature

enhancement occurs when the data set is differentiated.

Feature Removal

Occasionally we obtain a data set with an unwanted feature, such as an artifact or an extra peak from a contaminant. The moving average smoothing methods discussed previously are a form of integration of the data. If the averaging time exceeds the width of a feature, smoothing may completely remove the feature. We'll consider as an example, a signal that consists of the sum of two Gaussian curves as shown in Figure 16-7. Although we show only the first 32 rows, you should continue the calculations to row 108 (100 s).

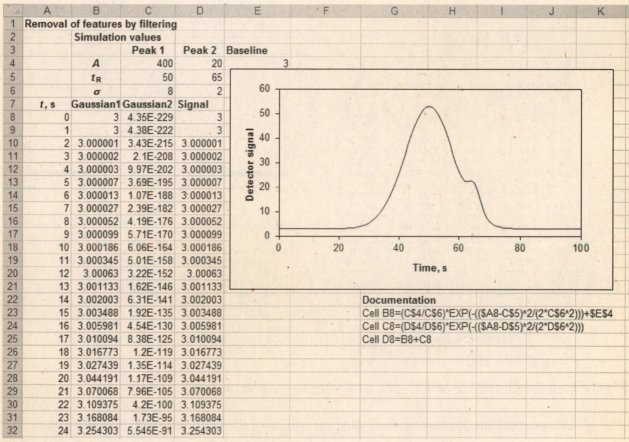

	A	B	C	D	E	F	G	H	I	J	K
1	Removal of features by filtering										
2		Simulation values									
3			Peak 1	Peak 2	Baseline						
4		*A*	400	20	3						
5		t_R	50	65							
6		σ	8	2							
7	*t*, s	Gaussian1	Gaussian2	Signal							
8	0	3	4.35E-229	3							
9	1	3	4.38E-222	3							
10	2	3.000001	3.43E-215	3.000001							
11	3	3.000002	2.1E-208	3.000002							
12	4	3.000003	9.97E-202	3.000003							
13	5	3.000007	3.69E-195	3.000007							
14	6	3.000013	1.07E-188	3.000013							
15	7	3.000027	2.39E-182	3.000027							
16	8	3.000052	4.19E-176	3.000052							
17	9	3.000099	5.71E-170	3.000099							
18	10	3.000186	6.06E-164	3.000186							
19	11	3.000345	5.01E-158	3.000345							
20	12	3.00063	3.22E-152	3.00063							
21	13	3.001133	1.62E-146	3.001133							
22	14	3.002003	6.31E-141	3.002003	Documentation						
23	15	3.003488	1.92E-135	3.003488	Cell B8=(C$4/C$6)*EXP(-(($A8-C$5)^2/(2*C$6^2)))+$E$4						
24	16	3.005981	4.54E-130	3.005981	Cell C8=(D$4/D$6)*EXP(-(($A8-D$5)^2/(2*D$6^2)))						
25	17	3.010094	8.38E-125	3.010094	Cell D8=B8+C8						
26	18	3.016773	1.2E-119	3.016773							
27	19	3.027439	1.35E-114	3.027439							
28	20	3.044191	1.17E-109	3.044191							
29	21	3.070068	7.96E-105	3.070068							
30	22	3.109375	4.2E-100	3.109375							
31	23	3.168084	1.73E-95	3.168084							
32	24	3.254303	5.545E-91	3.254303							

Figure 16-7 Generation of two Gaussians.

Let's consider that the second Gaussian that produces the shoulder at about 65 s is an undesired feature. Since the width of this feature is about 8 s (4σ), a moving average smooth with more than 9 points should be effective in removing the feature. We'll try a 15-point and a 21-point smooth. The results are shown in Figure 16-8. Note that the 15-point smooth completely removes the feature with only a small distortion of the first Gaussian. The 21-point smooth also removes the feature, but introduces more distortion of the first Gaussian. This occurs because the width of the smoothing window (21 s) is approaching the width of the first Gaussian ($4\sigma = 32$ s).

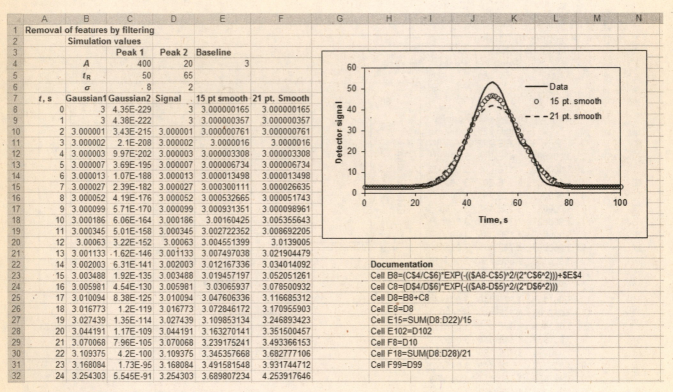

Figure 16-8 Results of 15- and 21-point smooths to remove unwanted feature.

As an extension of this exercise, vary the widths of the Gaussians and note the effect. You can also try smooths of different widths and see the effect. Note how critical it is to choose the appropriate smoothing width for this type of feature removal.

Feature Enhancement

Just as integration (smoothing) can lead to the removal of features, differentiation has the opposite effect. Differentiation enhances changing signals and can small reveal peaks and other features hidden by a large peak or even by noise in some cases.

To investigate this aspect, let's generate a single Gaussian curve as shown in Figure 16-9. Continue the simulation to at least row 68 (60 s). Next, we'll obtain the first and second derivatives of this signal. There are several ways to do this. The Savitzky-Golay approach obtains the derivatives while simultaneously smoothing the data. This is valuable because differentiation can enhance noise features since they change rapidly. In this case, our signal is not noisy, so we'll use a simple approach to compute the derivatives.

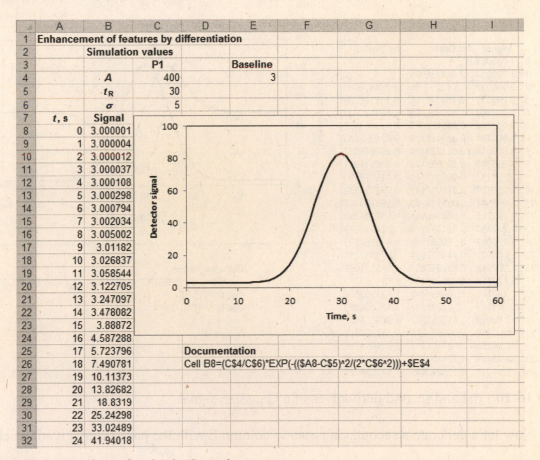

Figure 16-9 Generation of a single Gaussian curve.

In column C, let's calculate the first derivative by finding $\Delta S/\Delta t$, where S is the signal and t is time. We can do this by just finding the difference in successive signal values, ΔS, since we have equal time intervals of 1 s and thus $\Delta t = 1$. In column D, we'll find the second derivative by calculating $\Delta(\Delta S/\Delta t)/\Delta t$. Again we can do this by finding the differences in the first derivative. The resulting values and a chart are shown in Figure 16-10. We have plotted the two derivatives on a secondary axis as discussed in Chapter 7. Note that we have multiplied the second derivative by four to aid in plotting it on the same axis as the first derivative. Notice also the positions of the maxima and minima of the derivatives relative to the signal.

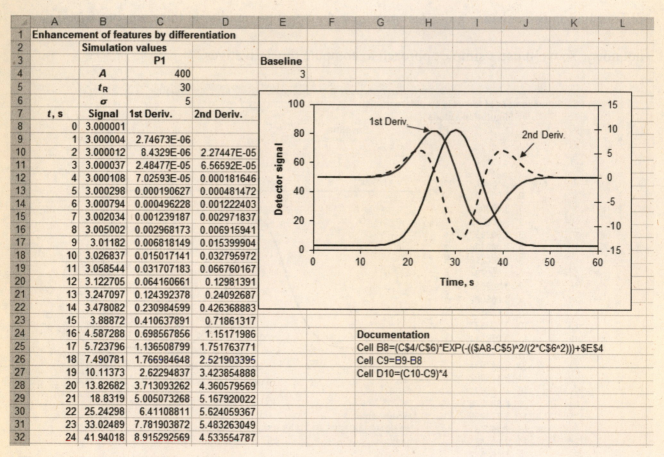

Figure 16-10 First and second derivatives of a Gaussian signal.

Now let's generate a second Gaussian in column E with the parameters given in cells D4:D6 of Figure 16-11. To make more room on our worksheet, let's change the widths of

Columns C and D to 4.00 (this number is the number of characters in the standard font that can be displayed in the cell). You can change the width by dragging the vertical line to the right of the column letter until **width: 4.00** is displayed, or you can select the columns and, on the Home tab, click Format in the Cells group. Under Cell size, select Column <u>W</u>idth and enter 4.00. In column F, add the Gaussian in column B to that in column E and adjust the chart to display only the Gaussian as shown in Figure 16-11. You can do this by right clicking on the chart, selecting S<u>e</u>lect Data… In the Select Data Source window, choose Edit and, in the Edit Series window show cells A8:A68 in the <u>X</u> Values: bar and cells F8:F68 in the <u>Y</u> Values bar. Note that it is very difficult to see the second Gaussian on the chart because it is hidden by the large first Gaussian.

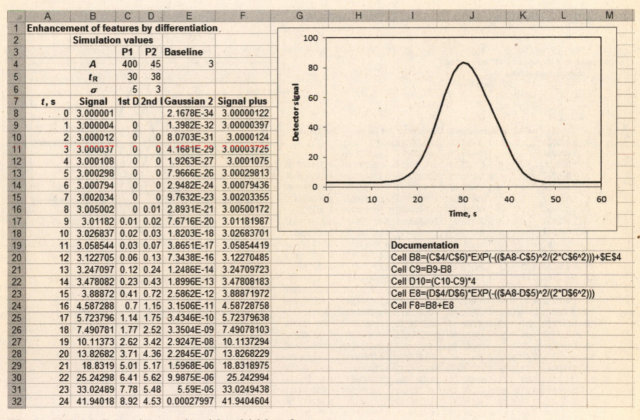

Figure 16-11 Gaussian peak with a hidden feature.

In columns G and H, let's calculate the first and second derivatives as we did for the pure Gaussian curve. We can add the two series to the chart and obtain the results shown in Figure 16-12. In this figure, we have changed some of the column widths so that we might better display the chart and spreadsheet documentation. Now note that the hidden feature (the small Gaussian curve) becomes apparent in both the first and second derivative plots. Feature enhancement by differentiation is commonly used in spectroscopy and chromatography to aid in interpreting the data.

You can extend this exercise by changing the parameters of the Gaussian peaks used to make up the composite peak. You may also want to add random noise as we did early in this chapter and note its effect on the first and second derivatives. You can also try moving average smoothing of the noisy data prior to differentiation.

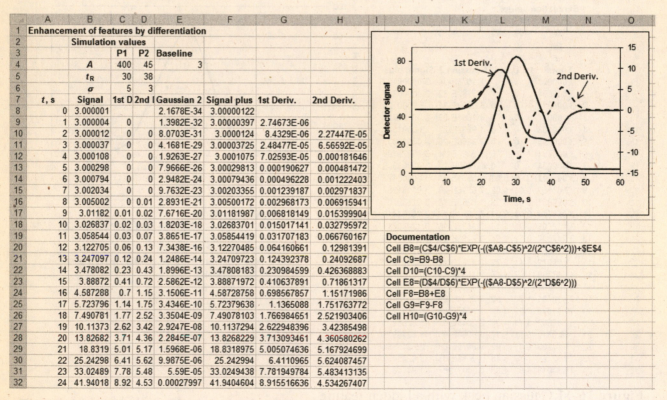

Figure 16-12 Feature enhancement by differentiation of the data

Recursive Estimation Methods

Most of our statistical calculations are done in a so-called batch mode. That is, the data set is collected and then such statistical parameters as the mean, the variance, the standard deviation, and others are estimated for the entire data set. Such batch methods are said to be nonrecursive. In recursive methods, by contrast, the statistical parameters are estimated as the data are received, one data point at a time. While this may not seem at first glance to be advantageous, recursive procedures allow us in more complex situations to calculate linear regression parameters, to carry out smoothing operations, and to estimate parameters of nonlinear systems while the data are received in real time. Such real-time procedures are very important in feedback control systems, in robotic systems, and in data processing from remote sites such as with telemetry data from spacecraft or satellites.

We'll consider here only the recursive calculation of the mean and standard deviation of a data set, but bear in mind that the principle is much more generally applicable. We'll take a data set with 23 points and a mean value of 55 as shown in Figure 16-13. To generate noisy data, we add a random number to 55 as shown in column C. We find the mean and standard deviation of the Data plus noise (column D) in the usual manner in cells D29:D31.

	A	B	C	D	E
1	**Recursive estimation of mean and standard deviation**				
2	23 data points with a mean of 55				
3					
4	Point No.	Data	Noise	Data plus noise	
5	1	55	0.06976837	55.06976837	
6	2	55	0.29531997	55.29531997	
7	3	55	-0.2155103	54.78448974	
8	4	55	-0.2727244	54.72727558	
9	5	55	-0.1783063	54.8216937	
10	6	55	0.234537	55.234537	
11	7	55	0.12621695	55.12621695	
12	8	55	-0.4995305	54.50046947	
13	9	55	-0.4423932	54.55760681	
14	10	55	0.33954841	55.33954841	
15	11	55	0.12109452	55.12109452	
16	12	55	0.46704459	55.46704459	
17	13	55	-0.4085048	54.59149521	
18	14	55	-0.4501166	54.54988345	
19	15	55	0.24685562	55.24685562	
20	16	55	0.43579235	55.43579235	
21	17	55	-0.4146468	54.58535322	
22	18	55	-0.0415367	54.95846335	
23	19	55	0.33313185	55.33313185	
24	20	55	-0.0303271	54.96967289	
25	21	55	0.19231169	55.19231169	
26	22	55	0.06970289	55.06970289	
27	23	55	-0.2848983	54.71510166	
28					
29			Batch Sum	1264.116129	
30			Batch mean	54.96157081	
31			Batch SD	0.28053611	
32	**Documentation**				
33	Cell C5=RAND()-0.5				
34	Cell D5=B5+C5				
35	Cell D29=SUM(D5:D27)				
36	Cell D30=D29/23				
37	Cell D31=STDEV(D5:D27)				

Figure 16-13 Data set with added random noise.

To calculate the mean in a recursive manner, let's see how we'd do this one number at a time. When the first number comes in, we call that the mean. When the second number is received, we calculate a new mean by taking one-half of the new number plus one-half of the first mean. Or in equation form

$$\bar{x}_1 = x_1$$
$$\bar{x}_2 = \frac{1}{2}\bar{x}_1 + \frac{1}{2}x_2$$

When the 3rd data point is received, the batch mean would be $\bar{x}_3 = (x_1 + x_2 + x_3)/3$. Since the second mean is $(x_1 + x_2)/2$, we can say that $(x_1 + x_2) = 2\bar{x}_2$ or

$$\bar{x}_3 = \frac{2}{3}\bar{x}_2 + \frac{1}{3}x_3$$

Generalizing this for the kth mean, we can write

$$\bar{x}_k = \left[\frac{k-1}{k}\right]\bar{x}_{k-1} + \frac{1}{k}x_k$$

Or, putting both right-hand terms over the common denominator and rearranging,

$$\bar{x}_k = \bar{x}_{k-1} + \frac{1}{k}(x_k - \bar{x}_{k-1})$$

This latter equation says that in order to calculate the kth mean, we take the previous mean and add to it a correction factor that varies with the number of observations according to a factor $(1/k)$ called the *gain factor*.

Let's now implement this in our worksheet. In column E, we'll put formulas corresponding to the above equation and obtain a running estimation of the mean as shown in Figure 16-14. Note that the 23rd mean in cell E27 is identical to that found from the usual batch mode equation (cell D30). The recursive equation is exactly equal to the batch equation and so the estimation after 10 data points would give identical results to the batch estimation for the same 10 data points. You may want to experiment to confirm that this is true. Also, recalculate the results several times by depressing F9. Each time a different random noise figure occurs, the results are slightly different. But, the recursive mean (E27) and batch mean (D30) are always identical.

	A	B	C	D	E
1	Recursive estimation of mean and standard deviation				
2	23 data points with a mean of 55				
3					
4	Point No.	Data	Noise	Data plus noise	Recursive mean
5	1	55	0.41093873	55.41093873	55.41093873
6	2	55	-0.2400787	54.75992133	55.08543003
7	3	55	-0.4235911	54.57640893	54.91575633
8	4	55	0.23173907	55.23173907	54.99475202
9	5	55	0.48113759	55.48113759	55.09202913
10	6	55	-0.2091136	54.79088639	55.04183867
11	7	55	-0.0324717	54.96752831	55.03122291
12	8	55	-0.0917572	54.9082428	55.01585039
13	9	55	-0.3694849	54.63051514	54.97303537
14	10	55	-0.1973475	54.80265252	54.95599708
15	11	55	-0.4198349	54.5801651	54.92183054
16	12	55	-0.466683	54.53331701	54.88945441
17	13	55	0.04652112	55.04652112	54.90153646
18	14	55	0.49483881	55.49483881	54.9439152
19	15	55	0.17102689	55.17102689	54.95905598
20	16	55	0.1536257	55.1536257	54.97121659
21	17	55	0.07097796	55.07097796	54.97708491
22	18	55	0.38402846	55.38402846	54.99969288
23	19	55	-0.0996612	54.90033881	54.99446372
24	20	55	-0.4904159	54.50958409	54.97021974
25	21	55	-0.0881952	54.91180485	54.96743808
26	22	55	-0.138631	54.861369	54.96261676
27	23	55	0.09394038	55.09394038	54.96832648
28					
29			Batch Sum	1264.271509	
30			Batch mean	54.96832648	
31			Batch SD	0.303722387	
32	Documentation				
33	Cell C5=RAND()-0.5			Cell E5=D5	
34	Cell D5=B5+C5			Cell E6=E5+(1/A6)*(D6-E5)	
35	Cell D29=SUM(D5:D27)				
36	Cell D30=D29/23				
37	Cell D31=STDEV(D5:D27)				

Figure 16-14 Recursive estimation of the mean.

Now we'll find the standard deviation in a recursive manner. We won't derive the relationship here, but the recursive mode variance of the kth data point is given by

$$s_k^2 = \left(\frac{1}{k-1}\right)\left[(k-2)s_{k-1}^2 + (k-1)(\bar{x}_{k-1}-\bar{x}_k)^2 + (\bar{x}_k-x_k)^2\right]$$

This equation can also be rearranged to give, on the right-hand side, the previous variance plus a correction factor. We'll use it, however, in the form shown above. Let's now find the recursive variance in column G and the standard deviation in column H as shown in Figure 16-15. Note once again that the recursive standard deviation is exactly equal to that obtained by a batch calculation. You can again depress F9 several times to change the calculation results by changing the random numbers.

Although we've only shown two basic examples of recursive calculations, many modern computational methods in analytical chemistry use such calculations. One of the most important examples is the Kalman filter which accomplishes filtering of the data and estimation of model parameters in a recursive mode. The principle is similar to that described above. A model equation is developed and used to predict the next data point. When the next data point is received, it is compared to the prediction. The difference is multiplied by a gain factor and model parameters are varied. The new model parameters are used to produce a new prediction. This procedure continues until an appropriate agreement between the model and the data is achieved or until an entire data set has been filtered.

	A	B	C	D	E	F	G
1	Recursive estimation of mean and standard deviation						
2	23 data points with a mean of 55						
3							
4	Point No.	Data	Noise	Data plus noise	Recursive mean	Recursive Var	Recursive SD
5	1	55	0.01751626	55.01751626	55.01751626	0	0
6	2	55	0.22078948	55.22078948	55.11915287	0.020660002	0.143735876
7	3	55	-0.2945894	54.70541062	54.98123879	0.067390884	0.259597542
8	4	55	0.39611764	55.39611764	55.0849585	0.087958372	0.296577767
9	5	55	-0.0834238	54.91657621	55.05128204	0.071639298	0.267655185
10	6	55	-0.1052231	54.89477691	55.02519785	0.061393748	0.247777618
11	7	55	0.48012427	55.48012427	55.09018734	0.080726891	0.284124781
12	8	55	-0.4831419	54.51685806	55.01852118	0.110282786	0.332088521
13	9	55	-0.3039443	54.69605572	54.98269169	0.108051213	0.328711443
14	10	55	-0.1395785	54.86042145	54.97046466	0.097540523	0.312314783
15	11	55	-0.0347522	54.96524783	54.96999041	0.087788945	0.296291993
16	12	55	-0.3913012	54.60869879	54.93988277	0.090685768	0.301140778
17	13	55	-0.2000292	54.79997085	54.92912032	0.084634416	0.290919949
18	14	55	0.49913611	55.49913611	54.96983573	0.101332505	0.31832767
19	15	55	-0.4576053	54.54239474	54.94133966	0.106274856	0.325998245
20	16	55	0.13034814	55.13034814	54.95315269	0.101422628	0.318469195
21	17	55	0.4622073	55.4622073	54.98309708	0.110327043	0.33215515
22	18	55	0.31711901	55.31711901	55.00165386	0.110035587	0.331716123
23	19	55	0.41511109	55.41511109	55.02341476	0.112919703	0.33603527
24	20	55	-0.2936319	54.7063681	55.00756243	0.11200249	0.334667731
25	21	55	0.02399766	55.02399766	55.00834506	0.106415228	0.32621347
26	22	55	-0.3971007	54.6028993	54.98991571	0.108819939	0.329878674
27	23	55	0.08538203	55.08538203	54.99406642	0.104269831	0.322908395
28							
29			Batch Sum	1264.863528			
30			Batch mean	54.99406642			
31			Batch SD	0.322908395			
32	Documentation						
33	Cell C5=RAND()-0.5			Cell E5=D5			
34	Cell D5=B5+C5			Cell E6=E5+(1/A6)*(D6-E5)			
35	Cell D29=SUM(D5:D27)			Cell F5=0			
36	Cell D30=D29/23			Cell F6=(1/(A6-1))*((A6-2)*F5+(A6-1)*((E5-E6)^2)+((E6-D6)^2))			
37	Cell D31=STDEV(D5:D27)			Cell G5=SQRT(F5)			

Figure 16-15 Recursive calculation of both mean and standard deviation.

Summary

In this our final chapter, we've investigated several ways in which Excel can be used in data processing. We've looked at smoothing of data, at feature enhancement and removal and at recursive estimation. Previous chapters have considered many additional data processing methods, such as curve fitting with Solver, the deconvolution of overlapping peaks, various statistical tests and procedures, and an array of graphical methods. We've shown Excel to be a very powerful and flexible tool for dealing with data collected in analytical chemistry.

Problems

1.   Generate a Gaussian chromatographic peak as in Figure 16-1. Add uniformly distributed random noise as was done in Figure 16-2. Vary the scaling factor from 0.3 to 1.0. Investigate the effects of smoothing with a moving average smooth of 5, 7, and 15 points. Use a 7-point Savitsky-Golay smooth. Comment on the efficiency of the smoothing functions and how much distortion is introduced.

2.   Set up a spreadsheet to study feature removal as in Figure 16-8. Vary the widths of the Gaussians and note the effect on the smoothed data. Also try a 7-point moving average smooth. Try a 7-point Savitsky-Golay smooth. Which functions are most effective in removing the unwanted feature?

3.   Construct a spreadsheet to study feature enhancement. Use a Gaussian peak with a hidden feature as in Figure 16-11. Add scaled random noise to the peak as we did in Figure 16-2. Take the first and second derivatives as was done in Figure 16-12. Now use a moving average or Savitsky-Golay smooth prior to differentiation. Is the smoothing effective here? Why or why not? You can also vary the widths of the Gaussian peaks and note the effect.

4.   The following data represent capillary electrophoresis results monitored by fluorescence detection. Try a 7-point Savitsky-Golay smooth on these data. Next use a 5-point moving average smooth. Then use a 7-point and a 15-point moving average smooth. Which

function gives better improvement in signal-to-noise with the least distortion of the peak

shape?

	A	B
1	**Smoothing of data**	
2	CE peak detected	
3	by fluorescence	
4	time, s	Fl. Intensity
5	100	0.0956
6	105	0.1195
7	110	0.1116
8	115	0.1275
9	120	0.0717
10	125	0.1036
11	130	0.0319
12	135	0.0717
13	140	0.1355
14	145	0.2231
15	150	0.1753
16	155	0.5817
17	160	1.8646
18	165	2.6535
19	170	2.8527
20	175	2.8846
21	180	2.8368
22	185	2.7890
23	190	2.7093
24	195	2.5180
25	200	2.3427
26	205	2.2312
27	210	1.9603
28	215	1.8248
29	220	1.6017
30	225	1.4901
31	230	1.2989
32	235	1.2590
33	240	1.1076
34	245	0.9642
35	250	0.9164
36	255	0.8845
37	260	0.7809
38	265	0.7172
39	270	0.6694
40	275	0.6215
41	280	0.6454
42	285	0.6454
43	290	0.5817
44	295	0.5259
45	300	0.5658
46	305	0.5259
47	310	0.5020
48	315	0.5419
49	320	0.5419
50	325	0.3426
51	330	0.4940
52	335	0.4383
53	340	0.4462
54	345	0.3984
55	350	0.4064
56	355	0.4383
57	360	0.3984
58	365	0.3905
59	370	0.4861
60	375	0.3984
61	380	0.3267
62	385	0.3905
63	390	0.3825
64	395	0.4303
65	400	0.3267
66	405	0.3586
67	410	0.3905
68	415	0.3745
69	420	0.3347
70	425	0.3426
71	430	0.2869
72	435	0.2550
73	440	0.3506
74	445	0.3028
75	450	0.3028
76	455	0.3108
77	460	0.2311
78	465	0.2709
79	470	0.3028
80	475	0.2789
81	480	0.3347
82	485	0.2311
83	490	0.1753
84	495	0.2869
85	500	0.2311

5.   Construct a spreadsheet as in Figure 16-15 to calculate the mean and standard deviation
by batch and recursive estimation methods. Use a mean value of 122.50 and add random
noise. Use 25 data points. Calculate the batch and recursive means. Calculate the batch
and recursive standard deviations. Use the F9 key to recalculate and change the random
noise figures. Are the means and standard deviations always identical? When might you
use a recursive procedure in place of the usual batch method? Try other mean values.